TECHNICAL GUIDE TO MANAGING GROUND WATER RESOURCES

United States Department of Agriculture

Forest Service

Minerals and Geology Management

Watershed, Fish, Wildlife, Air, and Rare Plants Engineering

FS-881

May 2007

Published by Books Express Publishing
Copyright © Books Express, 2011
ISBN 978-1-780391-83-0

Books Express publications are available from all good retail
and online booksellers. For publishing proposals and direct
ordering please contact us at: info@books-express.com

Contents

List of Figures

List of Tables

Part 1. Introduction

Ground water is the Nation's principal reserve of fresh water and represents much of its potential future water supply. Ground water on National Forest System (NFS) lands is a major contributor to flow in many streams and rivers and has a strong influence on the health and diversity of plant and animal species in forests, grasslands, riparian areas, lakes, wetlands, and cave systems. It also provides drinking water to hundreds of communities. Demands for safe drinking water and requirements to maintain healthy ecosystems are increasing, and complex social and scientific questions have arisen about how to assess and manage the water resources on NFS lands. This technical guide was developed to help address these issues. It describes the national ground water policy and provides management guidelines for the NFS.

Today, many of the concerns about ground water resources on or adjacent to public land involve questions about depletion of ground water storage, reductions in streamflow, potential loss of ground water-dependent ecosystems, land subsidence, saltwater intrusion, and changes in ground water quality. The effects of many human activities on ground water resources and on the broader environment need to be clearly understood in order to properly manage these systems. Throughout this technical guide, we emphasize that development, disruption, or contamination of ground water resources has consequences for hydrological systems and related environmental systems.

Ground water and surface water are interconnected and interdependent in almost all ecosystems. Ground water plays significant roles in sustaining the flow, chemistry, and temperature of streams, lakes, springs, wetlands, and cave systems in many settings, while surface waters provide recharge to ground water in other settings. Ground water has a major influence on rock weathering, streambank erosion, and the headward progression of stream channels. In steep terrain, it governs slope stability; in flat terrain, it limits soil compaction and land subsidence. Pumping of ground water can reduce river flows, lower lake levels, and reduce or eliminate discharges to wetlands and springs. It also can influence the sustainability of drinking-water supplies and maintenance of critical ground water-dependent habitats.

Increasingly, attention is being placed on how to manage ground water (and surface-water) resources on public lands in a sustainable manner. The potential for ground water resources to become contaminated from anthropogenic as well as natural sources is being scientifically assessed. Each ground water system and development situation is unique and requires a specific analysis to draw appropriate conclusions.

This technical guide begins by reviewing the legislative and policy framework, and the issues related to ground water inventory, monitoring, contamination, and development. Individual sections then focus on key concepts, principles and methods for managing ground water resources. Relevant special topics, case studies, and field examples are highlighted throughout the text. Additional information on some topics can be found in the appendixes.

Purpose and Objectives

This technical guide provides guidance for implementing the U.S. Department of Agriculture (USDA) Forest Service national ground water policy. It describes hydrological, geological, and ecological concepts, as well as the managerial responsibilities that must be considered to ensure the wise and sustainable use of ground water resources on NFS lands.

Scope and Organization

This document is one part of a four-part information system on ground water management on the national forests and grasslands. The other three parts are (1) Forest Service policy on ground water (Forest Service Manuals [FSM] 2543 and 2880); (2) a Forest Service sourcebook on State ground water laws, regulations, and case law for all 43 States with NFS land; and (3) a ground water inventory and monitoring technical guide. When complete, the four parts will provide line officers and technical specialists at all field levels with the science, policy, and legal framework for Forest Service ground water-resource management. Users of this document are strongly encouraged to refer to all of these documents when dealing with a ground water-resource issue.

This technical guide is intended for Forest Service line officers and managers and their technical-support staffs. Managers will be interested in Parts 1 and 2, in which information is presented on management considerations and on the importance of ground water issues. Part 3 and the appendixes provide more detailed information on basic hydrogeological principles and ground water investigation methods that may be most appropriate for technical support staffs.

Importance of Ground Water Resources on NFS Lands

Ground water is a valuable commodity and its use is growing nationwide. The NFS contains substantial ground water resources, for which stewardship and protection are mandated by congressional acts. Many other natural resources on NFS lands rely, directly or indirectly, on ground water and would be damaged or destroyed if that water were depleted or contaminated. Careful inventory of the quantity and quality of ground water on the NFS is needed to provide sufficient information to appraise the value and provide appropriate stewardship of these ground water resources. The following are the objectives of ground water inventory and monitoring:

- To ensure timely availability of hydrogeological resource information needed for the periodic assessment required by the Forest and Rangeland Renewable Resources Planning Act, as amended, and for land and resource management planning.
- To provide regionwide status and change data and to enhance the potential for combining data sets across geographic areas to address national trends.

- To classify aquifer types, establish baseline ground water quality, map flow systems and ground water-dependent ecosystems, and assess aquifer vulnerability based on a consistent standard throughout the NFS.

Ensuring sustainability of natural resources has become a fundamental requirement for Federal land management. In preparing to manage ground water resources within this framework, the following interdependent questions must be addressed:

- How much ground water is there, where is it, and what is its quality?
- What are the existing uses of ground water?
- What is the nature of the interconnections between the ground water and surface water systems?
- To what extent do other natural resources depend on ground water?
- How vulnerable are the aquifers to contamination or depletion?

To answer these questions, ground water resources need to be inventoried and assessed.

Overuse of ground water may impact streams, wetlands, riparian areas, forest stands, meadows, grasslands, seeps, springs, cave systems, and livestock and wildlife watering holes. It may lower lake and reservoir levels, and promote land subsidence, sinkhole formation, and cave collapse. Reduced water-table levels can impact biota that depend on ground water, particularly in riparian and wetland ecosystems.

When water is removed from saturated soils and deeper sediments, the soil, sediment, or rock structure that remains may partially collapse and result in visible slumping of soils, widespread subsidence of the land surface, or the formation of sinkholes. These changes in the land surface may damage highways, bridges, building foundations, and other structures. They also may damage natural resources. In addition, excessive well withdrawals can affect water quality in the aquifer. Saltwater may intrude into the aquifer, poor-quality or contaminated water may migrate from adjoining areas or surface water bodies, or chemical components of the desaturated aquifer may be mobilized. Ground water levels or pressures may drop, causing shallow wells to go dry and requiring deepening or replacement. Increased drawdown can impact ecological resources by depleting ground water that supports riparian vegetation, wetlands, or sensitive flora and fauna.

The list of elements and chemical compounds that may be accidentally or purposely released into the environment, and transported by ground water, is seemingly endless. The NFS contains thousands of public and private drinking-water supply systems located at campgrounds, rest areas, permittee sites, private in-holdings, and in-forest communities. The NFS also contains the headwaters of many streams that flow off-system lands and the recharge areas for many aquifers from which water is drawn for human use. The protection of all sources of public drinking water from contamination is a nationwide imperative, heralded by the Safe Drinking Water Act (SDWA) of 1974.

Many activities have the potential to contribute contamination to soils and ground water simply through the presence and use of fuels, oils, solvents, paints and detergents, and by the generation of solid or liquid wastes. Typical contamination sources on NFS lands include mines, oil and gas wells, landfills, and septic systems. Contamination of soils and ground water can be difficult, time-consuming, and expensive to address.

Although numerous Federal and State programs regulate activities that may release contaminants to soils and ground water, the implementation of these programs in rural areas generally lags behind that in urban areas. Because the release of even small amounts of stored chemicals or fuels may substantially damage soil and ground water resources, efforts must be made to ensure that all Forest Service activities and facilities comply with regulations for preventing soil and ground water contamination. Similarly, efforts must be made to collaborate or partner with States, permittees, owners of in-holdings, and forest-bounded communities to institute appropriate ground water protection measures.

Part 2. Managing Ground Water Resources

This section reviews the types of ground water issues that are important for all USDA Forest Service units, line officers, and staff to consider. Legal requirements and ground water-management strategies are discussed.

Federal Statutes

In addition to the Federal land management statutes cited in Forest Service Manual (FSM) 2501, the following Federal statutes provide pertinent direction to the Forest Service for its management of ground water resources in the National Forest System.

Safe Drinking Water Act of 1974, as amended. (42 U.S.C. §300f et seq). The intent of the SDWA is to ensure the safety of drinking-water supplies. Its authority is used to establish drinking-water standards and to protect surface- and ground water supplies from contamination.

Resource Conservation and Recovery Act of 1976, as amended. (42 U.S.C. §6901 et seq) The Resource Conservation and Recovery Act (RCRA) regulates the generation, transportation, treatment, storage and disposal of waste materials. It has very specific requirements for the protection and monitoring of ground water and surface water at operating facilities that may generate solid wastes or hazardous wastes.

Comprehensive Environmental Response, Compensation, and Liability Act of 1980, as amended. (42 U.S.C. §6901 et seq). Also known as "Superfund", the Comprehensive Environmental Response, Compensation, and Liability Act (CERCLA) regulates cleanup of existing environmental contamination at non-operating and abandoned sites (see also FSM 2160).

In addition, judicial doctrine and water-rights case law provide the legal interpretations of Federal and State statutes about usage and management of ground water (see FSM 2541.01 and Forest Service Handbook [FSH] 2509.16 for procedures to be followed for complying with Federal policy and State water-rights laws).

Overview of the National Ground Water Policy

The national ground water policy sets out the framework in which ground water resources are to be managed on NFS lands. The policy is designed to be located in two parts of the Forest Service Manual, FSM 2880, Geologic Resources, Hazards, and Services, and FSM 2543, Ground Water Resource Management. As of the publication date of this technical guide, FSM 2543 is in draft form and may change due to agency and public comment prior to finalization. Regional Foresters and Forest Supervisors are directed by the national ground water policy to perform the duties detailed below.

Land Management Planning

- Protection and sustainable development of ground water resources are appropriate components of land and resource management planning for NFS lands. Ground water inventories and monitoring data shall be integrated into the land and resource management process.
- When evaluating project alternatives or revising national forest plans, use the best available science, technology, models, information, and expertise to determine the location, extent, depths, amounts, flow paths, quality, and recharge and discharge areas of ground water resources and their hydrological connections with surface water.

Water Development

- Conduct appropriate National Environmental Policy Act (NEPA) analyses when evaluating applications for water wells or other activities that propose to test, study, monitor, modify, remediate, withdraw, or inject into ground water on NFS lands (see also FSH 2509).
- Always assume that hydrological connections exist between ground water and surface water in each watershed, unless it can be reasonably shown none exist in a local situation.
- Ensure that ground water that is needed to meet Forest Service and authorized purposes is used efficiently and, in water-scarce areas or time periods, frugally. Carefully evaluate alternative water sources, recognizing that the suitable and available ground water is often better than surface water for human consumption at administrative and public recreational sites.
- Prevent, if possible, or minimize the adverse impacts to streams, lakes, ponds, reservoirs, and other surface waters on NFS lands from ground water withdrawal.
- As applicable under State water-rights laws and adjudications, file water-use-permit applications and water-rights claims for beneficial uses of ground water by the Forest Service. Consult with the Office of General Counsel prior to filing (see also FSM 2541).
- Comply with wellhead protection (U.S. Environmental Protection Agency [EPA] 1994), sole-source aquifer, and underground injection control (UIC) requirements of Federal (40 Code of Federal Regulations [CFR] 144), State, and local agencies. Ensure that all public water systems (PWSs) on NFS lands that use ground water comply with EPA's ground water rules.
- Require all drinking-water systems that withdraw water from aquifers on NFS lands, and that are classified as community water systems (those that serve 25 year-round residents or have 15 or more service connections), to have flow meters installed and operating. Require wells on NFS lands that provide ground water that is later sold to consumers or used for industrial or commercial purposes to have flow meters installed and operating. Wells equipped with hand pumps are not required to have flow meters. Require injection wells with discharge pipes that are 4 inches inside diameter or larger to be metered.

Water Quality

- Identify the needs and opportunities for improving watersheds and improving ground water quality and quantity. Take appropriate steps to address the needs and take advantage of the opportunities.
- In areas where ground water on NFS land has become contaminated from human sources, evaluate the potential receptors, technical feasibility, costs, and likelihood of finding potentially responsible parties (PRPs), the risks of exacerbating the problem, and other relevant factors before making a decision to try to cleanup the ground water.
- Complete removal and/or remedial actions for ground water contamination at CERCLA/Superfund sites on NFS lands. Identify the PRPs and seek to have them perform the cleanup work, where possible, to minimize the cost of the cleanup to the Forest Service. At sites where the Forest Service is a PRP, the cleanup work should be aggressively performed in a timely manner to fulfill the agency's trustee responsibilities. Inform owners of non-federal property abutting NFS lands that overlie contaminated ground water of the existence of the contamination, the types of contaminants present, and the Forest Service plan for managing the contaminated ground water.

Ground Water-dependent Ecosystems

- Ecological processes and biodiversity of ground water-dependent ecosystems must be protected. Plan and implement appropriately to minimize adverse impacts on ground water-dependent ecosystems by (1) maintaining natural patterns of recharge and discharge, and minimizing disruption to ground water levels that are critical for ecosystems; (2) not polluting or causing significant changes in ground water quality; and (3) rehabilitating degraded ground water systems where possible.
- Manage ground water-dependent ecosystems to satisfy various legal mandates, including, but not limited to, those associated with floodplains, wetlands, water quality and quantity, dredge and fill material, endangered species, and cultural resources.
- Manage ground water-dependent ecosystems under the principles of multiple use and sustained yield, while emphasizing protection and improvement of soil, water, and vegetation, particularly because of effects upon aquatic and wildlife resources. Give preferential consideration to ground water-dependent resources when conflicts among land-use activities occur.
- Delineate and evaluate both ground water itself and ground water-dependent ecosystems before implementing any project activity with the potential to adversely affect those resources. Determine geographic boundaries of ground water-dependent ecosystems based on site-specific characteristics of water, geology, flora, and fauna.
- Establish maximum limits to which water levels can be drawn down at a specified distance from a ground water-dependent ecosystem in order to protect the character and function of that ecosystem.
- Establish a minimum distance from a connected river, stream, wetland, or other ground water-dependent ecosystem from which a ground water withdrawal may be sited.

Inventory and Monitoring	• Design inventory and monitoring programs to (1) gather enough information to develop management alternatives that will protect ground water resources, and (2) evaluate management concerns and issues expressed by the general public. Assign high priorities for survey, inventory, analysis, and monitoring to municipal water-supply aquifers, sensitive aquifers, unique ground water-dependent ecosystems, and high-value or intensively managed watersheds.
	• Develop estimates of the usable quantity of ground water in aquifers while protecting important NFS resources and monitor to detect excessive water withdrawal.
	• Define the present situation and detect spatial or temporal changes or trends in ground water quality or quantity and health of ground water-dependent ecosystems; detect impacts or changes over time and space, and quantify likely effects from human activities.
Data Management	• Establish guidelines and standards for the acquisition and reporting of ground water information to meet the specific needs of Forest Service programs. The storage of ground water data must conform to Forest Service Natural Resource Applications (FSNRA) standards and servicewide Geographic Information System (GIS) data standards. Storage will be in FSNRA databases upon availability.
Partnerships	• Close collaboration and partnership with other Federal Agencies and States/ Tribes, regional and local governments and other organizations is essential in gathering and analyzing information about ground water resources for which the Forest Service has stewardship.

Ground Water Uses

Some 83.8 billion gallons per day of fresh ground water were pumped in the United States in 2000 (Hutson and others 2004). This total was about 8 percent of the estimated daily natural recharge to the Nation's ground water. Much of this water was being withdrawn in excess of the recharge capabilities of local aquifers ("overpumping"). Withdrawals significantly in excess of natural recharge are located predominantly in coastal areas of California, Texas, Louisiana, Florida, and New York, in the Southwest, and in the Central Plains. In the United States, management of ground water is primarily the responsibility of State and local governments. The authority and responsibility for overseeing the allocation and development of water resources typically resides with the State's department of natural resources or water resources or the State engineer's office. The authority and responsibility to prevent undue contamination of ground water typically resides with the State's health department or department of environmental quality or environmental management and with local government (e.g., health department, county commissioners, city council). In addition on most Federal lands some overlapping responsibilities for both ground water and quantity resides with the management agency.

Management of water resources includes the management of land-use activities that include potential sources of contamination. As population density increases, an ever-increasing demand on water resources and an ever-increasing complexity of management issues are created. This complexity results from the uncertainties related to (1) how to manage water resources in a manner that achieves a sustainable annual supply, (2) how to prevent unplanned contamination of ground water, and (3) how to balance competing uses of interconnected water resources.

Pumping of ground water results in changes to the ground water system and, potentially, to the ecosystem of the region being developed. These changes may take many years to be observed because of the commonly slow movement of ground water. Some changes, such as the loss of aquifer storage capacity from land subsidence, may be irreversible. Some changes may not be readily observable because they are incrementally small with time, occur underground, or slowly affect the chemistry of the ground water or surface water. The consequences of pumping should be assessed for each level of development, and safe yield should be the maximum withdrawal for which the consequences are considered acceptable.

Management of Drinking-water Supplies

Ground water is one of the Nation's most important natural resources, providing about 40 percent of the Nation's public water supply (Alley and others 1999). In addition, more than 40 million people, including most of the rural population, supply their own drinking water from domestic wells. As a result, ground water is an important source of drinking water in every State and is also the source of much of the water used for agricultural irrigation. Therefore, protection of those water resources is an important goal of land-use planning and management nationwide. A valuable reference to use in assessing how much risk of contamination is associated with different land-use practices commonly occurring on NFS lands is titled "Drinking Water from Forests and Grasslands: A Synthesis of the Scientific Literature" published in 2000 by the Forest Service Southern Research Station as General Technical Report SRS-039 (http://www.srs.fs.fed.us/pubs/gtr/gtr_srs039).

The SDWA Amendments of 1996 (P.L. 104-182, 42 U.S.C. 300f et. seq.) revised the original 1974 Act by adopting a multiple barrier approach to the protection of drinking water from its source to the tap, creating a State revolving fund for financing water treatment improvements, and establishing reporting on the quality of water served to all water consumers by the water provider. All PWSs have to be assessed for vulnerability to current and potential sources of contamination and the source of water must be delineated. By definition, PWSs provide drinking water to at least 25 people or 15 service connections for at least 60 days a year. About 170,000 PWSs in the United States provide water to more than 250 million people. Of these, at least 3,500 communities and 60 million people get water directly from NFS lands. The EPA defines two main types of PWSs:

1. *Community water systems that provide drinking water to the same people yearlong.* All Federal drinking-water regulations apply to these systems.
2. *Noncommunity water systems that serve customers on less than a yearlong basis.* Such systems are considered to be transient if they serve people who are passing through the area and nontransient if they serve at least 25 of the same people more than 6 months in a year but not yearlong. Most federal drinking-water regulations apply to systems in the latter category, while only regulations concerning contaminants posing immediate health risks apply to systems in the transient category.

The Forest Service owns and operates about 6,000 water systems, most of which fall into the noncommunity, transient category. District Rangers and Forest Supervisors should make sure that these water systems are meeting all requirements of the law, are being tested in compliance with the law and regulations, and, if found to fail the bacteriological or any of the other standards, that the system will be immediately shut off and not reopened until all tests are in compliance. Exposure of the public or Forest Service employees to unsafe or SDWA non-compliant drinking water must not take place at Forest Service managed facilities. In addition, civil penalties can be imposed by States and EPA for violations of the SDWA.

Water systems owned and operated by the Forest Service should be maintained properly to ensure that the water provided to the public and employees is safe and meets all applicable standards. Systems nearing the end of their service life may need major overhaul or replacement. Line officers are responsible for requesting sufficient funding to maintain water systems or close down obsolete systems and switching those facilities to other water supplies, such as municipal water if available. Additional guidance can be found in FSM 7420 or through consultation with the regional environmental engineer.

Effective management of water resources in fractured-rock hydrogeological settings must be based on a sound conceptual understanding of the ground water flow system(s) that occur in the area to be managed. Because of the heterogeneous and anisotropic nature of fractured-rock settings, it has proven difficult to manage water resources in these settings. In fractured-rock settings, sustainable development is greatly complicated by uncertainties about actual watershed dimensions and annual water budgets in associated aquifer systems. The relationship between "deep" ground water in fractured rock and surface water is still not well understood at the watershed scale. At the watershed level, significant uncertainties also exist about whether water from non-consumptive uses returns to the deep ground water system or moves as interflow directly to a nearby stream.

Because, in part, of the relative ease of delineating and recognizing watershed boundaries in mountainous areas, the concept and practice of watershed-based resource management has evolved more rapidly in those regions of the country. Watershed-based management is a holistic approach that requires an understanding of ground water flow and the relationship between ground water and surface water. In fractured-rock settings such understandings, however, may be very difficult to achieve.

Source-water Protection

Another key provision of the 1996 Amendments to the SDWA was an increased focus on the prevention of contamination of drinking water at its source within a surface watershed or within a defined area surrounding a ground water extraction site, such as a well. States are required to do Source Water Assessments (U.S EPA 1997) for all public water supplies. For ground water, States commonly use one of two methods to define a well-head protection area: (1) "fixed radius," which is the area defined by a radius of set distance from a well, such as 1,000 feet or 1 mile; or (2) "time of travel," which is the area from which ground water flows horizontally to a well in a set time period, such as 1 year. Each method has its advantages and problems, and neither can provide 100 percent assurance that the ground water supply is really safe from contamination if appropriate land-use restrictions are applied within the deliineated area.

Coordination with Public Water-supply Purveyors

PWS utility operators near or within the NFS may request the Forest Service to add water-quality protective measures, including additional "best management practices" (BMPs), for many land uses and activities on NFS lands within delineated source watersheds and well-head protection areas. The Forest Service should work with water supply utilities and others to evaluate the likely effectiveness of such additional practices and the means for paying for their installation and maintenance. The Forest Service should also determine whether any reimbursement for revenues forgone to the U.S. Treasury expected from any contracts, leases or permits that are being ended or prevented should be required of the utility, municipality, or other entity (see also FSM 2542).

Protecting Public Water Supplies Through Forest Planning

During the first round of forest planning, a provision in the 1982 NFMA regulations required that municipal water-supply watersheds be identified as separate management areas. Many forests identified these watersheds and developed separate standards for them, but some did not. The new NFMA planning process does away with the requirement to delineate municipal watersheds, but it continues the emphasis on collaboration with stakeholders in the management of NFS lands, and the need to identify and quantify the amount and quality of water needed for multiple uses on and off these lands. As forest plans are revised, Forest Service units should invite participation from local water-utility managers and their staffs to help the agency make sure that forest plans recognize the importance of drinking-water sources on and under NFS lands and of developing and implementing sound water-quality and quantity protection strategies and measures.

11

Special-use Authorizations for Water Wells and Pipelines

This section addresses the authorization of water extraction or injection wells and water pipelines through special-use permits. Guidance in FSM 2729, FSM 2543, and FSH 2509, specifically addresses the authorization of water developments on NFS lands, and a decision tree summarizing the process is shown in figure 1. The basic laws authorizing water wells on NFS lands are the Organic Administration Act and the Federal Land Policy and Management Act (FLPMA). The permitting process for wells and pipelines is a discretionary activity; a permit for a well or pipeline may be denied if the agency's analysis indicates that NFS resources, including water, will not be adequately protected. Except when authorized by either U.S. Department of the Interior or USDA regulations for the management of mineral or energy exploration, development, or production, a special-use permit is required for all entities other than the Forest Service to drill water wells or construct water pipelines on NFS lands.

Where a State-based water right or State approval is needed for a water development, the process for securing State approvals should follow after or run concurrently with the Forest Service process for authorizing a water development. In all cases, State law must be observed when a State-based water right is involved. When a project proponent proposes to drill a well on NFS lands and/or to transport ground water across NFS lands through a pipeline, an analysis of the potential impacts of water removal from the aquifer along with the impacts of well drilling and/or pipeline construction is required (40 CFR 1508.25 Scope and 1508.7 Cumulative Impacts). For development of a water-injection well, the impact on ground water quality from the addition of non-native water and the impact that added volume would have on aquifer structure and function must be analyzed.

Laws and regulations governing wells include both State requirements for notification, drilling permits, well logs, well completion or abandonment procedures and documentation, and Federal requirements and recommendations for construction, sampling, and abandonment of monitoring wells. Under 36 CFR 251.51, the Forest Service has the authority to grant or deny a request for special-use authorization for a water diversion, extraction, or conveyance facility. No legal obligation exists to grant a special-use authorization for a water facility, even if the applicant controls a valid State-issued water right. When considering whether to grant a special-use authorization for a water diversion, the law requires the inclusion of terms and conditions necessary to protect national-forest resources as part of any decision granting a right-of-way across NFS lands (see Section 1765 of the FLPMA).

A solid administrative record must be developed to support decisions on special-use authorizations. Include the impacts to national-forest resources, details on the basis for mitigation measures required to protect those resources, and the reasons why the extraction or conveyance of ground water is consistent or inconsistent with the applicable land management plan.

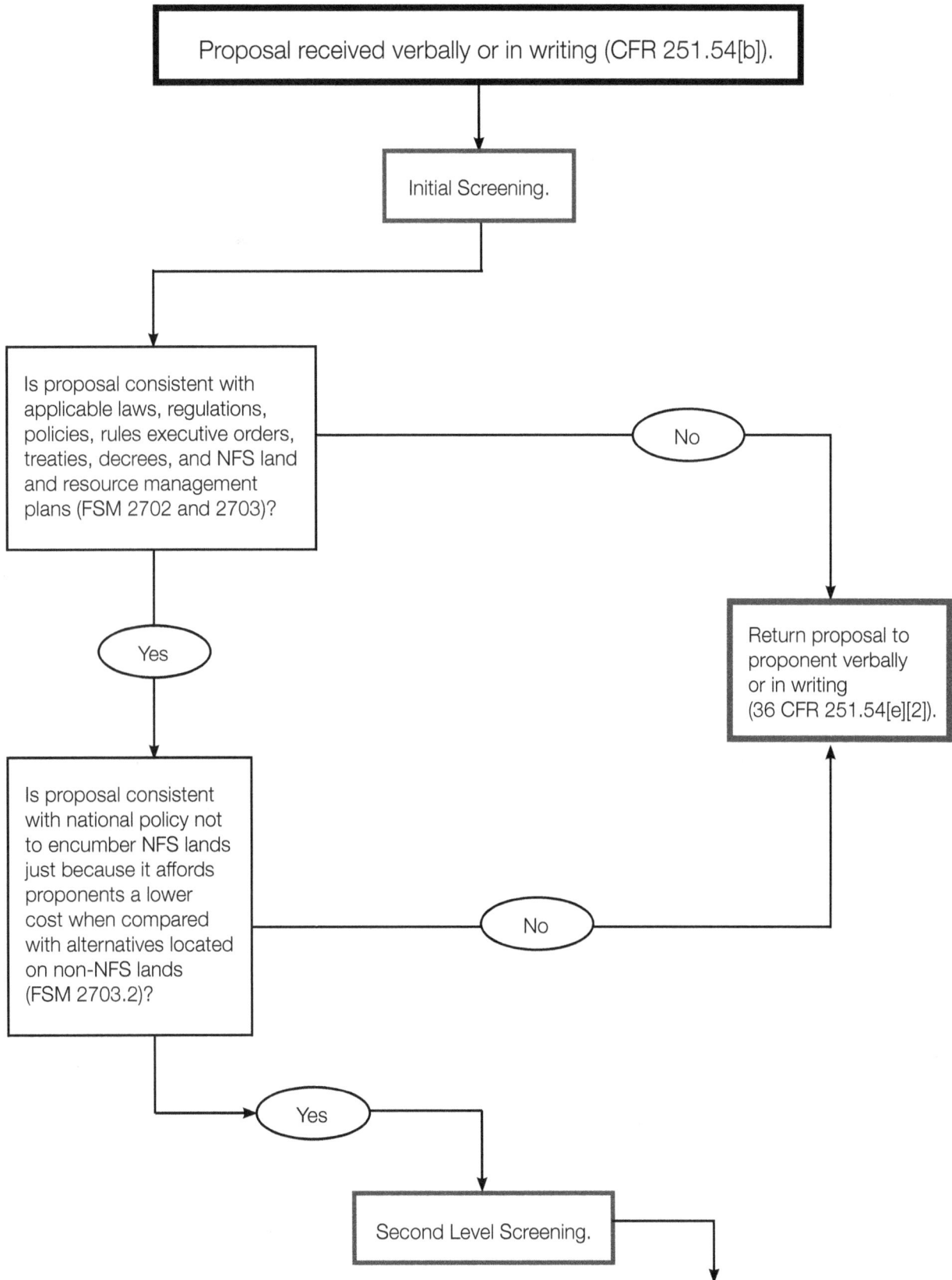

Figure 1. Decision tree for issuance of special-use permits for proposals to develop water supply wells on NFS land.

Proposal received verbally or in writing (CFR 251.54[b]).

Initial Screening.

Is proposal consistent with applicable laws, regulations, policies, rules executive orders, treaties, decrees, and NFS land and resource management plans (FSM 2702 and 2703)?

No

Yes

Is proposal consistent with national policy not to encumber NFS lands just because it affords proponents a lower cost when compared with alternatives located on non-NFS lands (FSM 2703.2)?

No

Return proposal to proponent verbally or in writing (36 CFR 251.54[e][2]).

Yes

Second Level Screening.

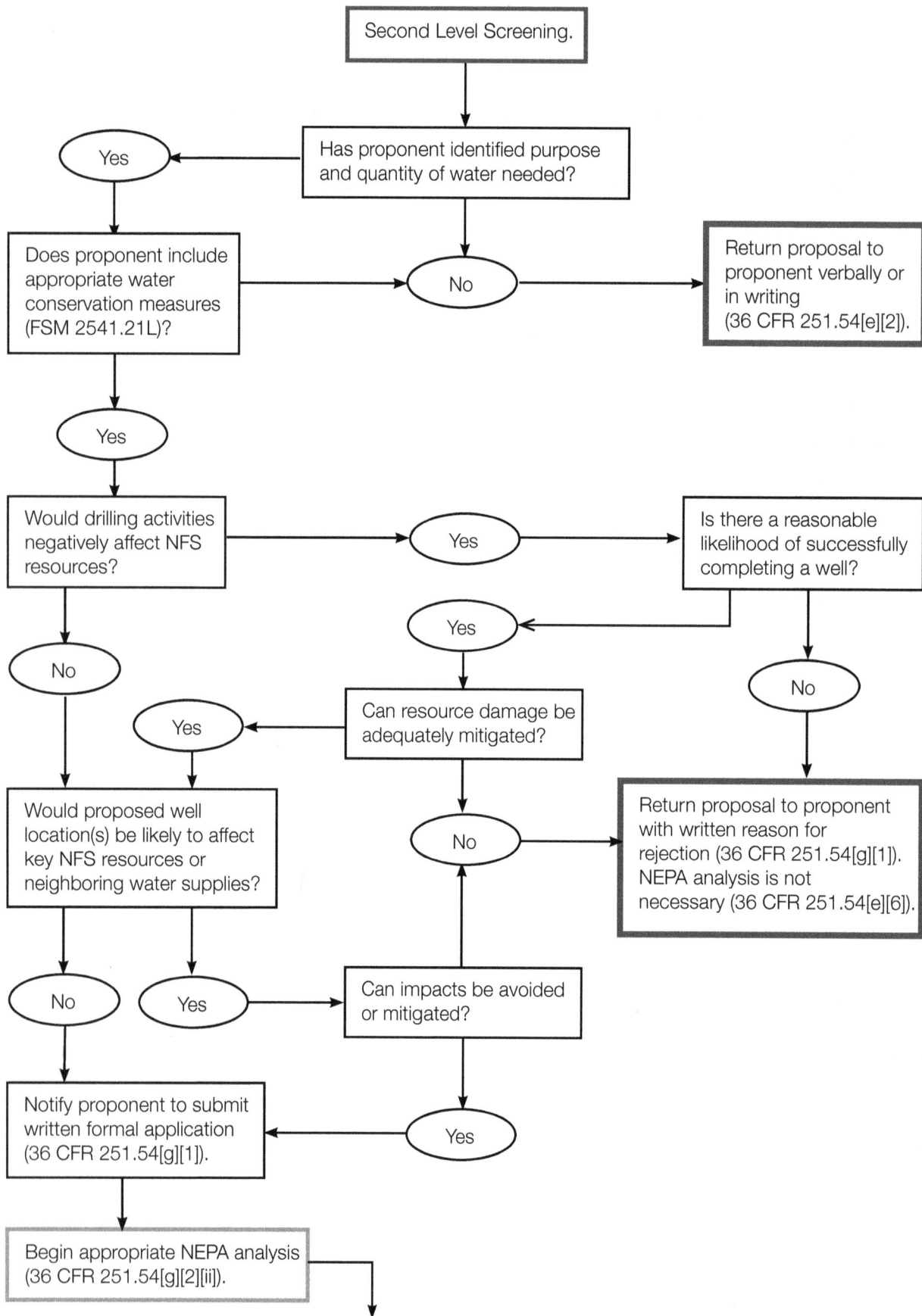

```
                        ┌─────────────────────────┐
                        │  Second Level Screening. │
                        └─────────────────────────┘
                                    │
                                    ▼
         ┌──────┐      ┌──────────────────────────────┐
         │ Yes  │◄─────│ Has proponent identified purpose│
         └──────┘      │ and quantity of water needed?  │
            │          └──────────────────────────────┘
            │                        │
            ▼                        ▼
 ┌────────────────────┐         ┌──────┐      ┌──────────────────────┐
 │ Does proponent     │         │  No  │─────►│ Return proposal to   │
 │ include            │────────►└──────┘      │ proponent verbally or│
 │ appropriate water  │                       │ in writing           │
 │ conservation       │                       │ (36 CFR 251.54[e][2])│
 │ measures           │                       └──────────────────────┘
 │ (FSM 2541.21L)?    │
 └────────────────────┘
            │
            ▼
         ┌──────┐
         │ Yes  │
         └──────┘
            │
            ▼
 ┌────────────────────┐      ┌──────┐      ┌──────────────────────┐
 │ Would drilling     │      │ Yes  │      │ Is there a reasonable│
 │ activities         │─────►└──────┘─────►│ likelihood of        │
 │ negatively affect  │                    │ successfully         │
 │ NFS resources?     │                    │ completing a well?   │
 └────────────────────┘                    └──────────────────────┘
            │                                        │
            ▼                   ┌──────┐             │
         ┌──────┐               │ Yes  │◄────────────┘
         │  No  │               └──────┘
         └──────┘                  │
            │        ┌──────┐      ▼
            │        │ Yes  │◄─┌──────────────────┐      ┌──────┐
            │        └──────┘  │ Can resource     │      │  No  │
            │           │      │ damage be        │      └──────┘
            │           │      │ adequately       │         │
            ▼           ▼      │ mitigated?       │         ▼
 ┌────────────────────┐└──────────────────┘  ┌──────────────────────┐
 │ Would proposed well│           │           │ Return proposal to   │
 │ location(s) be     │        ┌──────┐       │ proponent with       │
 │ likely to affect   │        │  No  │──────►│ written reason for   │
 │ key NFS resources  │        └──────┘       │ rejection            │
 │ or neighboring     │           ▲           │ (36 CFR 251.54[g][1])│
 │ water supplies?    │           │           │ NEPA analysis is not │
 └────────────────────┘           │           │ necessary            │
        │        │                │           │ (36 CFR 251.54[e][6])│
        ▼        ▼                │           └──────────────────────┘
    ┌──────┐ ┌──────┐    ┌──────────────────┐
    │  No  │ │ Yes  │───►│ Can impacts be   │
    └──────┘ └──────┘    │ avoided or       │
        │                │ mitigated?       │
        │                └──────────────────┘
        │                         │
        ▼                         ▼
 ┌────────────────────┐        ┌──────┐
 │ Notify proponent to│◄───────│ Yes  │
 │ submit written     │        └──────┘
 │ formal application │
 │ (36 CFR 251.54[g][1│
 │ ]).                │
 └────────────────────┘
            │
            ▼
 ┌────────────────────┐
 │ Begin appropriate  │──────┐
 │ NEPA analysis      │      │
 │ (36 CFR            │      │
 │ 251.54[g][2][ii]). │      ▼
 └────────────────────┘
```

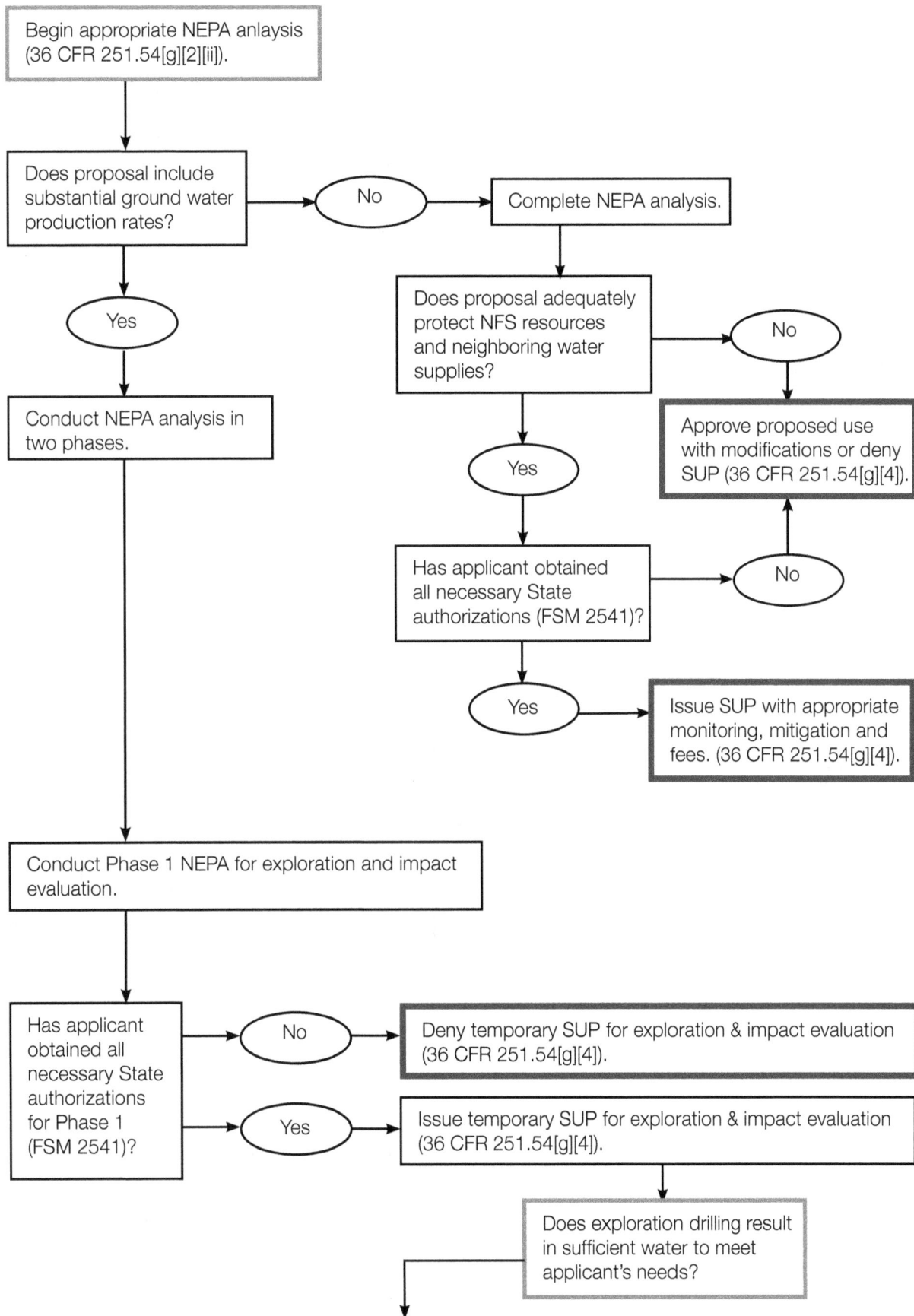

```
┌─────────────────────────┐
│ Begin appropriate NEPA  │
│ anlaysis                │
│ (36 CFR 251.54[g][2][ii]).│
└─────────────────────────┘
             │
             ▼
┌─────────────────────────┐         ┌──────┐      ┌──────────────────────┐
│ Does proposal include   │────────▶│  No  │─────▶│ Complete NEPA        │
│ substantial ground water│         └──────┘      │ analysis.            │
│ production rates?       │                        └──────────────────────┘
└─────────────────────────┘                                  │
             │                                                ▼
          ┌──────┐                        ┌──────────────────────┐      ┌──────┐
          │ Yes  │                        │ Does proposal        │─────▶│  No  │
          └──────┘                        │ adequately protect   │      └──────┘
             │                            │ NFS resources and    │         │
             ▼                            │ neighboring water    │         ▼
┌─────────────────────────┐              │ supplies?            │   ┌──────────────────────┐
│ Conduct NEPA analysis   │              └──────────────────────┘   │ Approve proposed use │
│ in two phases.          │                        │               │ with modifications or│
└─────────────────────────┘                     ┌──────┐           │ deny SUP             │
             │                                   │ Yes  │           │ (36 CFR 251.54[g][4]).│
             │                                   └──────┘           └──────────────────────┘
             │                                      │                        ▲
             │                                      ▼                        │
             │                            ┌──────────────────────┐      ┌──────┐
             │                            │ Has applicant obtained│────▶│  No  │
             │                            │ all necessary State   │      └──────┘
             │                            │ authorizations        │
             │                            │ (FSM 2541)?           │
             │                            └──────────────────────┘
             │                                      │
             │                                   ┌──────┐      ┌──────────────────────┐
             │                                   │ Yes  │─────▶│ Issue SUP with       │
             │                                   └──────┘      │ appropriate monitoring,│
             │                                                 │ mitigation and fees. │
             │                                                 │ (36 CFR 251.54[g][4]).│
             │                                                 └──────────────────────┘
             ▼
┌─────────────────────────────────────┐
│ Conduct Phase 1 NEPA for exploration │
│ and impact evaluation.               │
└─────────────────────────────────────┘
             │
             ▼
┌─────────────────┐      ┌──────┐      ┌──────────────────────────────────────┐
│ Has applicant   │─────▶│  No  │─────▶│ Deny temporary SUP for exploration & │
│ obtained all    │      └──────┘      │ impact evaluation                    │
│ necessary State │                    │ (36 CFR 251.54[g][4]).               │
│ authorizations  │                    └──────────────────────────────────────┘
│ for Phase 1     │
│ (FSM 2541)?     │      ┌──────┐      ┌──────────────────────────────────────┐
└─────────────────┘─────▶│ Yes  │─────▶│ Issue temporary SUP for exploration &│
                         └──────┘      │ impact evaluation                    │
                                       │ (36 CFR 251.54[g][4]).               │
                                       └──────────────────────────────────────┘
                                                        │
                                                        ▼
                                       ┌──────────────────────────────────────┐
                                       │ Does exploration drilling result in  │
                                       │ sufficient water to meet applicant's │
                                       │ needs?                               │
                                       └──────────────────────────────────────┘
                                                   │
                                                   ▼
```

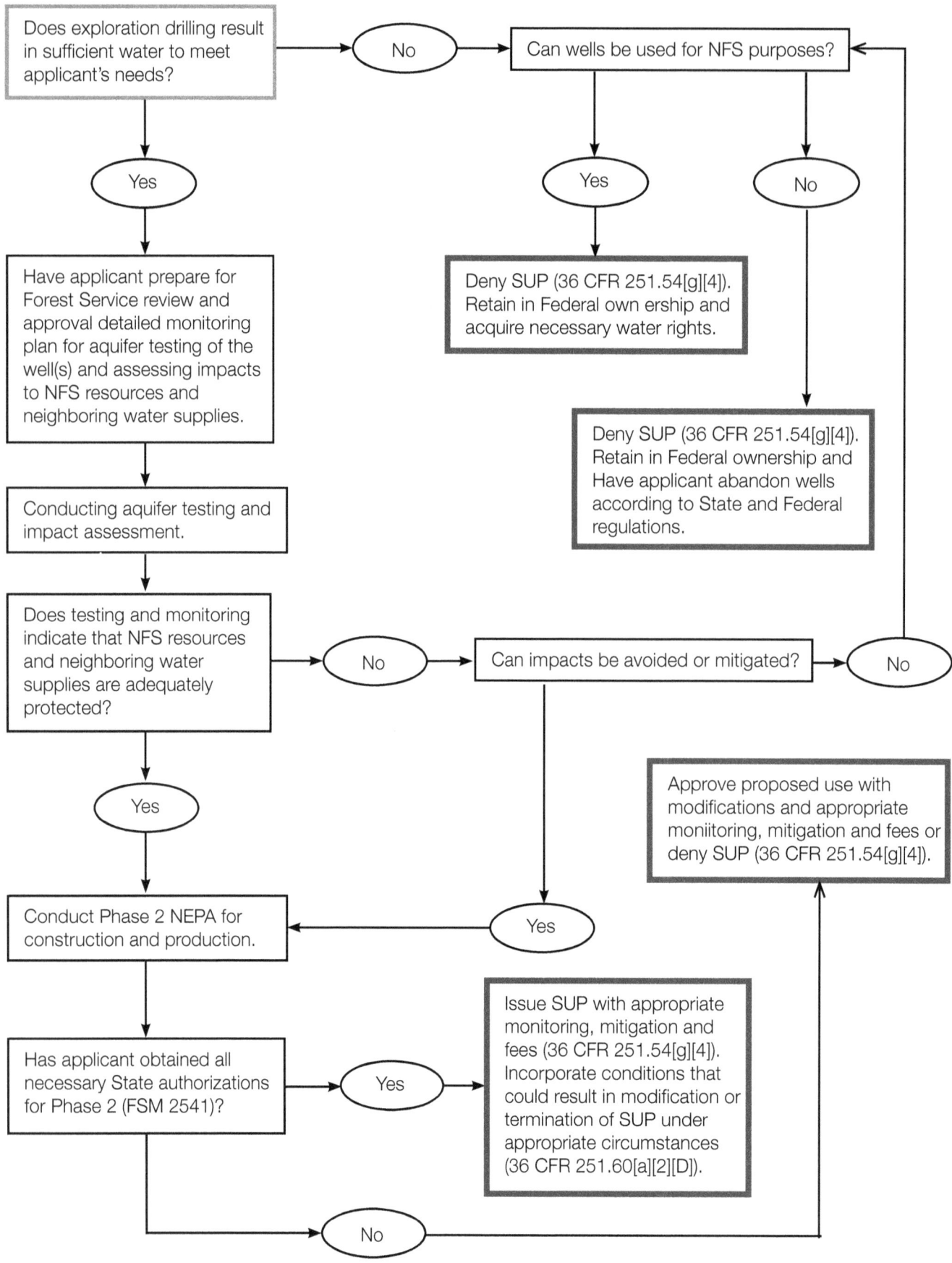

```
┌─────────────────────┐                    ┌───────────────────────────┐
│ Does exploration     │      ╭────╮        │ Can wells be used for     │◄───┐
│ drilling result      │─────►│ No │───────►│ NFS purposes?             │    │
│ in sufficient water  │      ╰────╯        └───────────────────────────┘    │
│ to meet applicant's  │                         │              │            │
│ needs?               │                    ╭────▼────╮    ╭─────▼────╮       │
└─────────────────────┘                    │   Yes   │    │    No    │       │
          │                                 ╰─────────╯    ╰──────────╯       │
     ╭────▼────╮                                 │              │             │
     │   Yes   │                                 │              │             │
     ╰─────────╯                    ┌────────────▼──────┐       │             │
          │                         │ Deny SUP (36 CFR  │       │             │
┌─────────▼───────────┐             │ 251.54[g][4]).    │       │             │
│ Have applicant       │            │ Retain in Federal │       │             │
│ prepare for Forest   │            │ own ership and    │       │             │
│ Service review and   │            │ acquire necessary │       │             │
│ approval detailed    │            │ water rights.     │       │             │
│ monitoring plan for  │            └───────────────────┘       │             │
│ aquifer testing of   │                                        │             │
│ the well(s) and      │                        ┌───────────────▼──────┐      │
│ assessing impacts    │                        │ Deny SUP (36 CFR     │      │
│ to NFS resources     │                        │ 251.54[g][4]).       │      │
│ and neighboring      │                        │ Retain in Federal    │      │
│ water supplies.      │                        │ ownership and        │      │
└─────────────────────┘                        │ Have applicant       │      │
          │                                     │ abandon wells        │      │
┌─────────▼───────────┐                         │ according to State   │      │
│ Conducting aquifer   │                        │ and Federal          │      │
│ testing and impact   │                        │ regulations.         │      │
│ assessment.          │                        └──────────────────────┘      │
└─────────────────────┘                                                       │
          │                                                                   │
┌─────────▼───────────┐                                                       │
│ Does testing and     │     ╭────╮     ┌──────────────────────────┐   ╭────╮ │
│ monitoring indicate  │────►│ No │────►│ Can impacts be avoided   │──►│ No │─┘
│ that NFS resources   │     ╰────╯     │ or mitigated?            │   ╰────╯
│ and neighboring      │               └──────────────────────────┘
│ water supplies are   │                       │
│ adequately           │                       │
│ protected?           │                       │
└─────────────────────┘                        │         ┌──────────────────────┐
          │                                     │         │ Approve proposed use │
     ╭────▼────╮                                │         │ with modifications   │
     │   Yes   │                                │         │ and appropriate      │
     ╰─────────╯                                │         │ moniitoring,         │
          │                                     │         │ mitigation and fees  │
          │                                     │         │ or deny SUP (36 CFR  │
          │                                     │         │ 251.54[g][4]).       │
          │                                     │         └──────────────────────┘
┌─────────▼───────────┐          ╭────╮         │                  ▲
│ Conduct Phase 2      │◄────────│ Yes│◄────────┘                  │
│ NEPA for             │         ╰────╯                            │
│ construction and     │                                           │
│ production.          │              ┌──────────────────────────┐ │
└─────────────────────┘              │ Issue SUP with            │ │
          │                          │ appropriate monitoring,   │ │
          │                          │ mitigation and fees (36   │ │
┌─────────▼───────────┐    ╭────╮    │ CFR 251.54[g][4]).        │ │
│ Has applicant        │───►│Yes │──►│ Incorporate conditions    │ │
│ obtained all         │    ╰────╯   │ that could result in      │ │
│ necessary State      │             │ modification or           │ │
│ authorizations for   │             │ termination of SUP under  │ │
│ Phase 2 (FSM 2541)?  │             │ appropriate circumstances │ │
└─────────────────────┘             │ (36 CFR 251.60[a][2][D]). │ │
          │                         └──────────────────────────┘ │
          │            ╭────╮                                     │
          └───────────►│ No │─────────────────────────────────────┘
                       ╰────╯
```

INITIAL SCREENING

As provided for in 36 CFR 251.54(b), initial proposals for ground water developments may be presented to the Forest Service either orally or in writing. Water developments related to a CERCLA response action are not subject to this initial NEPA screening, but are subject to CERCLA analysis in the engineering evaluation/cost analysis for removals or remedial investigation/feasibility study for remedial actions. To pass the initial information screening requirements, proposals to construct wells on NFS lands and/or pipelines across NFS lands must meet the following conditions:

1. The proposal to pump, inject, or transport water must be consistent with applicable laws, regulations, policies, rules, executive orders, treaties, compacts, and Forest Service land and resource management plans (FSM 2702 and 2703). Proposals are evaluated as specified in 36 CFR 251.54(e).

2. The proposal must be consistent with national policy not to encumber NFS lands just because it affords a proponent a lower cost when compared with alternatives located on non-NFS lands. If the intent of the proposal is to use ground water derived from NFS lands for a non-NFS purpose, the proponent must demonstrate that alternative water sources do not exist (FSM 2703.2).

Proposals that do not meet the minimal requirements of the initial information screening process are returned to the proponent as insufficient. The authorizing officer shall reply in writing if the proposal was presented in writing, or may reply orally if the proposal was presented orally (36 CFR 251.54[e][2]).

SECOND-LEVEL SCREENING

Additional information is required for proposals that pass initial information screening. In second-level screening, the proposal is evaluated as described in 36 CFR 251.54(e)(5) and as follows:

1. The quantity of water the proponent is seeking to pump from beneath NFS lands and the purpose of use of such water must be identified. If the proponent anticipates increased water needs in the future, those needs must be quantified. If the proponent seeks to inject water into the ground, the quantity, source(s), and quality of the injection water and the likely effects of this action must be identified.

2. Proposals to use ground water underlying NFS lands must include appropriate water conservation measures (FSM 2541.21h) and all community water system wells must be equipped with a flow metering device in good working order.

3. Drilling activities themselves can negatively impact NFS resources. In instances in which considerable disturbance may result from the drilling process, the proponent must demonstrate that a reasonable likelihood exists of successfully completing any water wells and adequately mitigating any resource damage.

4. Identify all anticipated facilities, such as roads, power lines, pipelines, water storage tanks, and pumps that could ultimately be needed to produce or inject, and convey water across NFS land. Proposals that involve construction and/or use of roads shall conform to the requirements of FLPMA, specifically Sections 502 and 505.
5. Identify key resources and existing water supplies to assist in evaluating the potential for the proposal to affect NFS resources and neighboring water supplies.
6. Return proposals that fail to pass second-level screening to the proponent with a written reason for rejection (36 CFR 251.54[g][1]). NEPA analysis is not required to make this determination (36 CFR 251.54[e][6]).

Where proposals pass second-level screening, notify the proponent that the Forest Service is prepared to accept a formal written application for a special-use authorization. Previously submitted information may be included in the application by reference. The Forest Service should begin the appropriate NEPA analysis on receipt of the formal application (36 CFR 251.54[g][2][ii]) and notify Federal, Tribal, State, and local entities involved in the management of water resources as early in the process as possible (FSM 1909.15, Conduct Scoping). Advise the proponent that any information provided will become public information once the formal application is received and the NEPA process initiated. Once the formal application is received, the proponent is referred to as the applicant.

ENVIRONMENTAL ANALYSIS

If information screening indicates that the proposal includes higher-than-average ground water production rates and/or potentially high-impact well(s) or transmission facilities, substantial additional analysis may be necessary. An application may be approved in two phases: (1) exploration, and (2) construction of water-production facilities. Using that approach, each phase requires separate NEPA analysis and documentation (refer to FSH 1909.15, chapters 20, 30, and 40). When the application uses existing wells, many of the evaluation procedures described here may still apply. The project applicants should be advised that obtaining approval for exploratory drilling and/or evaluation does not guarantee that construction of production phase facilities will be authorized. They should also be advised that there may be substantial mitigation measures required by the terms of a production authorization and that the scope of those measures may not be identified until the conclusion of the appropriate environmental analysis.

Where water supplies in sufficient quantities to meet the applicant's needs are located in existing wells or found through exploration, require a detailed plan to determine impacts. This plan must be site-specific and designed to identify potential impacts to NFS resources and neighboring water supplies, and must be approved before testing for impacts. In the absence of sufficient information to model impacts, an aquifer test with long-term pumping of existing and/or exploratory wells and monitoring of observation wells and surface water may be required. The purpose of the test is to evaluate the potential impacts of removing water at production levels from the well(s) under consideration. Where the proposal involves the transport of ground water pumped from nearby non-NFS lands across NFS lands, the above testing may still be required to evaluate impacts of the ground water withdrawal on NFS resources and neighboring water supplies (40 CFR 1508.25, Scope). When an injection well is proposed, a site-specific analysis of the impacts from the introduced water is required to determine potential impacts to NFS resources and neighboring water supplies. The results of testing, monitoring, and/or modeling should be shared with the appropriate State and local agencies.

EXPLORATORY DRILLING PROCEDURES

NEPA documentation must be completed, appropriate to the scale of operations, when screening indicates a reasonable likelihood of producing ground water or of injecting water without negative impacts to NFS resources or neighboring water supplies and all applicable State authorizations have been obtained. If the responsible official decides to allow exploration on NFS lands, a temporary permit may be issued for the exploration and impact-evaluation phase of the proposal. This temporary permit shall contain any conditions necessary to minimize impacts to NFS resources.

CONSTRUCTION AND PRODUCTION PERMITTING

The construction and production phase includes the construction of all infrastructure needed to pump, store, and convey water from its source to the location of use. Once a NEPA decision and all applicable State authorizations are in place, a special-use authorization is needed to occupy and use NFS lands for the purposes of constructing and operating facilities designed to produce, inject, and/or convey ground water (36 CFR 251.54 [g][5]). Refer to FSM 2711 for guidance on the type of permit or easement to issue. Refer to 36 CFR 251.56 for terms and conditions for permit issuance. Construction may be permitted separately from production. Once a permit is issued, the applicant is referred to as the holder. The Forest Service may amend the permit at any time, regardless of the length of time for which a permit is issued (FSM 2711.2). Continued monitoring of water developments is necessary to verify that their operation remains in the public interest.

MONITORING AND MITIGATION

Monitoring and/or mitigation measures necessary to ensure protection of NFS resources during the construction of water pumping, injection, storage, or transport facilities are included in annual plans of operation. Mitigation measures can include the cessation of pumping during critical times of the year or replacing water to streams and springs. If long-term monitoring detects additional or unforeseen adverse impacts to forest resources, or if mitigation measures do not adequately protect forest resources, the permit can be suspended or revoked (36 CFR 251.60[a][2][D]). To reverse or prevent a suspension, the holder shall undertake such efforts as are necessary to eliminate adverse impacts.

Case Study: Ground Water Development, Tonto National Forest, AZ

The Tonto National Forest's ground water policy evolved from experiences with ground water development projects on or adjacent to the forest. The discussion that follows briefly describes the Carlota Copper Company and Sunflower projects (fig. 2).

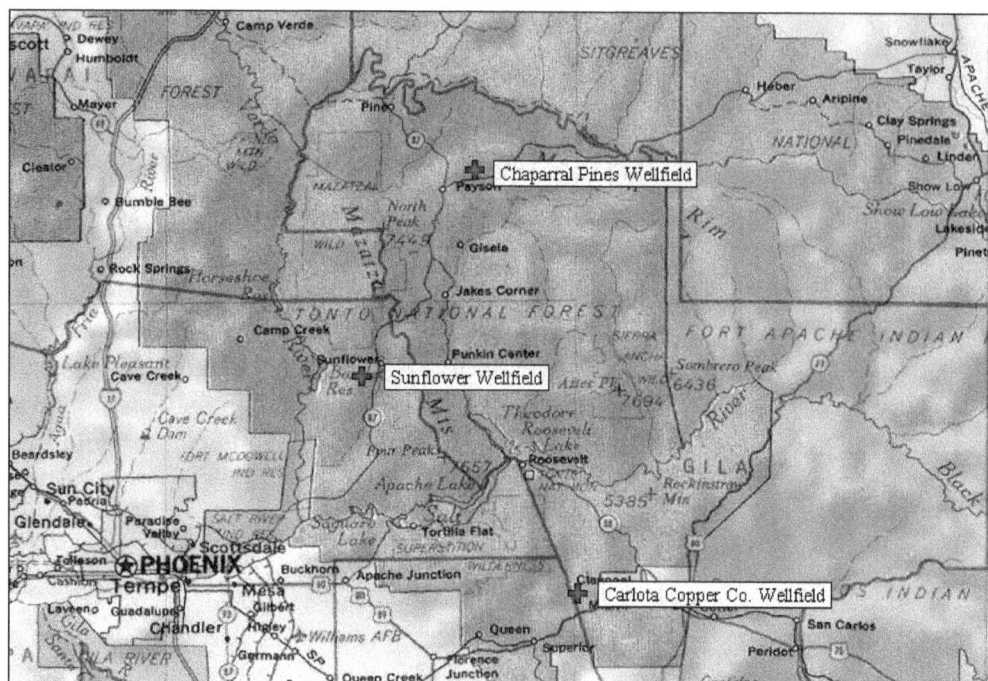

Figure 2. Ground water development project locations on the Tonto National Forest.

CARLOTA COPPER COMPANY WELLFIELD

The Carlota Copper Mine site is 6 miles west of Miami, AZ (fig. 2), at an elevation of approximately 3,700 feet above mean sea level in a rugged, mountainous, semiarid region. The Carlota Copper Company proposed to mine 100 million tons of ore from open pits over a 20-year period to produce 900 million pounds of copper. The ore would be leached with a sulfuric acid solution in a heap leach process. Predicted water requirements for the mine averaged 590 gallons per minute (gpm) with peak water requirements of 850 gpm during dry months.

The mine was proposed to be located in the Pinto Creek watershed (fig. 3), which drains into Roosevelt Lake, a major water supply reservoir for the Phoenix metropolitan area. Pinto Creek, which becomes perennial below the project area, is a valuable resource on the forest. The creek is a rare perennial stream in the Sonoran Desert and has been designated as an Aquatic Resource of National Importance by the EPA, studied for eligibility for inclusion in the National Wild and Scenic River System, nominated for unique waters status, named as one of the 10 most endangered rivers in the nation by American Rivers, and called a "jewel in the desert" by the late Senator Barry Goldwater. To protect the stream, the Tonto National Forest applied for and received an instream flow water right from the State that seeks to maintain existing median monthly flows along a 9-mile reach of the stream located approximately 4 miles below the Carlota project area. These flows range from 1 to 2.7 cubic feet per second (cfs).

Figure 3. Pinto Creek.

The mine conducted an extensive search for water and ultimately elected to use ground water from a wellfield approximately two miles downstream of the main project area in an area adjacent to the confluence of Pinto Creek and Haunted Canyon (fig. 4).

Three test wells ranging in depth from 755 feet to 1,220 feet were drilled at the site. All three wells experienced artesian flows with artesian discharge from the middle well (TW-2) flowing at 250 gpm. These wells were test pumped to evaluate the long-term yield potential of the aquifer, and the impact of pumping on surface-water resources and on water table elevations in alluvium. Well TW-2 was pumped for 25 days at a rate of 600 gpm. The monitoring network consisted of three shallow alluvial monitoring wells, four bedrock monitoring wells, weirs at two springs, and a weir or Parshall flume at two locations in Haunted Canyon and Pinto Creek.

During the 25-day pump test of TW-2, streamflow at a weir in Haunted Canyon approximately 2,300 feet south of the TW-2 well declined from approximately 45 gpm at the start of the test to 5 gpm at the end of the test (fig. 5). Flow increased progressively to approximately 27 gpm within a few days of shutting off the pump. The water level in an alluvial monitoring well in Haunted Canyon, located approximately 1,550 feet south of TW-2, declined approximately 1 foot during the 25-day test and recovered slowly following the test.

Based on these test results the Tonto National Forest sent a letter to the Arizona Department of Water Resources (ADWR) requesting an appropriability determination. In Arizona, water pumped from a well is considered to be appropriable if withdrawing that water tends to directly and appreciably reduce flow in a surface water source. ADWR concluded that the well was withdrawing appropriable water and would need a water right if it was to be

21

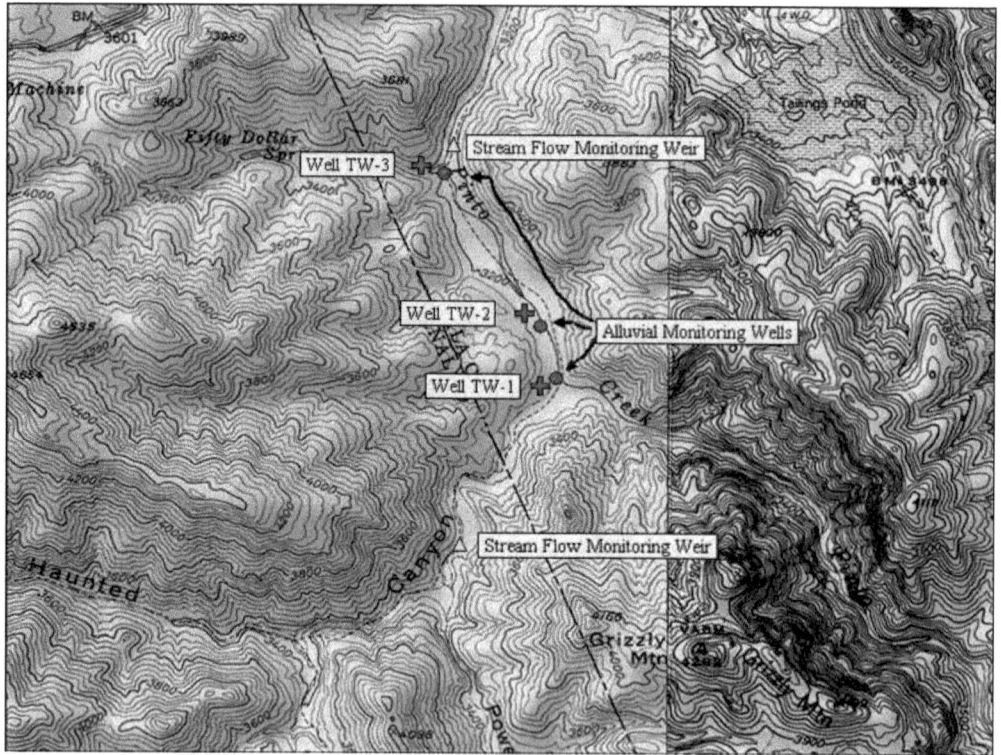

Figure 4. Map of the Carlota Wellfield and associated monitoring locations.

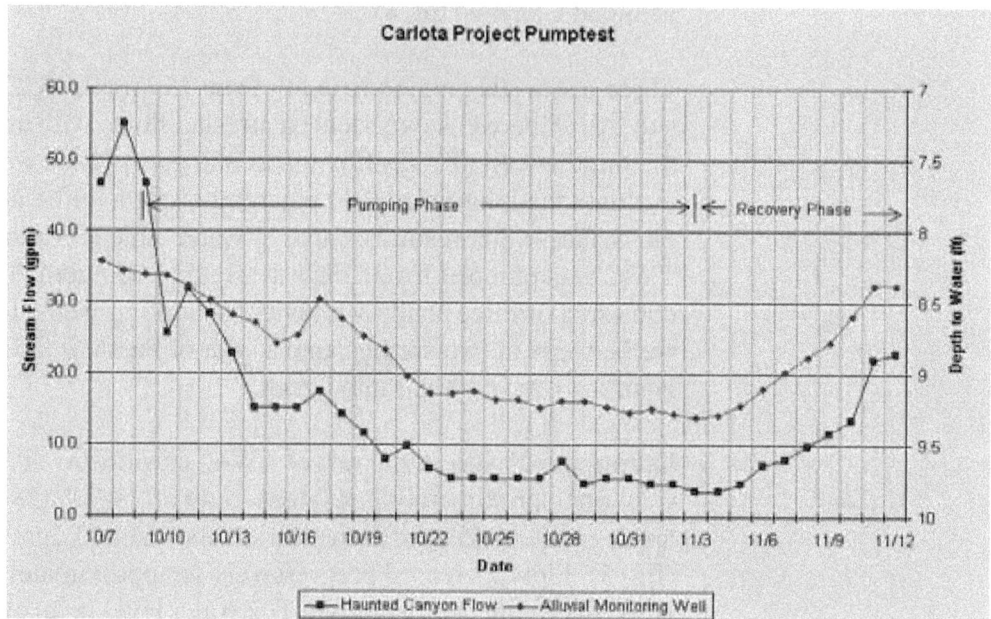

Figure 5. Hydrographs showing the decrease in stream flow and decline of the alluvial water table during the pump test.

used. The Carlota Copper Company subsequently submitted a water rights application. The Forest protested the application based on its instream flow water right downstream on Pinto Creek. The Forest negotiated a wellfield mitigation program with the mine that seeks to maintain median monthly flows in Haunted Canyon and Pinto Creek in exchange for the Forest's withdrawal of its protest.

The second ground water development project influencing the development of the Forest's ground water policy was the Sunflower Well. This well was proposed as a water supply source for upgrading a portion of State Highway 87 that carries heavy traffic from the Phoenix metropolitan area to summer recreation areas in the high country along the Mogollon Rim in north-central Arizona. Water requirements for highway construction were estimated to be about 200 gpm for compaction of fills and for dust control.

The Sunflower well was to be located on private land near Sycamore Creek (fig. 6), which has stream reaches with both intermittent and perennial flow near the well. Sycamore Creek, like Pinto Creek, is a stream with reaches of perennial flow in the Sonoran Desert. It supports valuable riparian vegetation, provides habitat for native fish, and is a popular recreation area. The Record of Decision for the Environmental Impact Statement (ROD) prepared for the highway upgrading project stated that construction water would not be withdrawn from Sycamore Creek. To evaluate the effects of the well on Sycamore Creek an aquifer test with observation wells and a streamflow monitoring flume was conducted.

The proposed production well (fig. 6) was completed to a depth of 240 feet in fractured basalt. Water rose under artesian pressure to a depth of about 20 feet below ground surface. The monitoring network consisted of four shallow observation wells in the alluvium bordering the creek, two deep observation wells in bedrock, and a Parshall flume in a perennial reach of Sycamore Creek just downstream of the well. The aquifer test was originally scheduled for 3 days with the production well pumping at an average rate of 250 gpm.

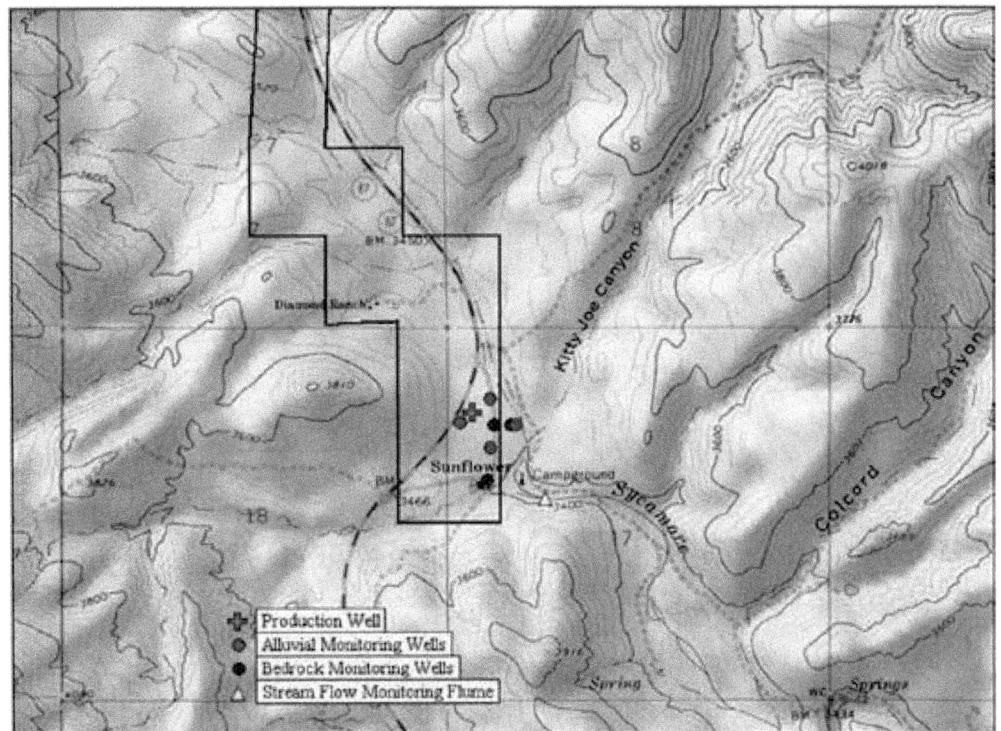

Figure 6. Map showing the Sunflower well and associated monitoring locations. The private lands are outlined.

23

Water levels in the shallow monitoring wells declined before, during, and after the test. Water levels declined at a slightly greater rate during the test. The majority of the decline is believed to be attributable to natural conditions. The impact of pumping on streamflow through the flume was dramatically different than the impact to the shallow observation wells. Prior to beginning the test, the flow rate through the flume was about 90 gpm (fig. 7). About 6 minutes after the pump in the production well was turned on, flow through the flume started to decline. Approximately 6 hours into the test, flow in Sycamore Creek declined to the point where there was no longer flow through the flume. One hour and twenty minutes after the pump was turned off, Sycamore Creek started flowing through the flume again. Two hours after the pump was turned off, flow through the flume was 37 gpm; 10 hours after turning the pump off, flow through the flume was 61 gpm.

Based on the results of this test, the contractor was not allowed to use the well for the highway upgrade project under the criteria of the ROD.

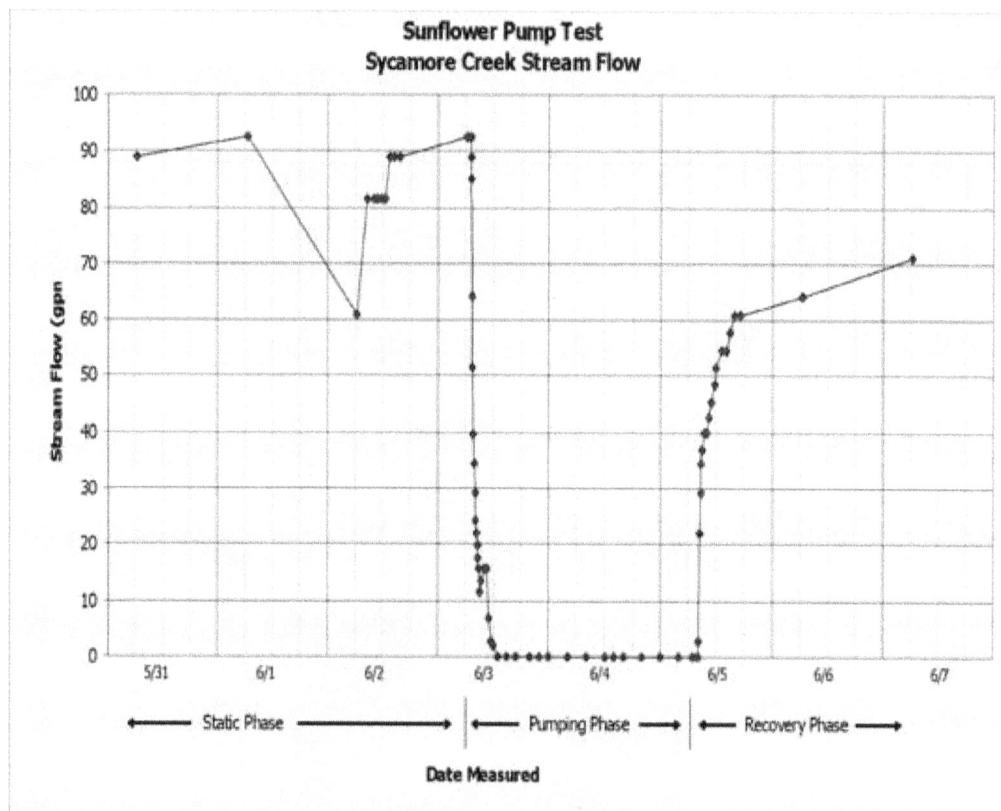

Figure 7. Hydrograph showing the decrease in stream flow during the pump test.

Stock Watering on Public Land

Springs for stock watering have been developed with little regard for the effects on ground water-dependent ecosystems that depend on springs (fig. 8). Spring development generally consists of excavation and conveying all water to a single discharge point. This type of spring development deprives the flora and fauna that depend on the spring water. A portion of the water from a spring should be allocated to protect the viability of the dependent ecosystem and the area should be fenced to eliminate trampling.

Figure 8. Example of poor management of a spring on NFS lands.

Managing Ground Water Quantity Problems

According to Galloway and others (2003), ground water management includes the engineering, economic, and political factors that affect the locations, rates, and timing of hydrological stresses to the ground water system (ground water withdrawals, artificial recharge, and so forth). These imposed stresses then affect the responses of the ground water system (ground water levels, discharge rates, and water quality), which in turn may affect streamflow rates, aquatic habitats, and other environmental conditions.

In managing withdrawals from a ground water system, it is important to understand the status of the system and the impacts of any withdrawals. To understand the status of a ground water system, basic information is needed on the geologic framework, boundary conditions, hydraulic-head distribution, water-transmitting and water-storage properties, and chemical distribution. Any quantitative analysis depends on the availability of data, the development of a conceptual model based on these data, and an understanding of the factors affecting the movement of ground water (Galloway and others 2003).

Detection and Monitoring

To monitor and evaluate changes in the ground water system, baseline conditions for the system must first be established. An inventory of existing wells or other sources of data is a first step in establishing baseline conditions. Such information may be obtained, often online, from the U.S. Geological Survey (USGS) water science center that covers the study area, or from the State engineer's or State geologist's offices. Once the status of existing data is established, areas where additional data are needed can be identified and new data can be obtained. Examples of needs may include new wells and water levels, new stream gages and stream flows, water-quality data, and water-use data. After baseline conditions are established, new data are collected from the monitoring network at a frequency appropriate for the problem. For many problems involving development of new wells or well fields, system response usually occurs quickly at first, particularly close to the new wells, then more slowly with time. Daily, or even hourly, observations may be needed close to the new wells at first. Weekly or monthly observations may be sufficient as transient effects of pumping begin to decrease.

Ground water systems are dynamic. They respond to short- and long-term changes in climate, ground water withdrawal, and land use (Taylor and Alley 2001). Monitoring of ground water conditions in response to these changes requires a monitoring network of observation wells, stream and spring gages, and meteorological and water-quality stations. Long-term, systematic measurements from such a network provide essential data needed to evaluate changes in the ground water system over time, to develop flow models and forecast trends, and to design, implement, and monitor the effectiveness of ground water management and protection programs (Taylor and Alley 2001).

Water-level measurements from observation wells (fig. 9) are the principal source of information about the hydrologic stresses acting on aquifers and how these stresses affect ground water recharge, storage, and discharge (Taylor and Alley 2001). The ideal observation network consists of wells drilled specifically for that network, as well as instrumentation to collect ancillary hydrological data such as rainfall and streamflow. Budgetary constraints may require the use of existing wells for all or part of the network, but care must be taken in the selection of existing wells for use in the network to enable correct interpretation of the data.

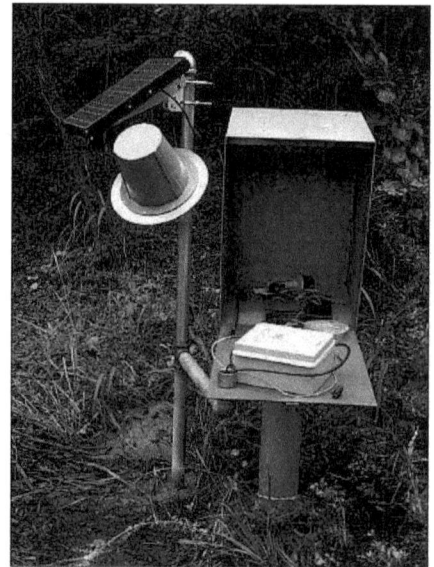

Figure 9. Water well instrumented for water-level data collection, satellite transmission, and real-time reporting on the Internet. (Photo by William Cunningham, USGS Circular 1217, 35.)

Particularly during low-flow conditions, measurement of stream and spring flow may also provide insights about the response of the ground water system to changing conditions. For example, decreasing flow in streams or from springs, despite average or above-average precipitation, may indicate adverse ground water response to pumping. Changes in ground- and surface-water quality may also be indicative of ground water responses to both natural and manmade system changes.

Monitoring of ground water use can be a critical component of a monitoring network. Ideally, pumping wells in the area of interest are equipped with meters to record the amount pumped, and these values are collected and documented.

Where metering is lacking, electric-power-consumption records or rated capacity of the well can be used as surrogates for actual pumpage data. Well-completion information is also crucial to understanding the impacts of ground water withdrawals. If undesirable effects because of new stresses on the ground water system are detected, informed management decisions can then be made. Some of the management options are described in the next sections. These options may be best evaluated through the use of numerical models, particularly in areas of complex hydrogeology.

Limiting Withdrawals

If groundwater withdrawals or springflow diversions are negatively affecting the ground water system, one management option is to limit the withdrawals to an established safe yield. Another is to specify the location of the new wells to minimize negative impacts (fig. 10). Although water levels near the pumping wells may recover relatively quickly, water-level declines (drawdown) may still occur at larger distances from the wells until new equilibrium conditions become established.

Increasing Recharge

Increasing recharge to the ground water system through the use of infiltration ponds, streamflow diversions, or injection wells can help to offset the effects of additional pumping by establishing new equilibrium conditions (fig. 11) (Galloway and others 2003). These methods, however, often require a high degree of maintenance to the recharge system facilities and equipment to keep it operating efficiently and may not result in a net positive effect on the targeted resources.

Conjunctive Management of Surface Water and Ground Water

Conjunctive use is the combined use of surface and ground water to optimize resource use and minimize adverse effects. Conjunctive use is often a cost-effective way to mitigate the negative impacts of excessive use of either resource (Galloway and others 2003). Moreover, the likelihood of more frequent surface water shortages, as urban and environmental demands on existing supplies increase, accentuates the differences in reliability between surface water and local ground water supplies. Conjunctive use can increase the yield of a water system by using existing resources more efficiently. By coordinating the use of surface- and ground water supplies at different times, in response to varying conditions, the overall use of water supplies can be improved in the short term and better sustained in the long term. Conjunctive use also can address ground water depletion problems, and help ensure the adequacy of ground water resources for periods of drought and surface water shortages.

Case Study: Conjunctive Use of Ground Water and Surface Water, Tonto National Forest, AZ

The Arizona Department of Transportation (ADOT) proposed to upgrade a 52-mile stretch of Highway 260 from Payson to Heber, AZ. Portions of the highway were to be realigned and the entire stretch was to be upgraded from two to four lanes. Construction would be completed in segments and plans called for an 8-year construction period.

Water was required for embankment compaction, dust control, and paving. Peak water requirements were estimated at approximately 180 gpm. ADOT investigated several water supply sources, including both ground- and surface-water supplies (fig. 12).

27

Natural conditions

Precipitation

Evaporation

Water table

Recharge

Riparian zone

Stream

Ground-water flow

Confining unit

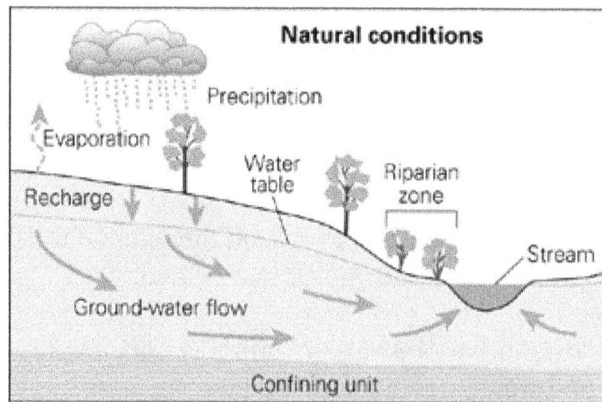

Under natural conditions, water that is recharged to the ground-water system flows toward the stream, where it discharges.

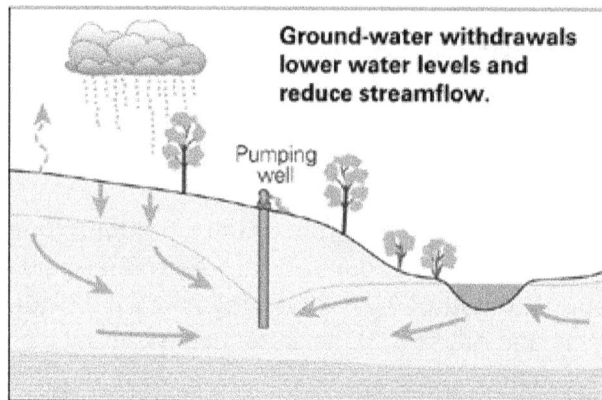

Ground-water withdrawals lower water levels and reduce streamflow.

Pumping well

Ground-water withdrawals at the well lower the water table and alter the direction of ground-water flow. Some water that flowed to the stream is now discharged by the well, and some of the streamflow may be drawn into the aquifer. Both processes reduce the amount of streamflow.

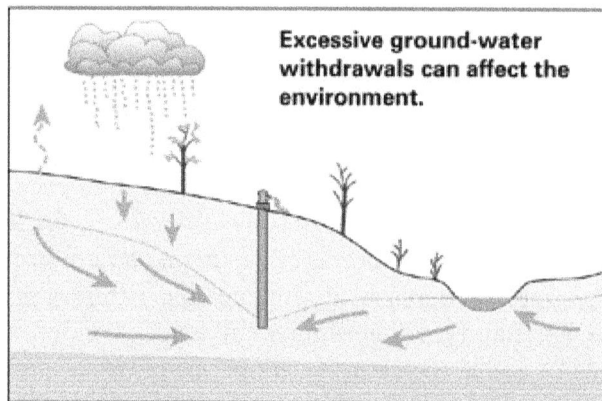

Excessive ground-water withdrawals can affect the environment.

Water-level declines may affect the environment for plants and animals. For example, plants in the riparian zone that grew because of the close proximity of the water table to the land surface may not survive as the depth to water increases. The environment for fish and other aquatic species also may be altered as the stream level drops.

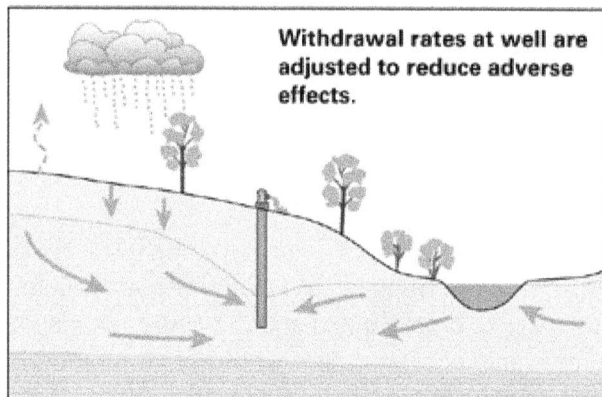

Withdrawal rates at well are adjusted to reduce adverse effects.

Various approaches could be implemented to reduce adverse effects during the critical summer months when streamflow is naturally low, including reducing withdrawal rates either seasonally or annually or artificial recharge during wet periods. A flow model of the ground-water system commonly is an integral part of evaluating these types of options.

Figure 10. Ground-water development near a stream can reduce streamflow and harm riparian vegetation, requiring management decisions to restore the system (last panel) (Galloway and others 2003).

28

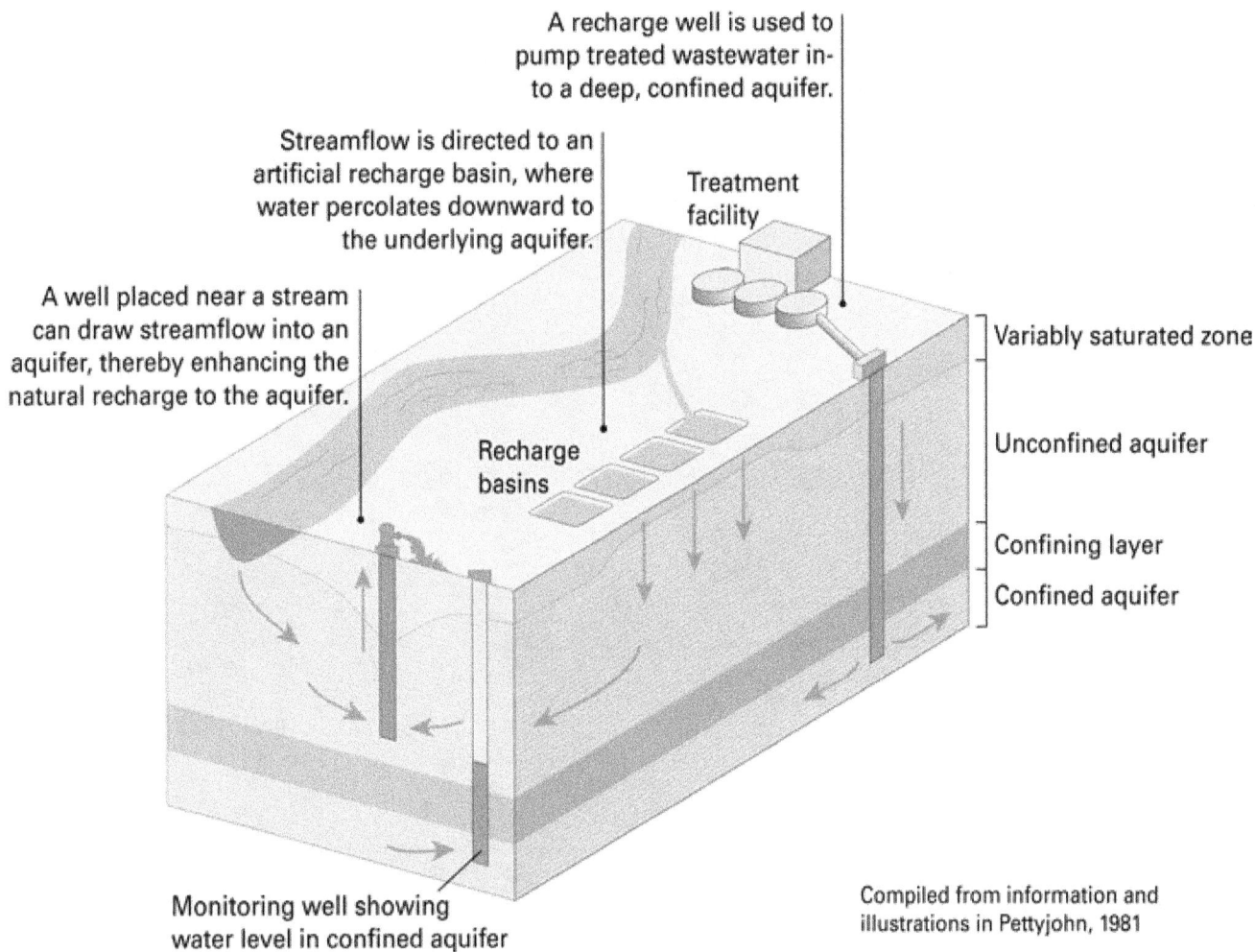

A recharge well is used to pump treated wastewater into a deep, confined aquifer.

Streamflow is directed to an artificial recharge basin, where water percolates downward to the underlying aquifer.

Treatment facility

A well placed near a stream can draw streamflow into an aquifer, thereby enhancing the natural recharge to the aquifer.

Variably saturated zone

Unconfined aquifer

Recharge basins

Confining layer

Confined aquifer

Monitoring well showing water level in confined aquifer

Compiled from information and illustrations in Pettyjohn, 1981

Figure 11. Schematic of artificial recharge processes (Galloway and others 2003).

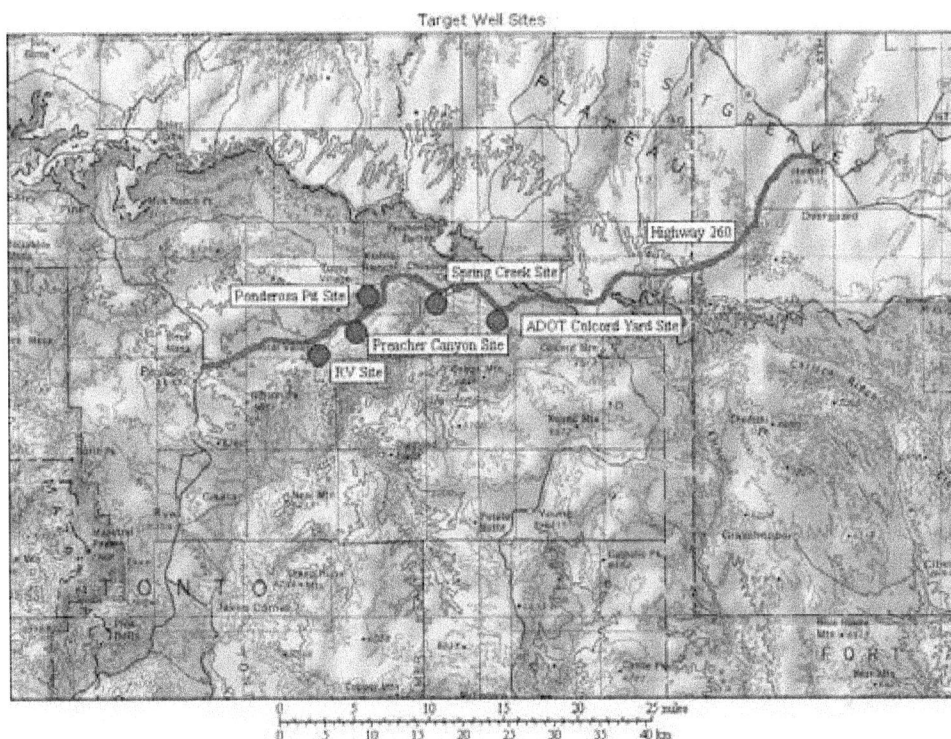

Target Well Sites

Highway 260

Spring Creek Site

Ponderosa Pit Site

ADOT Colcord Yard Site

Preacher Canyon Site

RV Site

Figure 12. Highway corridor and the location of ground-water sources investigated.

29

A recreational vehicle (RV) site on NFS land was selected for further study after investigations of the other sites suggested insufficient water supplies were available or the likelihood of adverse environmental impacts was too great. This site was located within a half mile of a stream with reaches of both perennial and intermittent flow (Little Green Valley Creek), within a mile of two private land subdivisions that relied on shallow wells for their water supply, and within a mile of three springs, two on NFS land and one on private land (fig. 13).

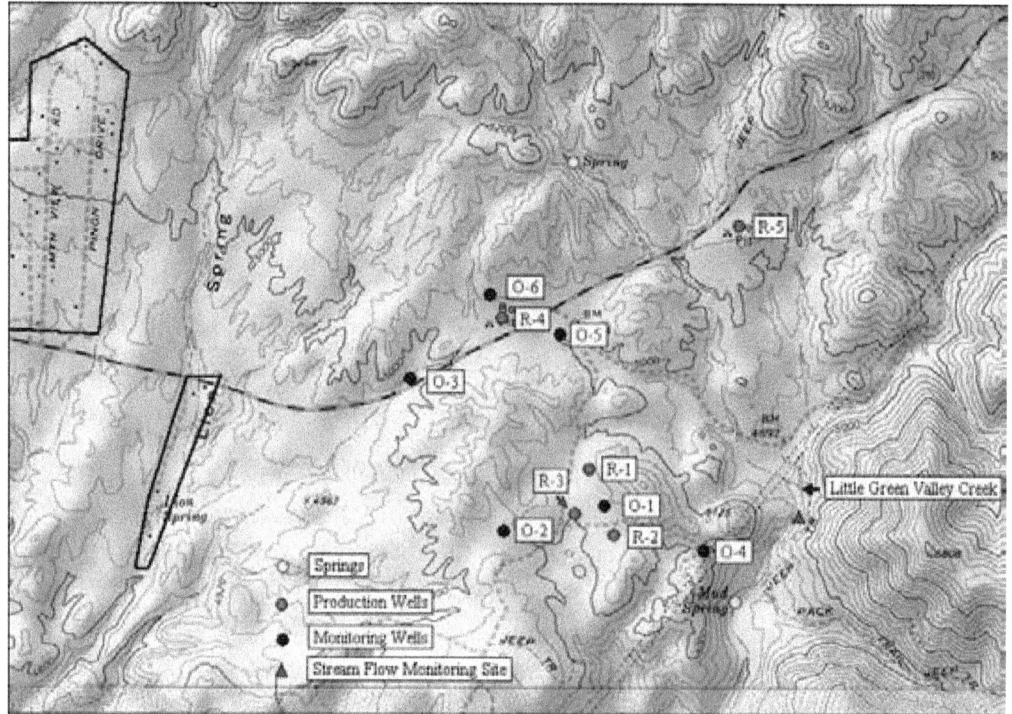

Figure 13. Locations of the production wells and associated monitoring locations. The private lands are outlined.

Before permitting ADOT to use this well field, the Forest Service required a long-term pump test (38 days) to assess impacts to the springs, Little Green Valley Creek, and wells on private land. Several wells were drilled at the site and selected wells were completed as potential production wells. Monitoring wells were installed at strategic locations around the production wells to monitor changes in water levels during the extended aquifer test. In addition, weirs were installed at springs on NFS land and in the channel of Little Green Valley Creek to monitor changes in flow during the test.

Flow in one of the springs stopped during the test, and flow in the other spring and in Little Green Valley Creek appeared to be declining, but impacts were unclear because of storm events during the pump test. Water table elevations declined in most production and monitoring wells during the test and recovered to varying degrees following the test. Aquifer test data indicated that a fracture-flow model should be used to analyze the data. A site-specific fracture-flow model was subsequently developed for the well field and calibrated against the pump test data. The model was then run to simulate well field operations for the life of the highway construction project (8 years). Model simulations suggested well field operations had the potential to lower water table elevations in the wells on private land.

To mitigate impacts to NFS resources and to wells on private land, the Forest Service required ADOT to discharge 10 gpm into Little Green Valley Creek and to install two monitoring wells next to the private land. When water table elevations in these monitoring wells drop more than 10 feet, ADOT is required to cease pumping.

ADOT was concerned that the aquifer at the RV site would not be able to supply the water needed for highway construction without dropping below the 10-foot mitigation threshold. After consultations with the Forest Service, the Arizona Game and Fish Department, the U.S. Fish and Wildlife Service, the ADWR, and the Salt River Project, a program to divert and use surface water during winter months and to inject surface waters into the RV site aquifer during the same time period was developed to prevent adverse water-table drawdowns in wells on private land. The well-field injection program consists of three wells with the capacity to inject from 225 to 450 gpm during periods when surface water is available for injection. The intent is to use the aquifer to store surface waters when surface waters are available, to restore water-table elevations in the aquifer, and then to withdraw stored water during periods when surface water is not available. Note that water-quality issues apparently were not a significant consideration in this case; careful consideration should be given to water-quality impacts on the aquifer system(s) used for storage of surface water prior to approval of any ground water-storage proposal.

ADOT is allowed to divert surface water from Tonto Creek, a tributary to Roosevelt Lake, from December to April, when riparian vegetation along Tonto Creek is dormant. Flows are allowed to be diverted from Tonto Creek when streamflows exceed threshold rates at two gages on Tonto Creek. No more than 10 percent of the flow in Tonto Creek up to a maximum of 1 cubic foot per second of stream flow can be diverted. Figure 14 displays the mitigation and conjunctive use measures incorporated into this project.

ADOT must repay the Salt River Project for the water diverted from Tonto Creek with water from the Central Arizona Project Canal. In Arizona, the ADWR reviews exchange agreements such as these to ensure that other water rights holders will not be injured.

Dowsing

In many parts of the world, particularly in rural areas, water-well locations may be determined by using the services of a "dowser" or "water witch." Dowsing is the action of a person who uses a rod, stick, or other device ("dowsing rod" or "divining rod") to locate ground water, metallic ores, oil, or other objects that may be hidden from sight. Dowsers may practice their art either in the field or over a map of the area of interest. The most common divining rods consist of either a forked stick or a pair of metallic rods. When the dowser crosses the

31

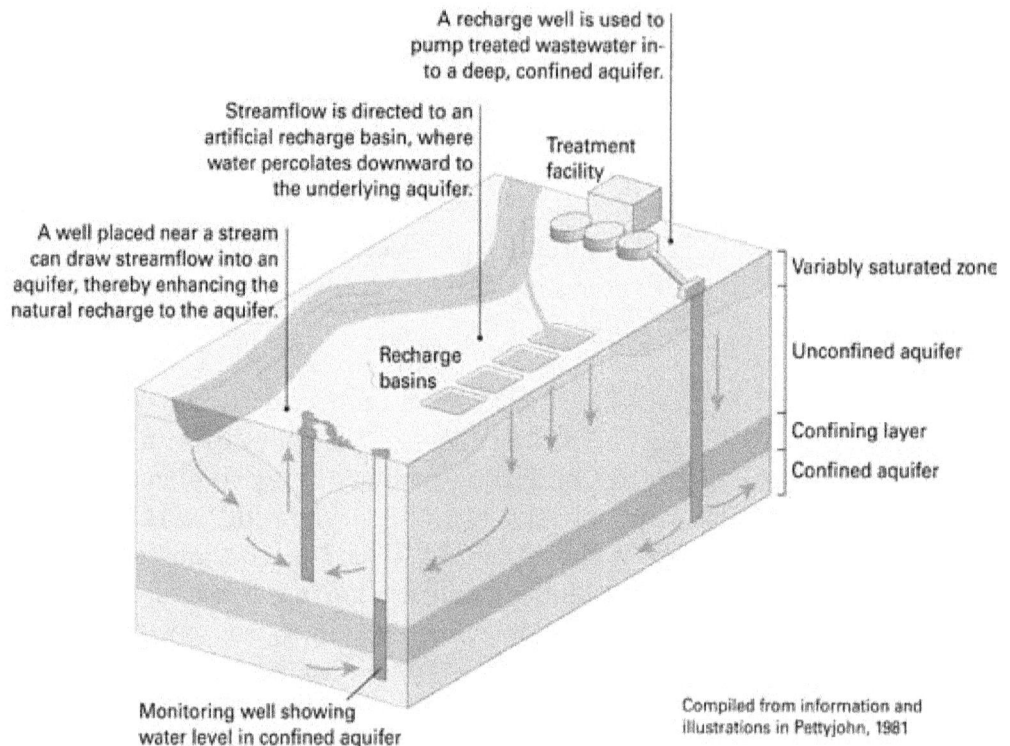

A recharge well is used to pump treated wastewater into a deep, confined aquifer.

Streamflow is directed to an artificial recharge basin, where water percolates downward to the underlying aquifer.

Treatment facility

A well placed near a stream can draw streamflow into an aquifer, thereby enhancing the natural recharge to the aquifer.

Variably saturated zone

Unconfined aquifer

Recharge basins

Confining layer

Confined aquifer

Monitoring well showing water level in confined aquifer

Compiled from information and illustrations in Pettyjohn, 1981

Figure 14. Mitigation and conjunctive use measures incorporated into the project. The private lands are outlined.

target (either in the field or over the location on a map) the forked stick bends downward or the pair of metal rods crosses. Many dowsers believe that water occurs in underground streams or rivers. Although such features are known to occur in Karst areas, they are relatively rare.

Dowsing has been practiced for hundreds, if not thousands, of years. Dowsers claim a high success rate, and many anecdotes of successful dowsing can be found; however, when subjected to scientifically controlled tests, the success rates of dowsers are no better than random chance (Carroll 2001). The natural explanation of the dowser's success is that in many areas water would be hard to miss. In a region of adequate rainfall and favorable geology, it is difficult not to drill and find water (U.S. Geological Survey 1993). In fact, some water exists under the Earth's surface almost everywhere. To accurately estimate the depth, quantity, and quality of ground water, a number of techniques must be used. Hydrological, geological, and geophysical knowledge is needed to determine the depths and extents of the different water-bearing strata and the quantity and quality of water found in each.

In response to many inquiries about dowsing, the USGS published a report on the subject in 1917, which was reprinted several times because of its popularity. They advised people "not to expend any money for the services of a 'water witch' or for the use or purchase of any machine or instrument devised for locating underground water or other minerals." Subsequent reprints (Ellis 1938), however, distinguished geophysical methods and equipment, which are commonly used to assist hydrologists in their search for ground water and minerals, from these types of "water finders." **Federal employees should never expend public funds on the services of a dowser.**

Ground Water Quality

Protection and management of ground water resources includes the establishment and implementation of water-quality standards that are designed to (1) protect public health, (2) maintain legally established designated uses, and (3) minimize impacts to ground water-dependent ecosystems.

One definition of water quality consists of the biological, chemical, and physical conditions of a water. Contamination can be defined as the introduction of substances into the hydrological environment that can adversely affect water quality as a result of human activities. Pollution then occurs when contaminant concentrations attain objectionable levels (in excess of applicable standards, health advisories, action limits, and so forth).

Certain land uses are known to cause ground water contamination. Specific types of contaminants are associated with specific types of land uses and industries. The Office of Technology Assessment of the US Congress (1984) identified the following six categories of major sources of ground water contamination:

1. Sources designed to discharge substances—septic tanks, injection wells, land application.
2. Sources designed to store, treat, or dispose of substances—landfills, surface impoundments, mine waste, storage tanks.
3. Sources designed to retain substances during transport—pipelines, material transport and transfer.
4. Sources discharging substances as a consequence of other planned activities—irrigation, pesticide and fertilizer application, road salt, urban runoff, mine drainage.
5. Sources providing a conduit for contaminated water to enter aquifers—wells, construction excavation.
6. Naturally occurring sources whose discharges are created or enhanced by human activity—ground water/surface-water interaction, natural leaching, saltwater intrusion.

Ground water quality is protected by Federal, State, local and tribal governments through rules and regulations aimed at managing these categories of contaminant sources. During the 1990s the EPA and State, local and tribal governments developed ground water protection strategies aimed at preventing ground water contamination. These strategies focus on proactive measures, including education, source-water protection, and utilization of public health authorities to prevent ground water contamination. Also, during the past 10 years the watershed management approach has proven to be effective as a way to manage water resources, including ground water.

Designated uses of water that are protected against water-quality degradation include domestic, municipal, agricultural, and industrial supply; power generation; recreation; aesthetic enjoyment; navigation; and preservation and enhancement of fish, wildlife, and other aquatic resources or preserves.

Application of Water Quality Regulations to Ground Water

Water-quality standards that are applied to ground water have been established by the EPA and State and tribal governments as authorized under the Safe Drinking Water Act (SDWA), the Clean Water Act (CWA), and State and tribal laws and regulations. Standards are derived for constituents that may be harmful to human health or the environment and for constituents that affect other designated uses. The promulgated values established for individual constituents are often based on toxicological studies. Applicable standards for a particular aquifer or ground water system are determined based on State requirements that may include ground water classification systems, ground water cleanup goals, or ground water discharge permit requirements. Most States also have primacy for enforcing the SDWA water-quality standards for drinking water systems. The EPA has direct implementation authority for Indian reservations and other selected lands.

Water-quality standards promulgated under the SDWA apply to public-water systems as defined in the SDWA. A PWS is a system for the provision to the public of water for human consumption through pipes or other constructed conveyances, if such system has at least 15 service connections or regularly serves at least 25 individuals. Numeric standards include the Maximum Contaminant Levels ("MCLs") and Secondary Maximum Contaminant Levels for public drinking-water supplies as established by the SDWA regulations. In some States, some numeric standards are set based on human-health risk assessment levels. In other States, the standards for some potential toxic pollutants, primarily pesticides, are set at laboratory detection limits (nondetectable levels).

Degradation has been defined as a change in water quality that lowers the quality of high-quality waters for a particular parameter. States determine whether a proposed activity may cause water-quality degradation based on information submitted by an applicant. Contaminants other than carcinogens are generally regulated under an anti-degradation policy for most natural waters. That policy allows for an increase in concentration of a contaminant in ground water, but does not allow for a standard to be exceeded. In at least one State (Montana), non-degradation rules apply to any activity resulting in a new or increased source that may cause water-quality degradation because of carcinogens.

Some States have applied some existing surface water-quality standards to ground water through statutes or rules administered under State ground water protection programs. Some States have established preventive action limits as an early warning of the presence of pollution before beneficial uses are adversely affected. The purpose is to achieve more stringent protection for higher quality ground water. For example, in Utah preventive levels are set in ground water discharge permits. The levels are set at 10 to 50 percent of the standard; if pollutant concentrations are detected that exceed the protection levels, then the source of the problem must be evaluated for potential correction.

Ground Water Classification

In many States, ground water classification schemes are used to help determine which standards may be applicable to selected aquifers or ground water beneath certain areas. Schemes are typically based on the current and/or potential future beneficial uses of the resource and existing water quality. Examples are drinking-water use, agricultural use, and industrial use. Boundary criteria for the classified areas may be physically based or otherwise determined, such as an aquifer or aquifer zone, a watershed, or a permitted discharge facility. Total dissolved solids (TDS) concentrations and specific conductance are commonly used to define the various classes of ground water (table 1). The classifications are used to establish in situ water-quality standards for implementing ground water protection programs, permitting discharges to ground water, and setting cleanup goals at contaminated sites.

Table 1. A common ground water classification system based on TDS concentration (mg/L).

TDS*	Classification
Less than 1,000	Fresh
1,000 to 3,000	Slightly brackish
3,000 to 10,000	Brackish
10,000 to 50,000	Saline
More than 50,000	Brine

*As a point of reference, the TDS concentration in seawater is approximately 35,000 mg/L.

Uncontaminated fresh ground water is generally suitable for human consumption, for livestock and other agricultural uses, and for most industrial uses. Slightly brackish water may not be suitable for those uses, depending on the relative amounts of the various major ions and trace elements. Brackish, saline, and brine waters are never suitable for human consumption. In some cases, brackish water can be used for livestock, but saline and brine waters never can (National Research Council 1981).

Most State ground water classification schemes are based on TDS. For example, in North Dakota and South Dakota ground water is classified as "potentially suitable" for drinking-water use if the TDS level is less than 10,000 parts per million (ppm), and suitable for no specific beneficial uses if the TDS level exceeds 10,000 ppm. In North Dakota, a second classification system based on aquifer sensitivity is also used to prioritize ground water monitoring to track the occurrence of agricultural chemicals and to help determine State activities in the Underground Injection Control (UIC) Class V Program.

In Colorado, a public hearing process in front of the State Water Quality Control Commission is required to classify specific ground water to set the applicable ground water-quality standards for protection and regulatory purposes. The classification scheme includes (1) quality for domestic use, (2) quality for agricultural use, (3) surface water-quality protection, (4) potentially usable quality, and (5) limited use and quality.

Ground Water Discharge Permits

In many States, facilities that discharge waste or pollutants directly or indirectly into ground water (other than those regulated under the UIC or National Pollution Discharge Elimination System [NPDES]) may be required to apply for a ground water discharge permit. The goal of this program is to allow economic development while maintaining ground water quality; in most cases, a limited zone of pollution (mixing zone) is permitted and quarterly compliance monitoring is instituted by the permittee. Ground water-quality standards and/or protection levels are used to determine the discharge requirements.

Facilities required to apply for ground water discharge permits are identified in the regulations. For example, Colorado requires all facilities under certain standard industrial classifications to apply for permits and some of these facilities are covered under a general permit for the UIC program. In Utah, facilities that pose little or no threat to ground water quality or that are permitted by other State ground water protection programs (such as septic tanks and discharges from permitted RCRA units) receive a permit by rule.

Generally, a facility needing a permit submits information to the State that describes the extent and quality of the ground water, the volume and composition of the discharge, how the discharge will be controlled or treated to meet standards and/or protection levels, and proposed inspection and monitoring plans to ensure compliance with the terms of the permit. In some States, the permitting process requires a contingency plan to bring the facility into compliance in the event of a significant release of contaminants to ground water from the facility. In South Dakota, a discharge plan includes three permits: (1) a ground water-quality variance, (2) a facility construction permit, and (3) a discharge permit from the Ground Water Quality Program.

Total Maximum Daily Loads

Regulations issued by EPA in 1985 and 1992, pursuant to the CWA, require the quantification of specific pollutants that impair the quality of surface-water bodies. The regulations require that total maximum daily loads (TMDLs) be established for selected streams or stream reaches that exceed water-quality standards because of contaminant loading. States typically have primacy for the TMDL program under their water-quality programs. The EPA is required to determine TMDLs if a State does not do so. While the TMDL programs typically focus on point-source loads to surface water, loading from ground water should be considered. It is not a requirement, however.

Water-quality Databases

The USGS, EPA, and many States maintain a number of water-related databases that contain water-quality information. Some of these systems are available on the Internet; however, access to some of them may necessitate a direct request to the right agency.

The USGS National Water Information System stores data on surface water stages and flows, ground water elevations, and water quality (http://waterdata.usgs.gov/nwis). The USGS National Water Quality Assessment Program is a commonly used source of information on ground water-quality available in the United States today. Under this program, USGS collects water-quality data in 60 special study regions of the country, conducts retrospective analyses of existing data (such as State data), and prepares national-scale syntheses of the results. Information from this program is also available separately on the Internet (http://water.usgs.gov//nawqa/data.html).

The EPA maintains the nationwide STORET database for water information, including water quality (http://www.epa.gov/storet/dbtop.html). The EPA is also developing a National Contaminant Occurrence Database (NCOD) to track contaminants in ground and surface sources of drinking water. The NCOD will aid in the identification and selection of contaminants for future drinking-water regulations, support regulation development or other appropriate actions, and assist in the review of existing regulations for possible modification. The NCOD will incorporate data of documented quality from existing Federal databases on regulated and unregulated physical, chemical, microbial, and radiological contaminants, as well as other contaminants that are known or are likely to occur in the source and finished waters of PWSs of the United States and its territories.

Ground Water-dependent Ecosystems

Ground water-dependent ecosystems are communities of plants, animals and other organisms whose extent and life processes depend on ground water. The following are examples of some ecosystems that may depend on ground water:

- Wetlands in areas of ground water discharge or shallow water table.
- Terrestrial vegetation and fauna, in areas with a shallow water table or in riparian zones.
- Aquatic ecosystems in ground water-fed streams and lakes.
- Cave and karst systems.
- Aquifer systems.
- Springs and seeps.

Ecological resources include sensitive fish, wildlife, plants, and habitats that are at risk from exposure to ground water contaminants or ground water depletion. Some examples are breeding, spawning, and nesting areas; early life-stage concentration and nursery areas; wintering or migratory areas; rare, threatened, and endangered species locations; and other types of concentrated-population or sensitive areas. These areas contain ecological resources that potentially are highly susceptible to permanent or long-term environmental damage from contaminated or depleted ground water.

Ground water-dependent ecosystems vary dramatically in how extensively they depend on ground water, from being entirely dependent to having occasional dependence. Unique ecosystems that depend on ground water, fens for example, can be entirely dependent on ground water, which makes them very vulnerable to local changes in ground water conditions. Ground water extraction by humans modifies the pre-existing hydrologic cycle. It can lower ground water levels and alter the natural variability of these levels. The result can be alteration of the timing, availability, and volume of ground water flow to dependent ecosystems.

Ground water-dependent ecosystems can be threatened by contamination and extraction. Particular threats include urban development, contamination from industry, intensive irrigation, clearing of vegetation, mining, and filling or draining of wetlands. In some caves and peatlands, scientific research into past environments relies upon the fossil record, and fluctuating water levels and changes in water quality can destroy this record.

The role ground water plays in controlling ecosystems on public land is poorly understood. Little information exists in the literature on this topic. Hatton and Evans (1998) provide an excellent discussion of ground water-dependent ecosystems in Australia, and the U.S. Bureau of Land Management (U.S. BLM 2001) discusses the occurrence, ecological values, and management of springs in the Western United States. Unseen and sometimes poorly understood, ground water nonetheless fundamentally controls the health of many ecosystems.

Ground water-dependent ecosystems have many values, including the following:

- *Water-quality benefits.* Microfauna in ground water help cleanup contaminants and may play an important, but not yet fully understood, role in maintaining the health of surface waters.
- *Biodiversity value.* Many species depend on habitats maintained by ground water discharge. They add to the ecological diversity of a region and can be indicators of the overall biological health of a system. Some plants and animals that depend on ground water are rare, unique, or threatened. The ecosystems in aquifers and caves may be among the oldest surviving on earth. They can be connected to other non-ground water-dependent ecosystems and thus integrated into many broader regional ecosystems.

- *Archeological and social value.* Some sites, such as springs, may have cultural significance, especially for Native Americans, and can have csocial, esthetic, and economic values.

Types of Ground Water-dependent Ecosystems

Terrestrial Vegetation and Fauna

Shallow ground water can support terrestrial vegetation, such as forests and woodlands, either permanently or seasonally (Baird and Wilby 1999). Examples occur in riparian areas along streams (Hayashi and Rosenberry 2002) and in upland areas that support forested wetland environments. Phreatophytes are plants whose roots generally extend downward to the water table and are common in these high-water-table areas. Some fauna depend on this vegetation and therefore indirectly depend on ground water. Terrestrial vegetation may depend to varying degrees on the diffuse discharge of shallow ground water, either to sustain transpiration and growth through a dry season or for the maintenance of perennially lush ecosystems in otherwise arid environments. Ground water-dependent terrestrial plant communities provide habitat for a variety of terrestrial, aquatic, and marine animals (U.S. BLM 2001), which by extension must also be considered ground water dependent. Some species are quite restricted to these habitats. For example, in Montana northern leopard frogs occur in fewer than six ponds or sloughs in the Flathead Lake watershed, and the northern bog lemming is known only from one fen complex in the Stillwater River watershed (Greenlee 1998).

An additional group of ground water-dependent fauna (including humans) rely on ground water as a source of drinking water. Ground water, as river baseflow or discharge to springs, is an important source of water across much of the country, particularly in the Southwestern United States and other areas with semiarid climate. Its significance is greater for larger mammals and birds, as many smaller animals can obtain most of their water requirements from other sources.

Ranchers in the West have made extensive use of ground water to supply drinking water to grazing stock. In addition to watering stock, ground water is also used by native fauna. Provision of water has allowed larger populations of both wildlife and pest animals to be sustained than would otherwise be the case. Ground water-dependent vegetation and wetlands may be used by terrestrial fauna as drought refuges. Access to ground water allows the vegetation to maintain its condition and normal phenology (for example, nectar production, new foliage initiation, seeding). Populations of some birds and mammals retreat to these areas during drought and then recolonize drier parts of the landscape following recovery. The long-term survival of such animal populations relies on maintaining the vegetation communities and ensuring that their water requirements are met.

Ecosystems in Streams and Lakes Fed by Ground Water

This category of ecosystem includes many ecosystems that are dependent on ground water-derived baseflow in streams and rivers (Gilbert and others 1998). Baseflow is that part of streamflow derived from ground water discharge and bank storage. River flow is often maintained largely by ground water, which provides baseflow long after rainfall or snow melt runoff ceases. On average, up to 40 percent of the flow of many rivers is estimated to be made up of ground water-fed baseflow. The baseflow typically emerges as springs or as diffuse flow from sediments underlying the river and banks. This water may be crucial for in-river and near-river ecosystems (Stanford and Gonser 1998). Localized areas of ground water discharge have a largely stable temperature and provide thermal refuges for fish in both winter and summer (Hayashi and Rosenberry 2002). Ground water also influences the spawning behavior of some fish. Reducing the baseflow to ground water-fed rivers could reduce upwelling or dry out riffles and reduce spawning success.

The ambient ground water flux is likely to be the key attribute influencing a surface-water ecosystem's dependency on ground water. The ground water level in riverine aquifers is important for maintaining a hydraulic gradient towards the stream that supports the necessary discharge flux. Sufficient discharge of ground water is needed to maintain the level of flow required by the various ecosystem components. Contamination of riverine aquifers by nutrients, pesticides, or other contaminants may adversely affect dependent ecosystems in baseflow-dominated streams.

Lakes, both natural and human made, can have complex ground water flow systems (Fetter 2000). Lakes interact with ground water in one of three basic ways: (1) some receive ground water inflow throughout their entire bed; (2) some have seepage loss to ground water throughout their entire bed; and (3) others, perhaps most, receive ground water inflow through part of their bed and have seepage loss to ground water through other parts (Winter and others 1998). Changes in flow patterns to lakes as a result of pumping may alter the natural fluxes to lakes of key constituents, such as nutrients. As a result, the distribution and composition of lake biota may be altered. Conversely, lakes perched well above local ground water year around may be immune to depletion of the underlying ground water system.

The chemistry of ground water and the direction and magnitude of exchange with surface water significantly affect the input of dissolved chemicals to lakes (Hayashi and Rosenberry 2002). In fact, ground water can be the principal source of dissolved chemicals to a lake, even in cases where ground water discharge is a small component of a lake's water budget. The importance of ground water is accentuated for dilute lakes (low TDS concentration), such as those in mountainous regions that rely on ground water as their primary source of dissolved solids and nutrients. In addition, a considerable proportion of the buffering capacity in many lakes is because of acid neutralizing capacity (ANC) contributed by influent ground water. ANC is particularly important for soft water lakes because of their extreme sensitivity to the adverse effects of acidic atmospheric deposition.

The transport of nutrients by ground water can be a significant source of water-quality degradation in lakes. Major sources of nutrient enrichment are inadequately designed and maintained household septic systems and nonpoint pollution sources, such as construction-project and agricultural runoff. The Lake Tahoe Basin Framework Study Groundwater Evaluation (U.S. Army Corps of Engineers 2003) was designed to enhance the understanding of the role ground water plays in eutrophication processes that reduce lake clarity. The study revealed that ground water contributed 12 percent of the nitrogen loading and 15 percent of the phosphorous loading to Lake Tahoe. While best management practices in the Lake Tahoe Basin represent an important step toward improving lake clarity, BMPs do not always take into account effects on ground water of either the original practice or the BMP itself.

Hyporheic and Hypolentic Zones

The interface between saturated ground water and surface water in streams and rivers is a zone of active mixing and interchange between the two and is known as the hyporheic zone (Jones and Mulholland 2000, Stanford and Ward 1988, 1993). In mountain streams with typical pool-and-riffle organization, ground water enters streams most readily at the upstream end of deep pools, and conversely, surface water moves into the subsurface beneath and to the sides of riffles (Harvey and Bencala 1993). The hyporheic zone is generally confined to the near stream area; however, in large alluvial or glacial outwash valleys (for example, Flathead River, MT) this zone may extend hundreds of feet away from the river channel. Hyporheic zones can be important for aquatic life (Gilbert and others 1998, Stanford and Ward 1993). In both gaining and losing streams, water and dissolved chemicals can move repeatedly over short distances between the stream and the shallow subsurface below the streambed. The spawning success of fish may be greater where flow from the stream brings oxygen into contact with eggs that were deposited within the coarse bottom sediment or where stream temperatures are controlled by ground water inflow. Upwelling of ground water provides stream organisms with nutrients, while downwelling stream water provides dissolved oxygen and organic matter to microbes and invertebrates in the hyporheic zone. This exchange zone is an important habitat for many invertebrates, and a refuge for some vertebrates during droughts and floods.

A similar mixing zone, called the hypolentic zone, occurs at the interface between saturated ground water and surface water in lakes and wetlands. In many lakes, the most active portion of the hypolentic zone is located in the littoral zone in close proximity to the shoreline (Hunt and others 2003, McBride and Pfannkuch 1975). Distinct vegetation and aquatic communities are likely to be associated with focused and diffuse discharge of ground water (Rosenberry and others 2000).

Springs

Springs typically are present where the water table intersects the land surface. In fractured-rock terrain, springs are fed through faults or fractures. Springs are important sources of water to streams and other surface-water features. They also may be important cultural and aesthetic features. The constant source of water at springs leads to the abundant growth of plants, and many times to unique habitats for endemic species like spring snails (U.S. BLM 2001). Ground water development can reduce spring flow, change springs from perennial to intermittent, or eliminate springs altogether. Springs typically represent points on the landscape where ground water flow paths from different sources converge. Ground water development may affect the amount of flow from these different sources to varying extents, thus affecting the chemical composition of the spring water.

Aquifer, Karst, and Cave Ecosystems

This category comprises the aquatic ecosystems that may be found in free water in cave and karst systems (Fetter 2000) and within aquifers themselves (Gilbert and others 1998). Aquifer ecosystems represent the most extended array of freshwater ecosystems across the entire planet (Gilbert 1996). Their fauna largely consists of invertebrates and microfauna. Very little is known about aquifer ecosystems under NFS lands, their importance for biodiversity, or their importance to the systems into which they discharge.

Some ecosystems, such as floodplains, exist along a continuum between fully aquatic communities and fully aquifer communities (Danielopol 1989). Aquifer ecosystems are not confined to near-surface environments. Stygofauna (animals occupying cave or aquifer habitats) have been identified at depths of up to 600 meters (Longley 1992). Aquifer ecosystems are characterized by darkness, consistency, persistence of habitat, and low energy and oxygen availability. The organisms that inhabit these environments are often specialized morphologically and physiologically. Their stable and confined environment results in high levels of endemism and high proportions of relict species compared with surface environments. Some cave fauna may have changed very little over the last hundreds of millions of years. Recent work in northwestern Australia has identified entire major lineages (orders or classes) of stygofauna that are thought not to have been represented in surface ecosystems since the Mesozoic Era.

Ground water level, flux, and quality are the three attributes of greatest significance to cave-karst and aquifer ecosystems. Ground water level and flux determine the amount of ground water available to support these ecosystems. Where the composition of aquifer ecosystems changes with depth, reductions in ground water levels may result in the loss of particular species or communities of aquatic organisms. Such aquifer ecosystems are highly specialized and may be lost entirely with changes in ground water level of only 1 to 2 meters (Humphreys 1999).

Many aquifer ecosystems have developed in very stable environments. Subtle changes in ground water quality because of contamination by agricultural chemicals, sediment, or septic tank effluent may alter ecosystem function. The potential sensitivity of aquifer ecosystems to changes in ground water quality can make them useful as bioindicators (Gilbert 1996).

Wetlands

Wetlands occur in widely diverse settings from coastal margins to floodplains to mountain valleys. Similar to streams and lakes, wetlands can receive inflow from ground water, recharge ground water, or do both. The persistence, size, and function of wetlands are controlled by hydrologic processes active at each site (Carter 1996). For example, the persistence of wetness for many wetlands depends on a relatively stable influx of ground water throughout seasonal and annual climatic cycles. Characterizing ground water discharge to wetlands and its relation to environmental factors such as moisture content and chemistry in the root zone of wetland plants is a critical but highly challenging aspect of wetlands hydrology (Hunt and others 1999).

Wetlands can be quite sensitive to the effects of ground water pumping. This pumping can affect wetlands not only by lowering the water table, but also by increasing seasonal changes in the elevation of the water table and exposing accumulated organic and inorganic material to oxidation. Some peat-forming wetlands are highly stable environments that may contain fossil material that provides insights into past environments. Overextraction of water, like the draining of wetlands for agriculture and other development, can destroy this valuable source of scientific data.

Fens are peat-forming wetlands that receive recharge and nutrients almost exclusively from ground water. The water table is at or just below the ground surface. Water moves into fens from upslope mineral soils, and flows through the fen at a low gradient. Fens differ from other peatlands because they are less acidic and have higher nutrient levels; therefore, they are able to support a much more diverse plant and animal community. Grasses, sedges, rushes, and wildflowers often cover these systems. Over time, peat may build up and separate the fen from its ground water supply. When this happens, the fen receives fewer nutrients and may become a bog. Patterned fens are characterized by a distribution of narrow, shrub-dominated ridges separated by wet depressions.

In North America, fens are common in the northeastern United States, the Great Lakes region, the Rocky Mountains, the Cascade and Siskiyou Mountains, and much of Canada. They are generally associated with low temperatures and short growing seasons. Slow decomposition of organic matter allows peat to accumulate. Fens provide important benefits in a watershed, including preventing or reducing the risk of floods, improving water quality, and providing habitat for unique plant and animal communities. Like most peatlands, fens have experienced a decline in acreage, mostly from mining and draining for cropland, fuel, and fertilizer. Because of the large historical loss of this ecosystem type, remaining fens are rare, and it is crucial to protect them. While mining and draining these ecosystems provide resources for people, up to 10,000 years are required to form a fen naturally.

Management Considerations and Protection Strategies

The Forest Service ground water policy is specifically designed to protect ground water-dependent ecosystems so that, wherever possible, the ecological processes and biodiversity of their dependent ecosystems are maintained, or restored, for the benefit of present and future generations. The general level of understanding of the role of ground water in maintaining ecosystems throughout the public lands is very low. Ground water resource managers and investigators tend to underestimate ecosystem vulnerability to ground water development, pollution, and land-use change. Planners must recognize ecosystem dependence on ground water and related processes. Perhaps such recognition can be best achieved by incorporating ground water resource inventory, monitoring, and protection into management plans.

The initial step in protecting ground water-dependent ecosystems is developing an inventory of those systems on NFS lands. Identify and describe their locations, ecological values, and degrees of dependence on ground water. Land management plans should then be reviewed and revised as necessary to incorporate ground water-level, ground water extraction-rate, ground water recharge-rate targets or other management rules that minimize localized impacts on dependent ecosystems. The degree of protection will vary according to the characteristics and dynamics of each ground water system and the significance of the ground water-dependent ecosystems. Protection may range from minimal where the aquifer is deep and has little connection to the surface, to significant where the connection is strong and the conservation value of dependent ecosystems is high. More localized measures for protecting ground water-dependent ecosystems may include the following steps:

- Establishing buffer zones around dependent ecosystems, within which ground water extraction is excluded or limited.
- Establishing maximum limits to which water levels can be drawn down at a specified distance from a dependent ecosystem.
- Establishing a minimum distance from a connected river, creek or other dependent ecosystem from which a well could be sited.
- Protecting ground water quality in areas that provide recharge to dependent ecosystems by limiting the types of activities that can take place there.

The social and economic costs of the recommended management prescriptions and protections, as well as the costs related to impacts from use, also need to be considered. Ground water extractions should be managed within the sustainable yield of aquifer systems so that the ecological processes and biodiversity of their dependent ecosystems are maintained or restored. In this process, threshold levels that are critical for ecosystem health should be estimated and considered. Planning, approval, and management of developments and land uses should aim to minimize adverse impacts on ground water systems by maintaining natural patterns of recharge and discharge, and by minimizing disruption to ground water levels that are critical for ecosystems.

Ground water-dependent ecosystems can have important values for ground water users, ecosystem managers, scientists, and the wider community. These values, and how threats to them may be avoided, should be identified in land management plans, and actions should be taken to ensure that the ecosystems are protected.

Case Study: Importance of Ground Water in Alpine-lake Ecosystems, Cabinet Mountains Wilderness, MT

An investigation in the Cabinet Mountains Wilderness, Kootenai National Forest, in northwestern Montana was conducted to assess the potential for adverse impacts from ground water withdrawal from a proposed underground mine. The study was prompted by concerns that mining under the wilderness could modify ground water hydraulics in the fractured bedrock aquifer and adversely impact the water balance, chemistry, and ecology of the overlying wilderness lakes.

Ground water plays an important role in the chemical composition of lakes, and the aquatic ecology of lakes is defined, in large part, by their hydrochemistry. The importance of ground water is accentuated for dilute lakes, like those in the Cabinet Mountains, which rely on ground water inputs as their primary source of dissolved solids and nutrients. Even though the volume of ground water inflow to these lakes is a small fraction of the annual hydrological budget, during the short ice-free period when peak biological activity takes place, ground water inflow can contribute considerable amounts of water and solutes.

Hydrological and chemical budget evaluations of Cliff Lake (fig. 15) and Rock Lake, two of the lakes overlying the Rock Creek ore body, were performed to help predict potential impacts from proposed mining. Nonsteady-state mass balances using naturally occurring tracers (solutes and stable isotopes) provided a means for estimating the quantity of ground water inflow into the lakes and evaluating the water balance over the short summer season.

Over the summer, the chemical composition of the lakes shifts toward that of local ground water, indicating a direct hydraulic connection to the ground water system. Compared with solute mass fluxes from precipitation or surface water, ground water is the principle source of dissolved solute load (fig. 16). For Rock Lake, ground water supplies about 59 percent of the ice-free season inflow but contributes 71 percent of the solute load. Similarly, for Cliff Lake, ground water supplies about 83 percent of the inflow and 96 percent of the solute load. In addition, a considerable proportion of the buffering capacity is a result of the ANC contributed by ground water. ANC is important for dilute lakes, such as these, because of their extreme sensitivity to the adverse effects of acid deposition.

Unless a surface-water body is directly connected to the underlying ground water system being affected by such mining, it will not experience significant disruptions in water or chemical budgets. This study established that Rock Lake and Cliff Lake are directly connected to the ground water system. Depletion of ground water inflow by mining-induced changes in hydraulic gradients and ground water flow paths could cause a shift in the hydrological, chemical, and consequently, the biological composition of these lakes.

For more information, see Montana Department of Environmental Quality and USDA Forest Service (2001) and Gurrieri and Furniss (2004).

Figure 15. Cliff Lake, Cabinet Mountains Wilderness, MT.

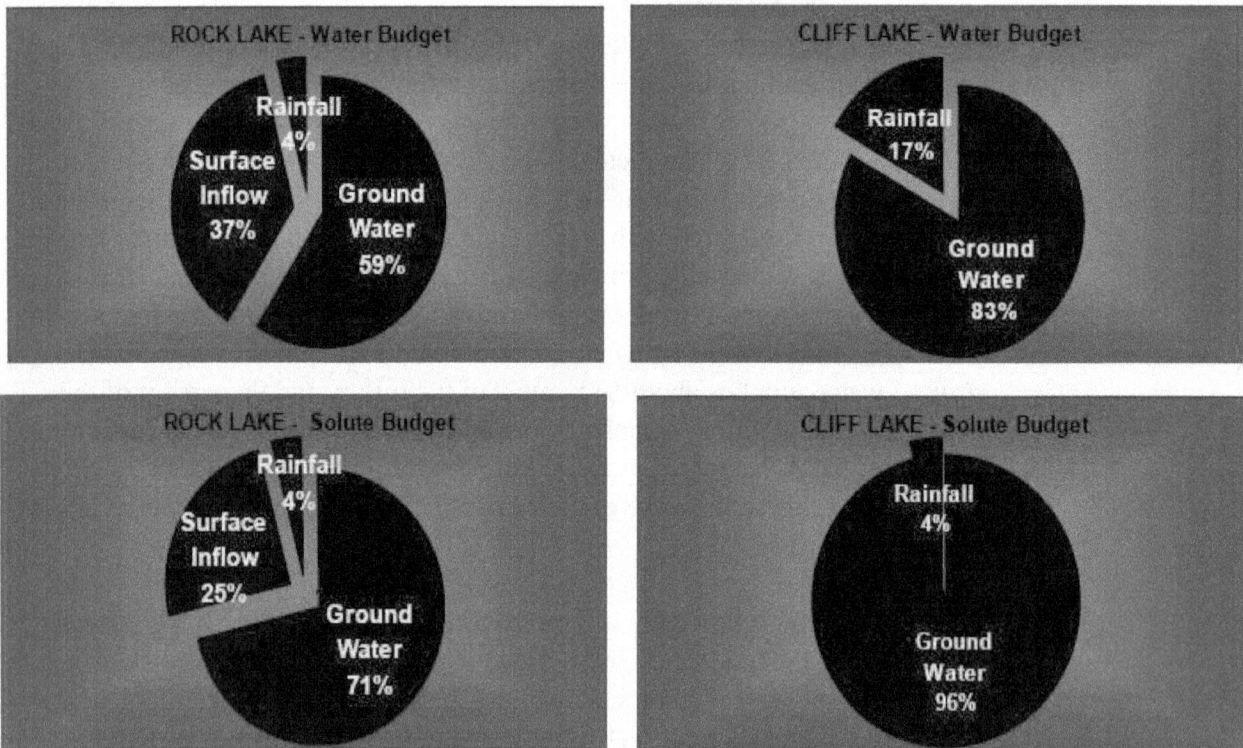

Figure 16. Water and solute budgets for Rock Lake and Cliff Lake in percent of ice-free season inflow.

46

Activities That Affect Ground Water

This section describes some of the activities that commonly cause ground water problems on NFS lands. See appendix IV for a discussion of possible techniques for remediating existing ground water contamination.

Mineral Development

Prospecting and developing mineral resources on NFS lands, including such materials as base and precious metals, oil and gas, coal, phosphate and gypsum, and aggregate and building stone, involve activities and land uses with the potential to significantly affect both the quantity and quality of the ground water resource associated with those lands. The primary issues associated with the major types of mineral development are presented below.

Hardrock Mining

Hardrock mining is defined as the extraction of precious and industrial metals and nonfuel minerals by surface and underground mining methods (Lyon and others 1993). In the United States, extensive hardrock mining started in the 1880s and, for the next 70 to 80 years, it was a major industry in many States. In 1992, more than 500 operating hardrock mines were located in the United States, of which more than 200 were gold mines. In 1997, approximately 60 mine sites in 26 States were on the Federal Superfund National Priorities List.

Many ore bodies and mines (both old and operating) are on public land administered by the Federal land management agencies. They are frequently in areas with relatively little other development. During the first half of the 20th century, environmental controls were very limited or nonexistent. As a result, numerous abandoned mines are currently causing serious environmental damage. Many thousand abandoned and inactive mines are on public land. The USDA Office of the Inspector General estimates that more than 38,000 abandoned and inactive hardrock mines are located on land administered by the Forest Service.

The two primary methods used to mine metals and minerals include surface, or open-pit, mining and underground mining. Surface mining methods are typically used for shallow ore bodies or ore bodies that have a low metal or mineral value per unit volume of rock, while underground mining methods are typically used when the ore body is deep or occurs in veins. Hardrock mining is a large-scale activity that typically disturbs large areas of land. The siting of a mine is largely dictated by the location of the ore body. Because of the high waste-to-product ratios associated with mining most ore bodies, large volumes of mining-related waste are generated. Mine waste includes all of the leftover material generated as a result of mining and processing the ore.

Ore Processing

Ore processing, or milling, refers to the altering of ore rock to (1) create a desired size of product, (2) remove unwanted constituents, and (3) concentrate or otherwise improve the quality of the product. Applicable milling processes are determined based on the physical and chemical properties of the target metal or mineral, the ore grade, and environmental considerations. Each method creates its own set of potential contaminants.

Amalgamation. In this process, metallic mercury is added to gold ore to separate the gold from the ore rock. When liquid mercury comes in contact with gold, it bonds with the surface of the gold particles (amalgamation). The mercury-coated gold particles coalesce or collect into a gray plastic mass. When this mass is heated, the mercury is driven off and the metallic gold remains.

Flotation. The physical and chemical properties of many minerals allow for separation and concentration by flotation. Finely crushed ore rock is added to water containing selected reagents. These reagents create a froth that selectively floats some minerals while others sink. Common reagents include copper, zinc, chromium, cyanide, nitrate and phenolic compounds, and sulfuric acid and lime for pH adjustment. The waste (tailings) and the wastewater are typically disposed of in large, constructed impoundments.

Leaching. Leaching typically involves spraying, pouring, or injecting an acid or cyanide solution over crushed and uncrushed ore to dissolve metals for later extraction. The main types include dump, heap, vat, and in situ leaching. For each type, a nearby holding area (typically a pond) is used to store the pregnant solution prior to recovery of the desired metal using chemical or electrical processes. Once the desired metal is recovered, the solution is reused in the leaching process.

In recent years, the most common and problematic technology has been cyanide heap leaching. In this process, the ore is usually crushed and is placed on a pad constructed of synthetic materials or clay. A leaching solution is sprayed or dripped over the top of the pile. Leaching can recover economic quantities of the desired mineral for months, years, or decades. When leaching no longer produces economical quantities of metals, the spent ore is typically rinsed to dilute or otherwise detoxify the reagent solution to meet environmental standards. If standards are met, the rinsing may be discontinued and the leached material may be allowed to drain. The spent ore is then typically left in place.

WATER MANAGEMENT

Management of water at large mine sites is a critical element of mine operation. At large mine sites that include a mill and a tailings impoundment, water management can be difficult and complex. The many management requirements include (1) the dewatering of open pits and/or underground mine workings, (2) the routing of surface runoff across mine sites, (3) the use and containment of water used for ore processing, and (4) the need to meet applicable water-quality standards for all discharges from the mine site. Historically, the management of water has not focused on prevention of environmental impacts. Nationwide, there have been numerous incidents in which contaminated water from a mine site has been improperly discharged to surface water and/or ground water.

Both surface and underground mines typically extend below the local or regional water table. The ground water that flows into the mine pit or underground workings must be removed to maintain acceptable working conditions. In open-pit mines, this water is typically pumped out and discharged to nearby surface waters or ephemeral drainages. In underground mines, the water can be pumped out and similarly discharged or, under certain conditions, drainage adits can be constructed at or below the lowest mine level to allow for free drainage of the water entering the workings. Many precious metal ore bodies occur in mountainous terrains or regions of continental shield where the host rock is commonly comprised of igneous and/or metamorphic rocks. In these types of rocks, ground water occurrence and flow are controlled by the distribution and orientation of geologic structures, such as fractures, joints, and faults. In these types of hydrogeological settings, ground water inflow into mine workings largely occurs only where the mine workings intersect water-bearing structures.

WASTE MANAGEMENT

Hardrock mining typically produces large volumes of solid waste, including overburden (spoil), development rock, waste rock, spent ore, and tailings. Waste rock, and in some cases development rock and spoil, can contain significant concentrations of metals, and therefore may present an environmental problem. In both surface and underground mining, extraction of ore waste materials requires the use of heavy equipment and explosives. The most commonly used explosive is ANFO, ammonium nitrate and fuel oil. Residual nitrogen in the waste rock and development rock can be leached out by precipitation and cause contamination of water resources.

Tailings are the waste solids remaining after ore processing. Commonly, tailings leave the mill as slurry consisting of 40 to 70 percent liquid and 30 to 60 percent fine-grained solids. Tailings and the associated carriage water (usually mill process water) can contain significant concentrations of heavy metals and other contaminants. Most tailings are disposed in onsite impoundments. Historically, tailings impoundments were not lined and were located without consideration of potential environmental impacts on streams and floodplains. Modern tailings impoundment design often includes low-permeability clay or synthetic liners designed to minimize seepage from the tailings, engineered caps designed to minimize infiltration of water into the tailings, and collection systems to capture leachate that collects within the impoundment. Some seepage from tailings impoundments is often unavoidable, and leachate may infiltrate to underlying ground water.

Spent ore is a waste material that is generated at mines that utilize a leaching process. The volume of spent ore can be very large and can contain environmentally significant residual amounts of leaching reagent and dissolved metals. Both spent ore and tailings need to be actively managed for years after mine closure to ensure that leachate does not contaminate underlying ground water.

MINE CLOSURE

Closure of a mining operation occurs during a temporary shutdown of operations or when the facilities are permanently decommissioned. Depending on the type of mine, the size and nature of the area of disturbance, and the type of ore processing used, active management of the mine site and water management may be necessary for years or even decades after closure. Until recently, reclamation was limited to grading and revegetating waste materials and pits to minimize erosion and improve the visual landscape. Permanent closure now routinely includes some or all of the following: removal/disposal of stored fuels and chemicals; structure demolition; removal of unnecessary roadways and ditches; shaft and adit plugging; waste detoxification; capping of tailings and waste rock; backfilling pits; and active water management, including assuring that all applicable water-quality standards are met. In numerous cases, this has meant operating and maintaining a water-treatment facility. At sites where acid drainage is a problem, post-closure water treatment may be necessary for decades.

POTENTIAL IMPACTS TO GROUND WATER RESOURCES

Information on potential environmental impacts related to hardrock mining has increased greatly in recent years. Numerous investigations and published reports have documented the release of toxic metals to ground water and surface water resulting from mobilization and transport of metals from mines and mine-related facilities.

In hardrock mines, adits and shafts, underground workings, open pits, overburden, development rock and waste rock dumps, tailings impoundments, leach pads, mills, and process water ponds are recognized as potential sources of acidity, metals, sulfate, cyanide, and nitrate. If released in environmentally harmful concentrations, these contaminants can significantly reduce the quality and usability of both ground and surface waters. Dissolved metals in ground waters can make it unsuitable for consumption. If contaminated ground water provides baseflow to a stream, the aquatic health of the stream and riparian ecosystems can be impacted. The impacts can be long term and large scale. They differ with the physical and geological setting of the ore body, type of ore extracted, the mining method, the method of ore processing, the effectiveness of water management, and the nature of mine closure.

A variety of complex geochemical and hydrogeological processes control the transport, attenuation, and ultimate distribution of metals and other mine-related contaminants in ground water (Drever 1997). Dissolved contaminants are transported to aquifers through complex overland and subsurface pathways. This complexity, combined with the large scale of many mining activities and the numerous mine-related sources of contaminants, makes water-quality assessments and restoration and remediation of mine sites very difficult.

Precious and heavy metal ore bodies are typically found in fractured-rock hydrogeologic settings. The extraction and processing of ore over the past 100 years has resulted in the release of heavy metals into the aquatic environment in mining districts across North America. During the past 10 years, research has shown that ground water flow can deliver significant metal loads to streams in mountainous areas. Adequate control of metal mobility at active and abandoned hardrock mine sites requires a good understanding of the local fractured-flow system and its geochemical conditions. The two major types of potential long-term quality impacts to ground water, acid drainage and dissolution and transport of contaminants, are discussed below.

Acid Drainage. A major problem at some hardrock mine sites is the formation of acid drainage, also known as acid rock drainage (ARD) or acid mine drainage (AMD), and the associated mobilization of toxic metals, iron, sulfate, and TDS. ARD results from the exposure of sulfide minerals (such as, pyrite, pyrrhotite, galena, sphalerite, and chalcopyrite) to air and water. Sulfide minerals are commonly associated with coal deposits and precious and heavy metal ore bodies. Pyrite (FeS_2), the most common sulfide mineral, reacts with water and oxygen to produce ferrous iron (Fe^{+2}), sulfate (SO_4), and acid (H^+). In oxygenated water with a pH greater than 3.5, ferrous iron will oxidize to ferric iron (Fe^{+3}), much of it will then precipitate as iron hydroxide ($Fe[OH]_3$), and additional acidity will be released. Some ferric iron remains in solution and continues to chemically accelerate the further oxidation of pyrite and subsequent generation of acidity. As the pH continues to decrease, the oxidation of ferrous iron and the precipitation of iron hydroxide decreases. The result is a greater dissolved concentration of ferric iron and therefore a greater rate of sulfide (pyrite) oxidation. The oxidation of sulfide minerals can be catalyzed by bacteria; *Thiobacillus ferrooxidans* is one example. This bacterium, which is common in the subsurface, can increase the rate of sulfide oxidation by 5 or 6 orders of magnitude. When low pH water comes in contact with metal-bearing rocks and minerals, a number of toxic metals go into solution and are transported by the water. Different metals dissolve over different ranges of pH. The most common metals associated with sulfide minerals include lead, zinc, copper, cadmium, and arsenic.

Water, oxygen, and sulfide minerals are necessary ingredients to generate acid drainage. Water serves as both a reactant and as a medium for the oxidation process. Water also transports the oxidation reaction products and the associated dissolved metals. Atmospheric oxygen is a very strong oxidizing agent and is important for bacterially catalyzed oxidation at pH values below 3.5. Surface water and shallow ground water typically have relatively high concentrations of dissolved oxygen.

Acid drainage can be discharged from underground mine workings, open-pit walls and floors, tailings impoundments, waste rock piles, and spent ore from leaching operations. It can also be released naturally from mineralized rock located at or near the surface; though, anthropogenic activity in such areas can enhance its release. It occurs at both active and abandoned mines. No easy or inexpensive solutions to acid drainage are currently available. The best approach is to avoid development of a problem through appropriate upfront planning and analysis. An appropriate management approach to possible acid generation is to isolate or otherwise segregate and specially handle wastes with acid generation potential. Oxygen contact and water contact with the isolated material should be minimized. Another approach is to ensure that an adequate amount of natural or introduced material is available to neutralize any acid produced. The neutralization approach, however, may not adequately address all of the solutes that could be released into solution during the oxidation process. Techniques used to isolate acid-generating materials include subaqueous disposal, barrier covers, waste blending, hydrologic controls, and bacterial control.

Transport of Dissolved Contaminants. Dissolved contaminants (primarily metals, sulfate, and nitrate) can migrate from mining operations to underlying ground water and surface water. Process water, mine water, and runoff and seepage from mine waste piles or impoundments can transport dissolved contaminants to ground water. The likelihood of contaminants dissolving and migrating from mine waste materials or mine workings to ground water depends on the nature and management of the waste materials and liquids, the local hydrogeological setting, and the geochemical conditions in the underlying vadose zone and aquifer.

Distinguishing between "natural" or background metal loadings and those resulting from mining is an issue that often arises at hardrock mine sites. A number of studies have attempted to separate "natural" loading from anthropogenic loading (Nimick and von Guerard 1998). Researchers have used water-quality data, including isotopes and tracers, to try to "fingerprint" water in an attempt to identify loading caused by leaching of unmined ore bodies from leaching of metals that is enhanced by mining activities. To date, however, no reliable technique has been developed to clearly separate these two general sources of loading.

At some locations, naturally occurring substances other than the target minerals can be a significant source of contaminants. The rocks that comprise ore bodies contain varying concentrations of nontarget minerals, often including radioactive minerals. Other minerals may be present at concentrations that can be toxic and can be mobilized by the same geochemical and hydrological processes that control transport of contaminants from mine sites. Nontarget substances that can pose a risk to ground water include aluminum, arsenic, asbestos, cadmium, chromium, copper, iron, lead, manganese, mercury, nickel, silver, selenium, sulfate, thallium, and zinc.

The impacts from mining can last for many years. As a result, environmental monitoring (including early warning, facility specific, and compliance monitoring), contingency planning and financial assurance have to be in place for many decades. Geochemical conditions within the ore body, waste rock, and tailings can change over time and must be tracked. Flexibility therefore is needed to make necessary changes in water control and water treatment after mine closure.

CASE STUDY:
DEPLETION OF STREAM-
FLOW BY UNDERGROUND
MINING, STILLWATER
MINE, CUSTER
NATIONAL FOREST, MT

Dewatering of shallow aquifers that are directly connected to surface water bodies can have a significant effect on the movement of water between these two water bodies. In mountainous terrain, the fracture-dominated ground water flow system adds complexity to predicting or monitoring the effects from mine dewatering. The disappearance of No Name Creek at the Stillwater Mine illustrates what can happen to surface water resources when mining disrupts the underlying ground water flow system.

The Stillwater Mine is an underground platinum and palladium mine located on the Custer National Forest in south-central Montana. The ore body is part of the Stillwater Complex, a vertically dipping, Precambrian-aged igneous rock unit. The mine began operations in 1986 and in July 1987 began developing the East Adit. After driving the adit about 4,000 feet, a large inflow of water was encountered that peaked at 884 gpm on May 25, 1988 (fig. 17). By July 1988, the inflow had decreased to its present steady-state rate of 200 gpm.

Overlying the adit is a 60-acre watershed that contained a perennial stream called No Name Creek. Baseflow of the stream was supported by a bedrock fracture spring located 830 feet vertically above the adit (fig. 18). At the same time the large adit inflow was encountered, No Name Creek began to dry up. By July 1988, the stream and another spring near the portal ceased to flow and have remained dry ever since.

Under predevelopment conditions, the ground water system was in a state of dynamic equilibrium and ground water discharging at the spring maintained the baseflow in No Name Creek. A new state of dynamic equilibrium was achieved after development of the adit and ground water that previously discharged to the spring was intercepted by the adit and now discharges out the portal (fig. 18). Tunneling activities induced a downward hydraulic gradient in the overlying fractured bedrock aquifer, and subsequent lowering of the potentiometric surface in the aquifer caused the spring to stop flowing. The enhanced vertical permeability along preexisting fractures created by the vertically dipping rocks likely contributed to the strong hydraulic connection between the adit and the overlying spring.

Interestingly, the flow of Nye Creek located adjacent to No Name Creek and also overlying the adit was not affected by the initial tunneling. In 1994, however, three springs in the upper Nye Creek basin were rendered dry by continued underground development of the ore body. Potential mitigation measures have been discussed. The most promising is grouting off the inflows in the underground adit. This effort could reestablish the spring as well as the baseflow of No Name Creek. This case illustrates consequences of ground water depletion and the difficulty of predicting the spatial as well as the temporal impacts from human activities on ground water/surface water interactions in a fractured bedrock aquifer.

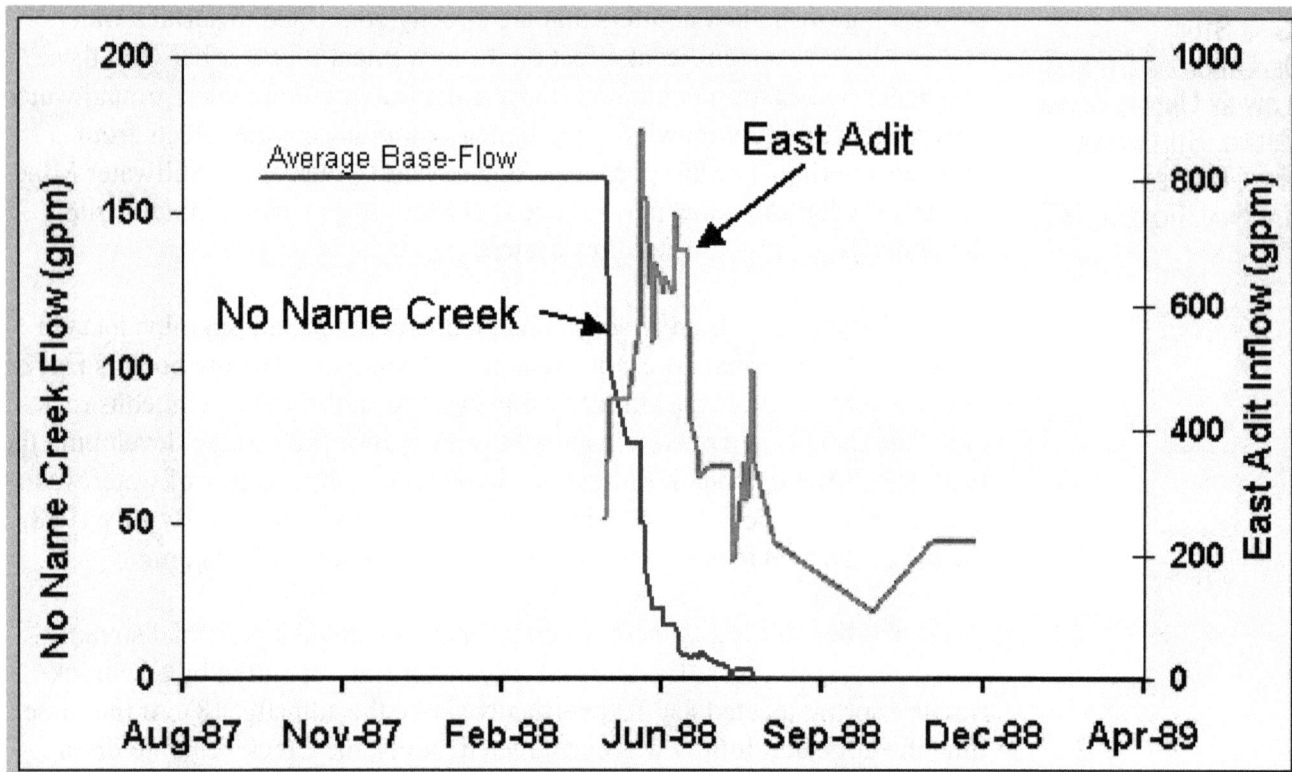

Figure 17. Hydrographs of No Name Creek and flow rate of the East Adit, Stillwater Mine, MT.

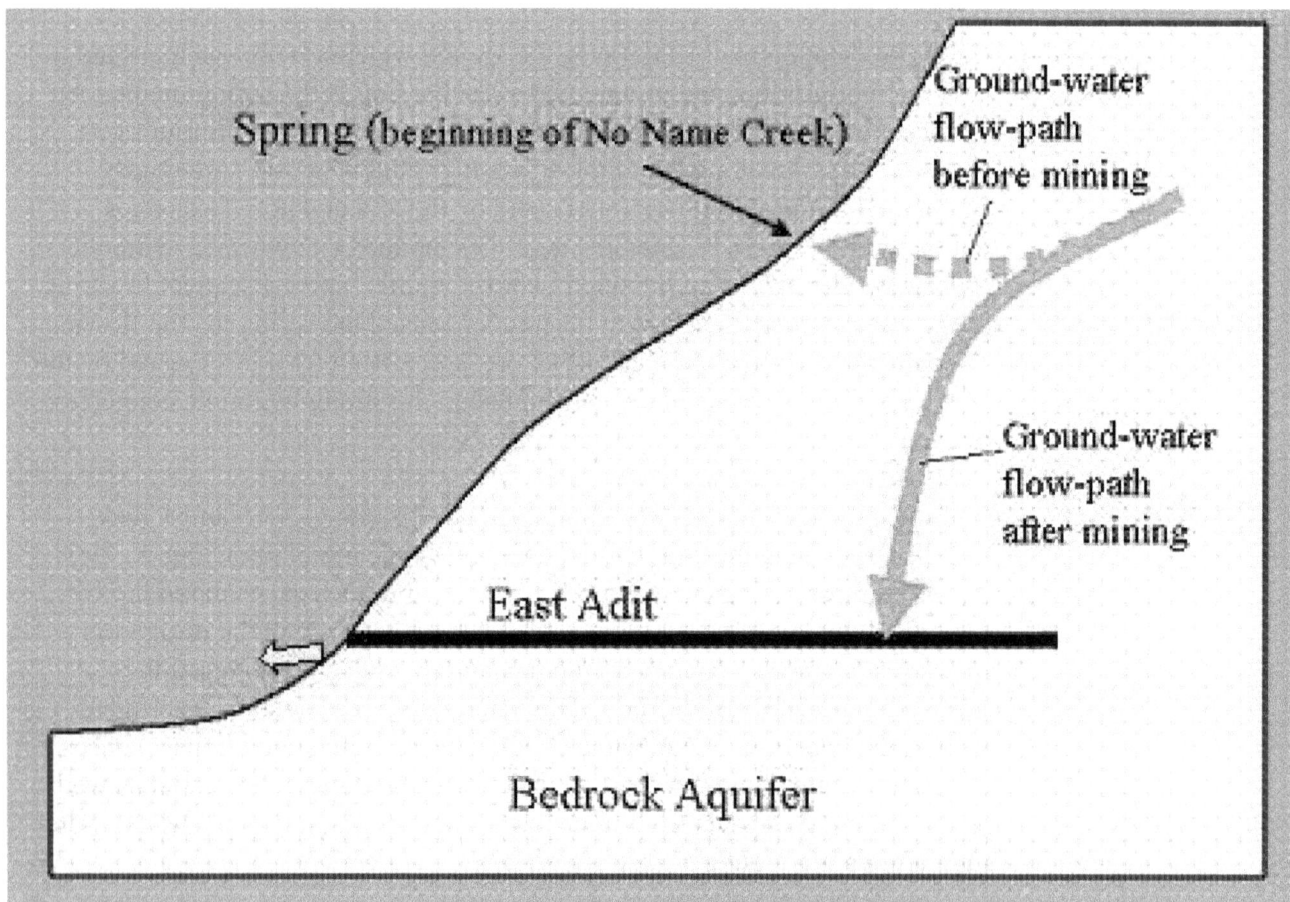

Figure 18. Before mining, ground water discharged to the spring and maintained the flow of No Name Creek. During development of the East Adit, ground water that would have discharged to the spring was intercepted and diverted into the adit.

Coal Mining

Coal accounts for one-third of the total energy usage and more than one-half of the electricity generated in the United States (USGS 1996). In 1998, total domestic production was 1.18 billion tons (National Mining Association 1999). Coal production in the West has almost doubled since 1991. Wyoming leads the nation in coal production; West Virginia and Kentucky are second and third, respectively. About 60 percent of domestic production is from surface mines and 40 percent from underground mines. On NFS lands, active coal mining occurs in Utah, Wyoming, Colorado, and West Virginia.

Strip mining is the most common method of producing coal from surface mines. Strip mining commonly includes the removal and storage of topsoil, the removal of any overburden material (spoil), and the subsequent excavation of the coal seam. As an individual "strip" advances across the land surface, only a relatively small area of the coal seam is actively mined. With this method, the spoil is removed from the advanced side of the active mine face and concurrently placed on the retreat side where the coal has been mined out.

Two methods of underground mining are commonly used: (1) room and pillar, and (2) longwall. In room and pillar mining, "entries" or adits are driven into the coal seam and crosscuts are driven at right angles to the adits at spacings dictated by the individual mine plan. The result is a checkerboard pattern of interconnected tunnels or "rooms" and unmined supports or "pillars." In longwall mining, numerous crosscuts are developed around a large block of coal. Once the crosscuts are fully developed the large block is completely excavated. Longwall mining results in fairly predictable subsidence of the overlying ground surface.

POTENTIAL IMPACTS TO GROUND WATER RESOURCES

In surface coal mines, dewatering may be required to lower the water table so that mining can proceed. Depending on the stratigraphic occurrence of the coal beds and the aerial extent of the economic coal seams, dewatering can result in a cone of depression that can extend for miles in the upgradient direction. Water levels can be lowered in ground water wells that are in the same hydrostratigraphic unit as the coal. Coal beds are often characterized by high hydraulic conductivity, and the associated high transmissivity often makes them attractive for accessing ground water for domestic use, livestock, and irrigation. It is not uncommon for coal-mining companies to enter into agreements with well owners to provide alternative water supplies if domestic, stock, or irrigation wells are impacted. Dewatering can also reduce ground water discharge to wetlands and springs, particularly if the coal beds to be mined occur in a confined hydrostratigraphic unit. In this type of hydrogeological setting, a small lowering of the potentiometric surface can cause a significant reduction in ground water discharge to surface waters.

Waste materials are generated from coal mining and coal preparation. Spoil materials removed for surface mining are often used to backfill the excavated area. Waste material from underground mining is disposed of in mined-out workings to the extent possible, but it often is placed in a designated waste rock disposal area on the surface. The waste material from the coal preparation process (both coarse material and fine-grained slurry) is typically disposed in disturbed portions of the mine site. The fine-slurry waste is commonly disposed of in an impoundment, where the slurry solid settles and the water can be reclaimed.

As with precious metal mining, coal mining can expose sulfide minerals to oxygen, water, and bacteria. Pyrite and, less commonly, marcasite (FeS_2) and greigite (Fe_3S_4) are the primary sulfide minerals found in coal and adjacent rock. Oxidation of these minerals can generate acidic water and mobilize and transport heavy metals to ground water and surface water. Mine waste and coal preparation waste can contain significant amounts of pyrite and metals, including cadmium, chromium, mercury, nickel, lead, and zinc. These metals and sulfur can be concentrated in waste materials by factors of 3 to 10 compared to raw coal (National Research Council 1979). Therefore, acid drainage and associated mobilization of metals and sulfate are potentially significant threats to ground water resources from coal mining.

Underground coal mining using the longwall extraction method directly leads the overlying strata to break and fracture as subsidence occurs. Room and pillar extraction can also lead to fracturing and subsidence. This fracturing of the overlying strata changes the intrinsic permeability of the strata, and can alter ground water flow paths, create areas of increased permeability, and cause fluctuations in the water table. Any changes to the ground water can take years to establish a new equilibrium. Where the overlying rock strata are thin (less than about 600 feet) between the mined coal seam and the land surface, rock fracturing associated with longwall mine subsidence can also directly affect surface water. With respect to ground water, shallow aquifers could drain into subsidence fractures, or surface waters and recharge could be diverted into fractures. Sometimes, underground mining can encounter faults in the subsurface. The faults can sometimes contain ground water that discharges into the underground mine. The effects discussed above, however, do not occur everywhere, and the local geology, occurrence of ground water and surface water, and mining scenario must be evaluated carefully to ensure an adequate understanding of a particular site.

Oil and Gas Exploration and Development

Although geophysical and geological investigations are useful for oil and gas exploration, only exploratory drilling can confirm the presence of commercially valuable oil and gas reserves. Tens of thousands of exploration holes are completed every year. The majority of these wells are "dry" holes, meaning that no commercially significant quantities of oil and gas are encountered. Oil and gas companies are required to properly plug and abandon "dry" holes as well as exploration and production wells and injection and disposal wells that are no longer in use. Plugging and abandonment must be completed in accordance with State law. By 1993, more than 3.3 million wells had been drilled in the United States by the petroleum industry, and approximately 1.2 million wells had been plugged and 1 million had been abandoned or were inactive (American Petroleum Institute 1993).

Oil and gas contained in geologic formations is often not under sufficient hydraulic pressure to flow freely to a production well. The formation may have low permeability or the area immediately surrounding the well may become packed with cuttings. A number of techniques are used to increase or enhance the flow. They include hydraulic fracturing and acid introduction to dissolve formation matrix and create larger void space. The use of these flow-enhancement techniques and secondary recovery methods result in physical changes to the geologic formation that will affect the hydraulic properties of the formation. Typically, the effects of these techniques and methods are localized to the area immediately surrounding the individual well, are limited to the specific oil and gas reservoir, and do not impact adjacent aquifers.

POTENTIAL IMPACTS TO GROUND WATER RESOURCES

The Forest Service plays only a partial role in the regulation of oil and gas production activities on NFS lands under lease for oil and gas. The BLM oversees oil and gas drilling on NFS lands and is the formal leasing agency. The Forest Service only has responsibility for surface activities and surface impact evaluation. The EPA and State agencies regulate the disposal of wastes generated by the development and production of oil and gas. Underground waste disposal is regulated under the UIC program, which was authorized under the SDWA. RCRA conditionally exempted wastes associated with exploration, development, and production of oil and gas from regulation as a hazardous waste. Exempted wastes include well completion, treatment and stimulation fluids, workover wastes, packing fluids, and constituents removed from produced water before disposal.

Exploration, development, and production of traditional oil and gas resources typically do not significantly deplete ground water. Oil and gas resources are often developed from geological reservoirs that do not contain significant amounts of freshwater; however, the development and production of oil and gas can affect adjacent or nearby aquifers. Potential impacts result from the creation of artificial pathways between oil and gas reservoirs and adjacent aquifers. Modification of ground water flow paths may cause fresh ground water to come in contact with oil or gas. In addition, improper disposal of waste waters (brine, storm runoff), drilling fluids, and other wastes can impact the quality of underlying ground water (U.S EPA 1987).

A high risk of fluid migration exists along the vertical pathways created by inadequately constructed wells and unplugged inactive wells. Brine or hydrocarbons can migrate to overlying or underlying aquifers in such wells. This problem is well known in the oil fields around Midland, TX. Since the 1930s, most States have required that multiple barriers be included in well construction and abandonment to prevent migration of injected water, formation fluids, and produced fluids. These barriers include (1) setting surface casing below all known aquifers and cementing the casing to the surface, and (2) extending the casing from the surface to the production or injection interval and cementing the interval. Barriers that can be used to prevent fluid migration in abandoned wells include cement or mechanical plugs. They should be installed (1) at points where the casing has been cut, (2) at the base of the lowermost aquifer, (3) across the surface casing shoe, and (4) at the surface. Individual States and the BLM have casing programs for oil and gas wells to limit cross contamination of aquifers.

Coal-bed Methane

Coal-bed methane is natural methane gas that is produced during the transformation of plant and other organic material to coal (a process called coalification) and subsequently trapped in coal beds (DeBruin and others 2001). As the coalification process proceeds and lignite, sub-bituminous, and bituminous coal are formed, various gases, including methane, carbon dioxide, and nitrogen are released. These gases can then be trapped in the coal beds by ground water pressure. Two types of methane gas can be created during the coalification process: (1) biogenic methane, which is produced by bacterial activity; and (2) thermogenic methane, which is produced by heating, usually during burial. Coal-bed methane can be stored in four different ways within coal beds: (1) as a free gas within micropores, (2) as dissolved gas in ground water that occurs within the coal beds, (3) as adsorbed gas, and (4) as absorbed gas.

Economically viable coal-bed methane resources can occur in coal fields that include shallow, thick, laterally continuous coal beds. Historically, coal-bed methane production focused on high-rank, high-gas-content coal beds. Recently a new production technique has been developed that makes it more economical to produce methane from shallow, low-gas-content coal beds. Using this technique, coal-bed methane well casings are set to the top of the target coal bed, and the underlying target zone is reamed. A submersible pump is then used to pump water up the tubing, and the methane gas separates from the water and flows up the annulus. The flow of methane gas up the annulus is facilitated by a decrease in hydraulic head because of dewatering. At the wellhead, gas is piped to a compressor and the "produced" water is discarded. Coal-bed methane wells go through three stages of production: (1) dewatering stage—water production exceeds gas production, (2) stable production stage—maximum methane production and stable water production, and (3) declining stage—methane production declines until it becomes uneconomic.

In some locales, the production of coal-bed methane requires that large volumes of ground water be pumped out of the coal seams to recover the gas. These amounts can vary widely depending on the local hydrogeological regime. The depletion and disposal of the "produced" ground water is a significant water-management issue. Because the annual amount of ground water produced from a coal-bed aquifer can easily exceed the annual recharge, removing large volumes of ground water can lower local and even regional aquifer water levels. The result can be reduced yields and increased pumping costs for wells developed in these aquifers. It is fairly common for companies that produce coal-bed methane to enter into agreements to provide water to owners of impacted wells. In most coal-bed methane production areas great uncertainty exists as to how long it will take to recharge the depleted aquifers after methane production has ceased. Depending on the disposal method, the use of ground water resources in coal-bed methane production areas may be severely limited for years or decades into the future.

The Western Governors' Association has published the handbook *Coal Bed Methane Best Management Practices* (Western Governors' Association 2004). The reader is advised to refer to it.

DISPOSAL OF PRODUCTION WATER

The quality of coal-bed methane "produced" water can vary significantly. The quality of some ground water contained in coal beds is very good and is sometimes used for domestic consumption. Ground water that occurs in coal bed aquifers can contain significant concentrations of cations such as sodium, calcium, and magnesium. Many cations are readily sorbed to clay particles and can be easily exchanged for other cations. Excess sodium sorbed to clay soil will cause the soil to swell and reduce the soil permeability. The sodium adsorption ratio (SAR) is a measure of the ratio of sodium to calcium plus magnesium and is used to provide an indication of the degree to which free sodium ions could occupy exchange sites on clay particles. High SAR values can indicate that the use of water for irrigation purposes should be limited. This is an important issue where "produced" water is discharged to streams above locations where stream water is diverted for irrigation of crops.

Ground water from coal-bed methane wells is most often disposed of by direct discharge to ephemeral or intermittent streams. Other disposal methods include direct discharge to perennial streams, disposal through shallow or deep injection wells, and recharge to the subsurface through infiltration from recharge basins (Wireman 2002). It is important to adequately evaluate the technical and environmental issues associated with the disposal of ground water produced as a result of coal-bed methane production. Disposal of "produced" water through injection wells or via infiltration from recharge basins or spray-irrigation areas facilitates ground water recharge and can result in lower net loss of the resource. Disposal to perennial streams is more legally complicated and may require a National Pollution Discharge Elimination System (NPDES) permit or the State equivalent.

Whether recharge to the subsurface via infiltration or injection is a viable disposal option for "produced" water will depend on a number of legal, engineering, and hydrogeological factors. Legal factors that need to be considered include permitting requirements and the potential for infiltrated or injected water to resurface in nearby drainage channels. Engineering factors include cost, geotechnical considerations, and operation and maintenance requirements. Hydrogeological factors that need to be considered include (1) the volume, rate and quality of water to be disposed, and (2) the hydraulic and chemical characteristics of the soils/rock to receive the water. In addition, for recharge basins or irrigation areas the thickness and hydraulic properties of the unsaturated zone beneath the recharge basin are important. The construction and use of recharge basins or injection wells may need to be permitted. The need for a permit and the permit conditions will depend, in part, on whether or not the infiltrated water will discharge back to the land surface at some distance from the recharge basin or injection well or discharge directly to a nearby stream or lake.

It is important to site recharge basins in locations where hydrogeological conditions will prevent or minimize the local discharge of infiltrated water. The rate of infiltration depends on the infiltration capacity of the soil or sediment underlying the recharge basin. The rate of infiltration will decline from an initial faster rate to an approximately constant rate for water with low suspended solids and low to moderate dissolved solids. For any given soil or sediment type a limiting curve defines the maximum possible rates of infiltration versus time (Horton 1933). The final constant rate is lower for clay soils with fine pores than for open-textured sandy soils or sediments. The final constant infiltration rate is numerically equivalent to the saturated hydraulic conductivity of the soil or sediment (Rubin and Steinhardt 1963, 1964). The latter can be determined from saturated hydraulic conductivity data, which are more readily available.

Ground Water Pumping

As surface water resources become fully developed and appropriated, ground water commonly offers the only available source for new development. In many areas of the United States, however, pumping of ground water has resulted in significant depletion of ground water storage (Alley and others 1999). These ground water depletions can result in lowered water levels in wells, hydraulic interference between pumping wells, reduced surface water discharge, land subsidence, and adverse changes in water quality.

Declining Water Levels

It is useful to consider three terms that have long been associated with ground water sustainability: (1) safe yield, (2) ground water mining, and (3) overdraft. The term "safe yield" commonly is used in efforts to quantify sustainable ground water development. The term should be used with respect to specific effects of pumping, such as water-level declines, reduced streamflow, and degradation of water quality. The consequences of pumping should be assessed for each level of development, and safe yield should be taken as the maximum pumpage for which the consequences are considered acceptable. The term "ground water mining" typically refers to a prolonged, progressive, and, in many cases, permanent decrease in the amount of water stored in a ground water system. This phenomenon may occur, for example, in heavily pumped aquifers in arid and semiarid regions. Ground water mining is a hydrologic term without connotations about water-management practices (U.S. Water Resources Council 1980). The term "overdraft" refers to withdrawals of ground water from an aquifer at rates considered to be excessive and therefore carries the value judgment of overdevelopment. Thus, overdraft may refer to ground water mining that is considered excessive as well as to other undesirable effects of ground water withdrawals.

Pumping ground water from a well always causes (1) a decline in ground water levels at and near the well; and (2) a diversion of ground water to the pumping well that was moving slowly to its natural, possibly distant, area of discharge (fig. 19). Pumping of a single low-capacity well typically has a local effect on the ground water flow system. Pumping of high-capacity wells or many wells (sometimes hundreds or thousands of wells) in large areas can have regionally significant effects on ground water systems. Where a new pumping well is installed near an existing pumping well and both are tapping the same aquifer, overlapping cones of depression (well interference) can result (Fetter 2000). The effect on the existing well from pumping the new well is lowered water levels, an increased rate of decline, and reduced yield. In addition, changes in water chemistry at the existing well can result. The new well likewise has a lower yield than if it had been placed farther from the existing pumping well.

Ground water heads respond to pumping to markedly different degrees in unconfined and confined aquifers. Pumping the same quantity of water from wells in confined and in unconfined aquifers initially results in much larger declines in heads over much larger areas for the confined aquifers. This is because less water is available from confined aquifers for a given loss of head compared to similar unconfined aquifers.

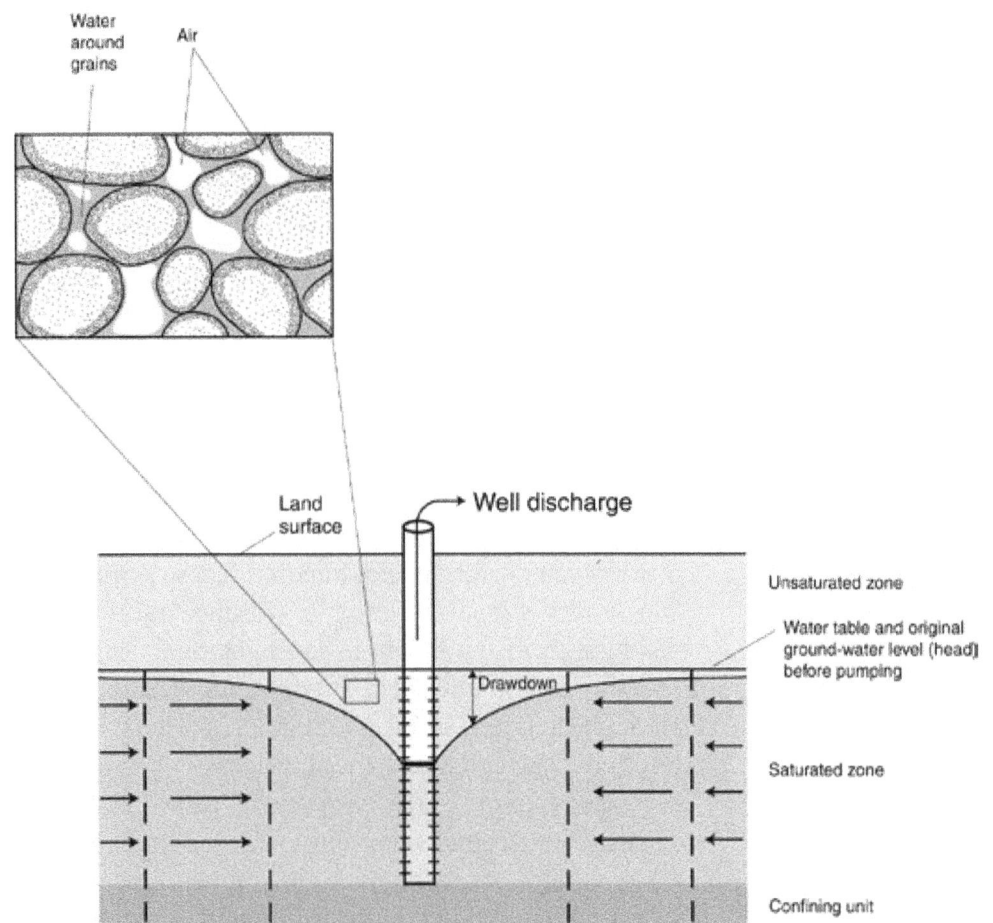

Figure 19. Pumping a single well in an idealized unconfined aquifer. Dewatering occurs in a cone of depression of unconfined aquifers during pumping by wells (Alley and others 1999).

61

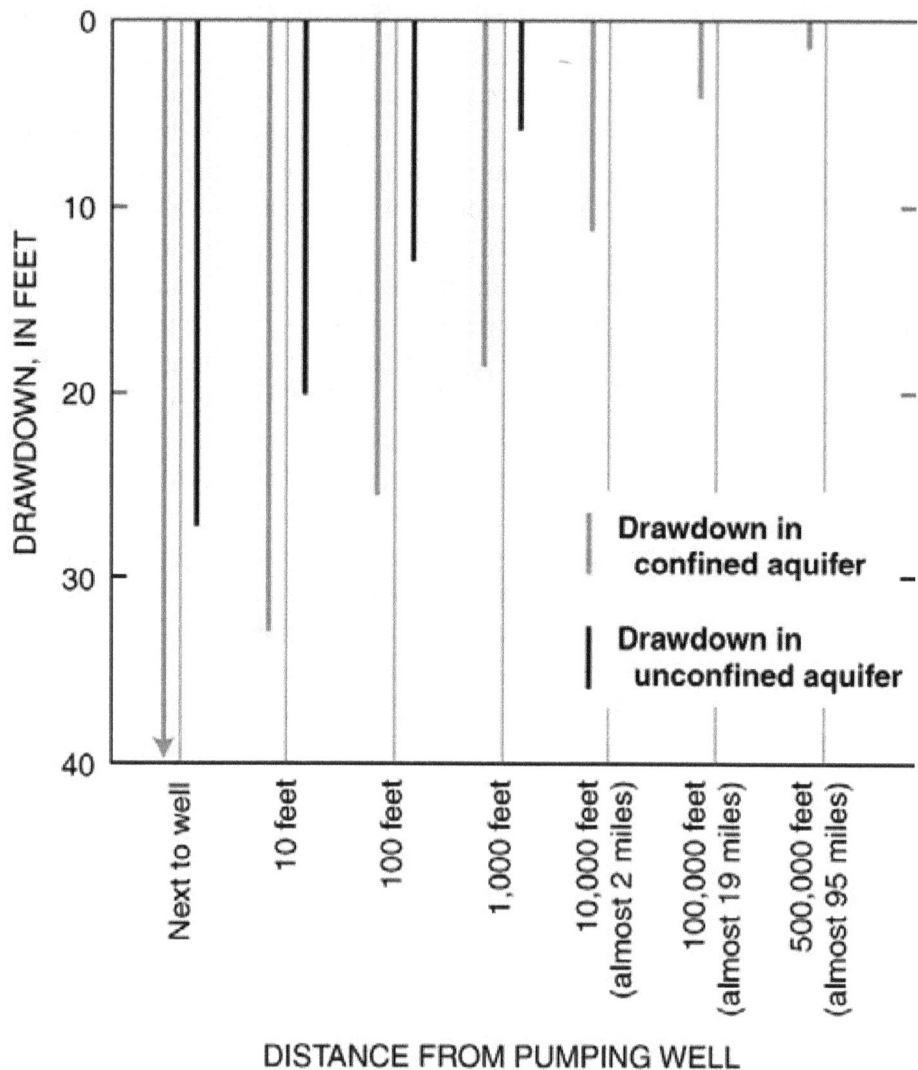

Figure 20. Comparison of drawdowns after 1 year at selected distances from single wells that are pumped at the same rate in idealized confined and unconfined aquifers (Alley and others 1999).

As might be expected, declines in heads and associated reductions in storage in response to pumping can be large compared to changes in unstressed ground water systems. For example, declines in heads as a result of intense pumping can reach several hundred feet in some hydrogeological settings. Drawdown is typically larger in confined aquifers (fig. 20). Widespread pumping that is sufficient to cause regional declines in aquifer heads can result in several unwanted effects: (1) substantially decreased aquifer storage, particularly in unconfined aquifers; (2) dried up wells in places because the lowered heads are below the screened or open intervals of these wells; (3) decreased pumping efficiency and increased pumping costs because the vertical distance that ground water must be lifted to the land surface increases; (4) changed rates of movement of low quality or contaminated ground water and increased likelihood that the low quality or contaminated ground water will be intercepted by a pumping well; and (5) land subsidence or intrusion of saline ground water may result in some hydrogeologic settings.

Perennially flowing springs can be adversely affected by too much water well pumping. Flows may diminish or cease if too much pumping occurs in an aquifer where a hydrologic connection exists between a spring and a well. Many examples of this phenomenon can be found in all parts of the United States. The same holds true for surface streamflows, especially during baseflow periods and in times of drought when all of the streamflow comes from ground water discharge.

Depletion of ground water also can lower water levels in lakes, ponds, wetlands, and riparian areas. Water temperatures can rise from solar heating of smaller volumes of water and depletion of cooler ground water inflows. In turn, geochemical reaction rates may increase and affect the organisms in those waters, possibly to their detriment. Algae blooms are more likely in these lakes, ponds, and reservoirs, and when the algae die, fall to the bottom, and decompose, dissolved oxygen is consumed in the water body, causing stress to or killing fish and other aquatic species.

Where the depletion of ground water causes a decline in surface water or even total stream dewatering, terrestrial species may be adversely affected similarly to aquatic species. If any species so affected are listed under the Endangered Species Act of 1973, the Forest Service line officer has a duty to consult with the appropriate agency responsible for administering that act (U.S. Fish and Wildlife Service or National Marine Fisheries Service) and implement its recommendations for species protection or recovery. Recommendations can include modifying or canceling an authorization for water extraction from NFS land.

Land Subsidence

Land subsidence is a gradual settling or sudden sinking of the Earth's surface because of subsurface movement of earth materials. More than 80 percent of the identified subsidence in the United States is a consequence of human impact on subsurface water. This effect is an often-overlooked environmental consequence of our water-use practices. Impacts from land subsidence include damage to manmade structures, such as buildings and roads, as well as irrecoverable damage to aquifers. Subsidence is a global problem. In the United States, more than 17,000 square miles in 45 States have been directly affected by subsidence. In the late 1980s, the estimated annual costs in the United States from flooding and structural damage caused by land subsidence exceeded $125 million (National Research Council 1991). This section provides an overview of land subsidence principles and impacts. Detailed information on subsidence is provided by Galloway and others (2003).

In some areas, excessive pumping can cause the collapse of the framework of aquifer materials. The result is aquifer compaction and subsidence at the land surface (fig. 21). This compaction results in the permanent loss of aquifer storage, even if the water table should later recover when pumping stops. Although the water table may recover to prepumping levels, resumption of pumping will result in rapid drawdown because of the loss of aquifer storage capacity. In some parts of Florida, the lowering of the water table from

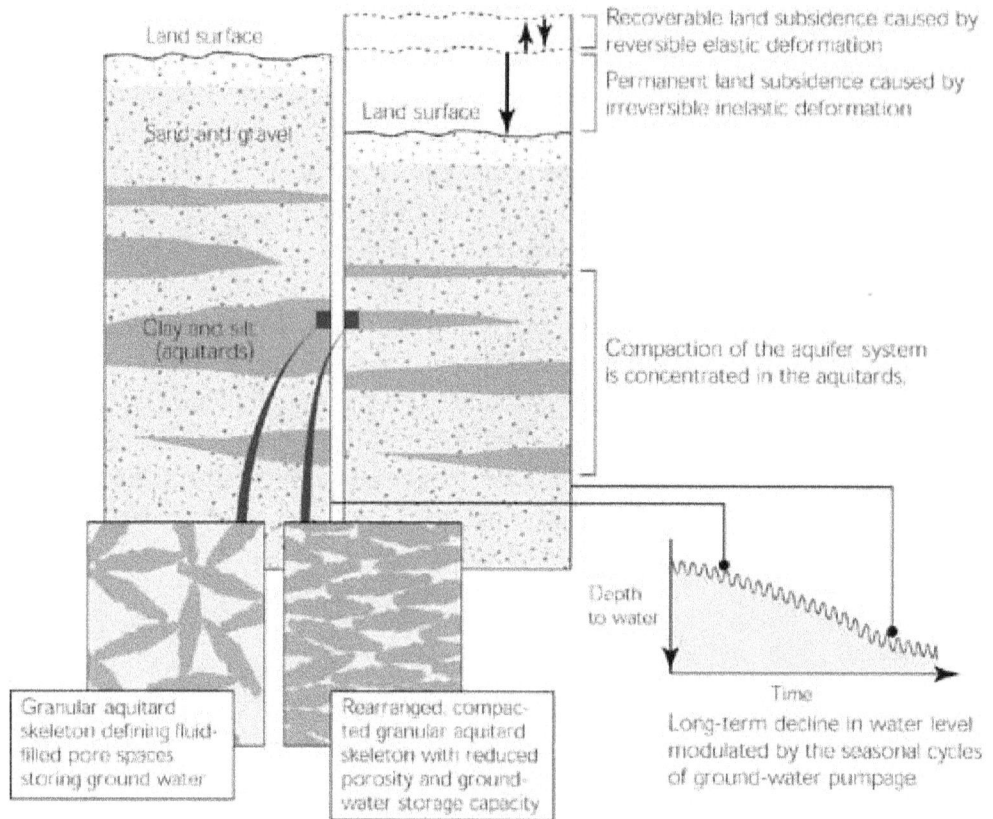

Recoverable land subsidence caused by reversible elastic deformation

Permanent land subsidence caused by irreversible inelastic deformation

Land surface

Sand and gravel

Land surface

Clay and silt (aquitards)

Compaction of the aquifer system is concentrated in the aquitards.

Granular aquitard skeleton defining fluid-filled pore spaces storing ground water

Rearranged, compacted granular aquitard skeleton with reduced porosity and ground-water storage capacity

Depth to water

Time

Long-term decline in water level modulated by the seasonal cycles of ground-water pumpage

When long-term pumping lowers ground-water levels and raises stresses on the aquitards beyond the preconsolidation-stress thresholds, the aquitards compact and the land surface subsides permanently.

Figure 21. A reduction in the total storage capacity of the aquifer system can occur if pumping of water causes an unrecoverable reduction in the pore volumes of compacted aquitards because of a collapse of the sediment structure (Galloway and others 1999).

pumping has resulted in sinkhole development. Subsidence resulting from drainage of organic soils is a problem in wetland areas, and such changes can adversely affect wetland ecosystems. Subsidence also can severely damage building foundations, roads, and buried pipelines, and can increase the frequency of flooding in low-lying areas.

A time lag often occurs between the dewatering of an aquifer and subsidence because much of the compaction results from the slow draining of water from confining units adjacent to the aquifer (Galloway and others 1999). This phenomenon is called "hydrodynamic consolidation." It is also responsible for residual compaction, which may continue long after water levels are initially lowered or even after pumping stops.

Three distinct processes account for most water-related subsidence: (1) compaction of aquifer systems, (2) drainage and subsequent oxidation of organic soils, and (3) dissolution and collapse of susceptible rocks. Other causes of subsidence include underground mining (particularly coal mining), removal of oil and gas reserves from the subsurface, thawing of permafrost, consolidation of sedimentary deposits over geologic time, and tectonism.

Examples of subsidence caused by overdraft of ground water and aquifer compaction include the San Joaquin Valley in California (fig. 22), agricultural areas in south-central Arizona, the Houston-Galveston area of Texas, and Las Vegas, NV. Subsidence because of drainage and subsequent oxidation of organic soils is a major problem in the Florida Everglades. The causes are conversion of marshland to urban areas and farmland, periodic droughts, and associated wildfires. Subsidence exceeds 5 feet in the agricultural areas (fig. 23). This amount of subsidence is especially significant to this near-sea-level wetlands system in which flow is driven by less than 20 feet of relief.

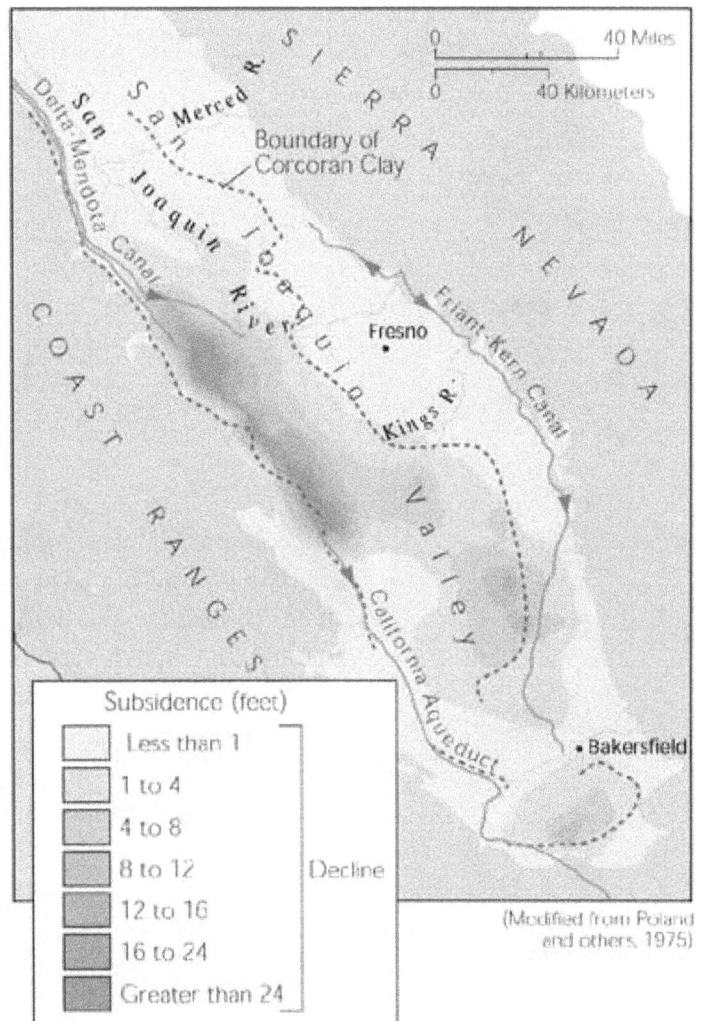

Figure 22. Land subsidence in the San Joaquin Valley, 1926-70 (modified from Poland and others 1975). The approximate location of maximum subsidence (28 feet) in the United States is Mendota, San Joaquin Valley, CA. Signs on the pole show approximate altitude of the land surface in 1925, 1955, and 1977 (Galloway and others 1999).

65

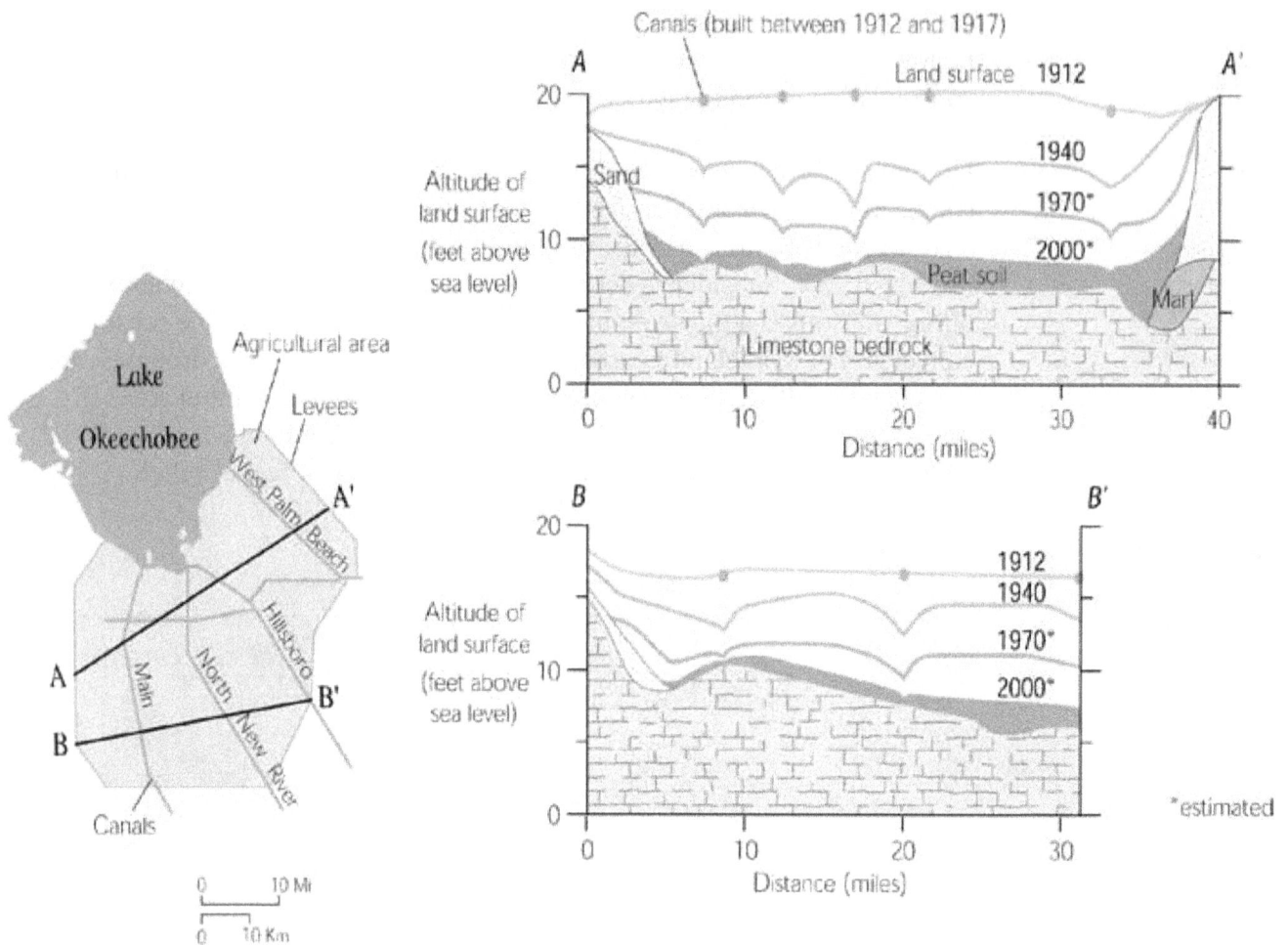

Figure 23. Cross sections through the agricultural area of a portion of former Everglades area, central Florida, showing the decrease in land-surface elevation (Galloway and others 1999).

Discrete collapse features (sinkholes and cavities) tend to be associated with specific rock types, such as evaporites (salt, gypsum, and anhydrite) and carbonates (limestone and dolomite). These rocks are susceptible to dissolution in water and the formation of cavities. This process can occur naturally where these rocks are present near the surface. Evaporite and carbonate rocks underlie about 35 to 40 percent of the United States, but they are buried at great depths in many areas. Collapse of the land surface above cavities can be triggered by ground water-level declines caused by pumping and by enhanced percolation of ground water (fig. 24). It often results in some of the most visually spectacular examples of subsidence. Large-scale pumping can induce sinkholes by abruptly changing ground water levels and disturbing the equilibrium between a buried cavity and the overlying earth materials (Newton 1986). Rapid water-level declines can cause a loss of fluid-pressure support, bringing more weight to bear on the soils and rocks that span the buried voids. As stresses on these supporting materials increase, the cavity roof may fail and the ground suddenly collapse. Although the collapses tend to be highly localized, their effects can extend beyond the collapse zone through the potential facilitation of contaminant movement into and through the ground water system.

Sediments spall into a cavity. As spalling continues, the cohesive covering sediments form a structural arch. The cavity migrates upward by progressive roof collapse. The cavity eventually breaches the ground surface, creating sudden and dramatic sinkholes.

Overburden (mostly clay)

Carbonate bedrock

Figure 24. Cover-collapse sinkholes may develop abruptly and cause catastrophic damage. They occur where the covering sediments contain a significant amount of clay (Galloway and others 1999).

IMPACTS OF SUBSIDENCE

Localized surface impacts of subsidence include earth fissures and sinkholes. Earth fissures occur as a result of ground failure in areas of uneven or differential compaction. Most fissures occur near the margins of alluvial basins or near exposed or shallow buried bedrock in regions where differential land subsidence has occurred. They tend to be concentrated where the thickness of alluvium changes markedly. When they first open, fissures are usually narrow vertical cracks, less than an inch wide and up to hundreds of feet long. They can subsequently lengthen to many thousands of feet and widen to more than 10 feet as a result of erosion and collapse. Vertical offset along the fissure is usually no more than a few inches, but a fissure in central Arizona has a vertical offset of more than 2 feet. Apparent depths of fissures range from a few feet to more than 30 feet.

The large-scale and differential settling of the ground surface that accompanies subsidence can have a profound impact on manmade structures. The cost of damage caused by subsidence is estimated to be millions of dollars each year (National Research Council 1991). Types of potential damage to manmade structures caused by subsidence include the following:

- Damaged roads.
- Broken foundations.
- Severed utilities and pipelines.
- Damaged underground and above-ground storage tanks.
- Damaged storage reservoirs and treatment lagoons.
- Cracked canals and aqueducts.
- Broken well casings and damaged pumps.
- Damaged railroad tracks, bridges and tunnels.
- Flood damage in low-lying areas and damage to flood-control dikes.
- Damage to irrigated fields.
- Loss of property because of catastrophic sinkhole collapse.

Ground Water and Slope Stability

Ground water can play an important role in slope movements because its presence in soil pores reduces slope stability. Slope movements often occur during the wet season, or following major rainfall or snowmelt events (Terzaghi 1950). They are quite common in the forests of the Western United

67

States. Intense rainfall associated with hurricanes and large frontal systems also can produce landslides in forests of the Southeast and on the Caribbean National Forest in Puerto Rico (Neary and Swift 1987). Heavy rainfall on January 24, 1997, triggered the Mill Creek landslide on the Eldorado National Forest in the Sierra Nevada of California. This landslide damaged or destroyed three cabins and dammed the South Fork of the American River for 5 hours; 4 weeks and $4.5 million were required to remove the slide from U.S. Highway 50 (fig. 25) (Reid and LaHusen 1998). In addition to the potential loss of life and property associated with landslides, they result in other environmental impacts such as soil erosion and increased sediment concentrations in streams.

Figure 25. Aerial view of the Mill Creek landslide blocking U.S. Highway 50 (Photo courtesy of Lynn Harrison, CalTrans).

Slopes move when gravitational forces exceed the strength of earth material making up the slopes. These movements, landslides, involve both rock and soil (Cruden and Varnes 1996). Movement of rock or soil masses on a slope is resisted along contacts between rock block surfaces or between individual soil particles, and any fluids present in the voids (spaces) between them tend to decrease such resistance (Kenney 1984). Rock blocks and soil particles derive their resisting strength mainly from friction at the points of contact with surrounding blocks or particles. Increased pore pressure within the voids reduces this resisting strength. As the pore pressure increases, the potential for slope movement increases along planes where gravitational force and resisting force becomes nearly equal (Keppeler and Brown 1998).

In crystalline rock masses, pore pressure changes are rapid and they can lead to slope movement in highly fractured masses (fig. 26). Horizontally bedded rock masses, in which the principal direction of possible movement is horizontal, do not develop pore pressures as great as in rock masses where bedding and the principal direction of movement are parallel to the slope face (fig. 26) (Freeze and Cherry 1979).

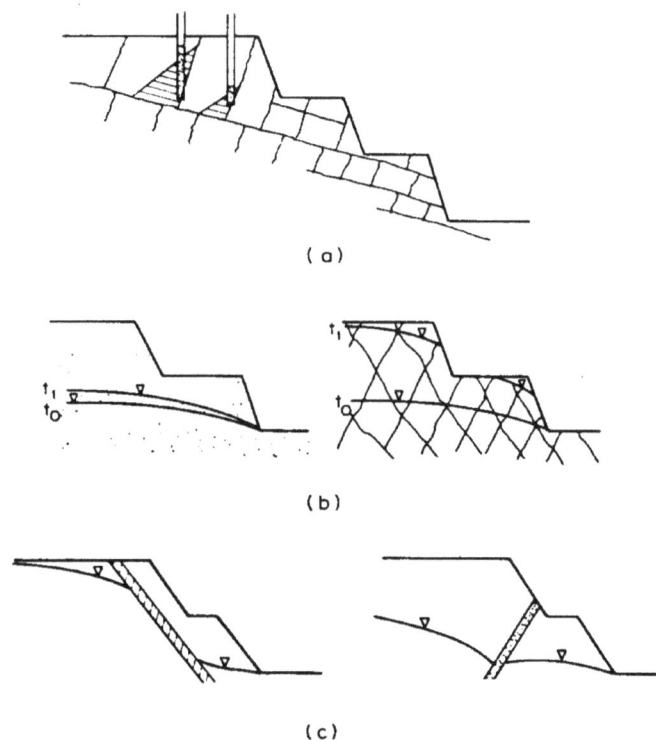

Figure 10.8 Some aspects of groundwater flow in rock slopes. (a) Possible large differences in fluid pressures in adjacent rock joints; (b) comparison of transient water-table fluctuations in porous soil slopes and low-porosity rock slopes; (c) fault as a low-permeability groundwater barrier and as a high-permeability subsurface drain (after Patton and Deere, 1971).

Figure 26. Some aspects of ground water flow in rock slopes: (1) possible large differences in fluid pressure in adjacent rock joints; (2) comparison of transient water-table fluctuations in porous soil slopes and low-porosity rock slopes; (3) fault as a low-permeability ground water barrier, and as a high-permeability subsurface drain (after Patton and Deere 1971).

69

Similarly, granular soils experience a decrease in strength as pore pressure in the intergranular spaces reduces the contact between adjacent particles. In soil with a significant proportion of clay-size particles, the effect of pore pressure is somewhat more complex. Such a soil has resistant strength because of frictional contact between particles and interparticle attraction, called cohesion, between the finer sized particles. Pore pressure increases, however, lead to strength decreases in these soils, too. A detailed presentation of ground water and slope stability can be found in both "*Slope Stability Guide for the National Forests in the United States*" by Prellwitz and others (1994) and "*Landslides— Investigation and Mitigation*" edited by Turner and Schuster (1996).

It has long been recognized in the fields of soil mechanics, agronomy, geological engineering, and environmental geology that soil erosion on a slope depends greatly on the amount of water the slope contains. If ground water recharge is sufficient in a given location to bring the ground water level near to the land surface, the erosion potential of the surface soil in response to runoff events will be much higher than with lower ground water levels. Little infiltration capacity is available when ground water levels are high, so that virtually all of the rainfall or snowmelt that occurs becomes runoff. In addition, the surface soil grains may be nearly buoyant and easily dislodged by runoff water. On even modest hillslopes, bare soils may be eroded rapidly under such circumstances. Gully formation occurs most rapidly under such circumstances, with saturated soils yielding high runoff and offering little resistance to erosion.

High ground water levels adjacent to streams lead to unstable, easily erodible streambanks and streambeds. Just as dry garden soils may be so hard that it is difficult to push in a shovel blade, but when saturated with water are easily worked with a shovel, so too are streambanks much softer and more erodible when saturated than when dry. Streambeds produce sediment much more easily when high ground water levels are providing a buoyant effect (by pushing up through the streambed to discharge into the stream) than when low ground water levels allow water to escape the stream by seeping through the streambed to recharge the ground water that lies below. In addition to increasing particle buoyancy and reducing particle friction, ground water may play a significant role in weathering of streambank materials into smaller, more transportable particles. Pore pressures within soils exposed on the free face of the streambank can detach particles and lead to differences in erosion on the bank face. This seepage pressure often leads to part of the bank being undermined either because of a difference in permeability between the layers of soil exposed or because of the height of the saturated zone. Sufficient undermining then results in a part of the streambank moving as a small landslide or slump. As long as the basic conditions persist, this action continues to modify the streambank.

Where ground water is found at shallow depths, it may dominate and accelerate the headward progression of stream channels and the formation and shape of tributaries. Several field observations, laboratory flume experiments, and computer modeling efforts have provided evidence that the presence of ground water may dominate the initiation and rate of headcut progression. Laboratory flume experiments of headcut migration under hydrogeological conditions similar to those on the Colorado Plateau yielded patterns of stream networks (long valleys, short tributaries, and amphitheater heads) that compared well with field descriptions. Various computer models indicated that headcuts formed spontaneously with the introduction of ground water seepage and that headward erosion rates increased by as much as 60 times.

The presence of water in the subsurface offers several mechanisms to influence geomorphic processes and rock weathering. These mechanisms include freezing and thawing cycles, wetting and drying cycles, chemical dissolution, and particle transport and piping (Higgins and Coates 1990). Often, geomorphic development of gullies, streambank erosion, and sediment production may involve more than one of the mechanisms described below.

Freeze-Thaw Cycles. Water that infiltrates the voids between soil particles (grains) expands as it freezes and exerts sufficient force to disrupt the existing order of the soil particles. At the ground surface, this is often evident as frost heaving, which is an upward swelling of the soil surface during freezing. Similarly, water that percolates into the fractures of otherwise impermeable rock expands as it freezes, widening, deepening, and lengthening the fractures, and initiating other fractures. As many fractures propagate and many others are initiated, this weathering process cleaves pieces of rock from the parent body. Water also enters fine fractures in the cleaved pieces of rock. When it freezes, it leads to additional cleaving and the creation of smaller pieces of rock.

Wet-Dry Cycles. As in the freeze-thaw cycle, water in this process infiltrates into the voids between particles in soils that have a significant proportion of clay particles (Dunne 1990). These cohesive soils tend to crack as they dry. These shrinkage cracks significantly influence the development of gullies. The effect of seasonal wetting and drying can be accentuated when the clays present in the soil have great shrinking and swelling potential. In massive bedrock, this wetting and drying promotes chemical weathering of the rock minerals exposed along cracks penetrated by the ground water. Over time, the zone of weathered minerals associated with soil development on either side of the crack becomes wider.

Chemical Dissolution. Some kinds of rock, such as limestone and gypsum, are somewhat soluble in water. Given sufficient contact time, appreciable masses of these rocks may be dissolved away by the presence of even small volumes of water. Given large volumes of water, the rate of dissolution may be dramatic and the ultimate impacts catastrophic. Karstic limestone, for example, is so riddled with solution cavities and widened fractures that extensive cave systems may form and evolve quickly; the collapse of cavern ceilings and the formation of sinkholes are major adverse repercussions.

Seepage Erosion and Piping. In subsurface strata that consist of uncemented and unconsolidated sediments such as boulders, cobbles, gravels, sands, silts, and clays, it is possible for very fine particles to be transported by water through the voids between larger particles. This process, along with animal burrowing and decaying of plant roots, creates major conduits for water flow. The phenomenon, called "piping," can be a major concern (fig. 27). Some research has focused on the effects of logging on piping (Ziemer 1992).

Civil engineers have spent a great deal of time studying the piping of small particles from the sediments at the bases of dams; leakage through the base of a dam must not be allowed to become large enough to exert buoyant forces at the toe or immediately downstream of the dam. Otherwise, a disaster like the Teton Dam collapse may result. In karst terrain, the solution cavities and passageways open to flow may be large enough to permit very large sediment sizes, such as gravels and cobbles, to be transported. This phenomenon can be a major concern (fig. 27).

Figure 27. Ground water seeps are evident in a roadcut located 30 miles from Lohman, ID. It is clear that ground water seepage has been eroding the rocky face of the roadcut. Note the headward progression of erosion, which may continue until the gully becomes a permanent stream channel fed by ground water.

Midslope and valley-bottom roads in mountainous terrain often intercept and redistribute shallow ground water flow. Effects on ground water-dependent ecosystems and streamflow timing and duration can be significant. Roads also may aid contaminant and hazardous waste migration.

Effects of Vegetation Management on Ground Water

Manipulation of forest vegetation, including both trees and shrubs, can directly and indirectly affect ground water. Vegetation influences the water budget through its effects on water inputs to the basin and more directly through plant water use. By intercepting rain and snow, the vegetation canopy can facilitate water loss to sublimation and evaporation. This interception loss may affect the amount of water available for ground water recharge. By shading ground and water surfaces, vegetation can also influence the rate and timing of snowmelt and evaporation from those surfaces. Plants with access to ground water (phreatophytes) also influence ground water quantity. They take up ground water directly for transpiration. Management activities that intentionally or unintentionally influence the density, structure, and species composition of vegetation may have measurable effects on the quantity and quality of ground water.

Phreatophyte Management

Plants growing in valley bottoms and along river margins generally have better access to water than plants growing in upland areas. Although most phreatophytic plants utilize soil water when available, phreatophytes primarily use ground water (Smith and others 1998). This use may cause quite dramatic diurnal fluctuations in shallow alluvial aquifers in areas near streams. Because of higher water availability in areas adjacent to stream channels and on floodplains, plants growing in these areas generally transpire at higher rates than vegetation in uplands where water is limiting. As a consequence of these high rates of water use by plants with access to ground water, attempts have been made to estimate potential water salvage through the removal of phreatophytes. Although the volumes of salvaged water proposed in these studies are often quite impressive (U.S. Bureau of Reclamation 1963), very few studies have actually demonstrated that removal of even extensive areas of vegetation have resulted in measurable increases in streamflow (Muckel 1966). Most studies have indicated that clearing of phreatophytes results in no measurable change in streamflow (Culler and others 1982, Welder 1988). Removal of phreatophytes, however, does often result in increases in water table elevations in shallow aquifers (Welder 1988) and destabilization of streambanks. Water salvage from removing such vegetation is often significantly less than expected and sometimes results in higher water loss from an area than before removal (Welder 1988). Depending on the depth from the soil surface to the water table, an elevated water table may result in increased evaporative losses from the site if the capillary fringe comes into contact with the atmosphere. Furthermore, water is used by the vegetation that replaces the phreatophytes.

Evapotranspiration in stands dominated by phreatophytes has been estimated to be from 1.1 to 9 acre-feet of water per acre per year in arid areas of the Southwestern United States (Anderson and others 1976). Robinson (1967) reported that annual savings in areas of dense vegetation may amount to 2 to 3 feet of water, depending on depth to the water table. Years of effort and tens of millions of Federal dollars were spent in the 1970s to eliminate or thin phreatophytes in New Mexico, Arizona, Texas, and elsewhere, when phreatophytes were viewed only as "water thieves." The benefits of riparian vegetation to fish, wildlife, and humans are now recognized and far fewer projects to eliminate them are being undertaken (Campbell 1970). The recent drought throughout the Western United States, however, has stimulated a new push for control of nonnative phreatophytes (mainly *Tamarix* spp.[tamarisk]) as well as native species such as mesquite (*Prosopis* spp.) and willow (*Salix* spp.). Two recently (2004) signed bills will commit $100 million to removal of tamarisk from western rivers over several years for the purpose of water salvage.

The presence, density, and composition of phreatophytes can affect the quality of ground water through uptake of nutrients and pollutants. Phreatophytic vegetation has been used for bioremediation of soil and ground water toxicity caused by mining and solid waste disposal. Certain species can take up and store particular ions, heavy metals, and other pollutants. Phreatophytic vegetation may be very effective in removing nitrate from ground water as well as phosphorous and other nutrients (Griffiths and others 1997, Dosskey 2001).

Upland Forest Management

Removal of the forest canopy affects the amount of interception of snow and rain by the canopy, as well as the infiltration rate of the precipitation that reaches the forest floor. Both of these processes can affect ground water recharge and the rate of ground water movement at a local scale. Anderson and others (1976) summarized interception in rain-dominated areas as ranging from about 8 percent of annual precipitation in hardwoods to about 20 percent for conifers. In snow-dominated regions, interception losses ranged from about 10 to 30 percent for conifers. Intercepted water is not available for ground water recharge; however, if the forest canopy is reduced or removed, this water can become available as long as the forest floor has not been compacted by heavy machinery such as log skidders or removed by erosion. Under certain conditions, forest fires can form impermeable layers (hydrophobicity), which hinder or even prevent infiltration of water on the forest floor, limiting water on the ground surface from recharging shallow aquifers. Slow drainage of soil moisture in the range of field capacity is the source of a large proportion of the baseflow of forested headwaters streams, where organic matter content of forest soils tends to be high. Depth of forest soils throughout the country varies widely but generally ranges from 2 to 8 feet before parent material or impermeable layers are found. Some areas like the Midsouth and the Pacific Northwest have deeper forest soils and, hence, deeper rooting zones with probable larger effects on ground water if the tree roots are killed by logging, fire, or other means.

Studies have shown that management of upland forests can increase total annual water yield in a basin, particularly if total annual precipitation in the watershed exceeds 450 millimeters (118 inches) and deep rooted plants can be replaced by shallow-rooted species (Woods 1966, Hibbert 1983). Increases in water yield can be accomplished through mechanical thinning and removal of existing trees and deep rooted shrubs through use of herbicides. The use of herbicides and pesticides, as well as fertilizers, to treat forest stands or selected understory species can affect the quality of ground water and surface water. The fate and transport of these chemicals is reported on elsewhere in this technical guide. The human health aspect of these chemicals in the forested environment is covered extensively in Dissmeyer (2000).

Case Study: Effects of Acidic Precipitation on Ground Water, Hubbard Brook Experimental Forest, NH

Interdisciplinary studies of the northern hardwoods ecosystem at Hubbard Brook Experimental Forest in the White Mountains of New Hampshire began in 1955 by the Northeastern Research Station of the Forest Service. Much of the early focus was on the effects on small watersheds of clear-cutting and herbicide spraying to prevent regrowth of vegetation with respect to streamflow quantity and quality. Over time, the studies expanded to include the effects of acid rain on soil chemistry, nutrient cycling, and ground water chemistry.

Nilsson and others (1982) found that acid deposition could include mobile anions that fall directly on stream surfaces or on soils where they are routed quickly to streams and ground water. They concluded that, over time, acid deposition can be expected to lead to stream-water acidification. Where a hydrologic connection exists between surface water and ground water, it is likely that the ground water also will become increasingly acidic. Surface streams in that part of New Hampshire typically have pH values between 4.0 and 6.0, while the long-term average pH of the precipitation falling at Hubbard Brook is 4.4, which is typical for much of New England, according to Hornbeck and Leak (1992).

Studies at Mirror Lake in the Hubbard Brook Experimental Forest by the U.S. Geological Survey and others found that the extent of ground water flow systems tributary to surface-water bodies were much larger than the surface watershed divides would indicate. Also, multiple vertical layers of ground ground water flows with deeper layers coming from increasingly larger recharge areas. Therefore, the deeper layers would be expected to be fed by larger quantities of acidic precipitation, which was contaminated primarily by sulfate in that area. This conclusion illustrates the need to do a thorough study before significant land- or water-use decisions are made in areas known to experience high levels of acidic deposition.

For more information about effects of acidic deposition on surface and ground water and northern hardwood ecosystems, visit http://www.lternet.edu/ or http://www.hubbardbrook.org The USGS Circular 1139, "Ground Water and Surface Water: A Single Resource" by T.C. Winter and others (1998), also contains information on the effects of acidic deposition on ground water (http://pubs.usgs.gov/circ/circ1139/).

Impacts from Fires

Young and others (2003) present the latest research on the effects of fire on aquatic ecosystems. The effects of fire on ground water have not been well established or researched. The ultimate impacts of fire on ground water are generally manifested as slope failures and increased baseflow in streams and springs. Burned areas typically yield more runoff to streams and more infiltration to ground water, compared to preburn conditions. Although the soil surface is typically rendered slightly to highly hydrophobic by fires, with more intense burns and higher loads of vegetation yielding more hydrophobic character, the hydrophobic soil surface is easily disturbed by differential solar heating and frost heaving. Rapid infiltration of precipitation may then occur through discontinuities in the hydrophobic soil surface. Normally, the forest canopy and ground cover afforded by living vegetation intercept a fair amount of precipitation and much of that is returned to the atmosphere by evaporation. When fire destroys the above-ground tree and plant structures, far less interception of precipitation and its subsequent evaporation occur. The increased infiltration that occurs may result in slope failures, as the moisture contents of the soil and subsoil increase.

Since transpiration and interception of precipitation decrease after a fire (at least until substantial new vegetation growth takes place), ground water discharges to streams and springs often increase after a fire. Year-round baseflow and the severity of floods also increase. Gaining streams exhibit higher flows and losing streams may become gaining streams, depending on the increased magnitude of ground water recharge versus the increases overland flow and runoff from the burned areas. Fire-induced vegetation changes can alter the water-holding capacity of soils, the rate of snow melt, and local water tables, and these factors can lead to changes in the timing of peak and low-water events and the formation of small forest pools (Pilliod and others 2003). Small pools often form in areas of gentle slope after loss of vegetation from logging or fire, because decreased evapotranspiration results in elevated water tables and increased soil saturation.

Ground water is a factor in the ecology of the forest. Its decline in dense forests may be a factor in the decline of species diversity, and its increase in burned areas may cause shifts away from naturally occurring forest species to those that are more competitive in habitats with wetter soils and ponded water. Ground water discharges to wetlands and riparian areas may significantly increase after fires and result in shifts in amphibian populations. Small isolated wetlands are particularly important amphibian habitats, and their formation in burned forests may benefit some amphibians.

Increasing ground water levels may significantly increase the potential for slope failures and landslides. The loss of vegetation during a fire causes soil moisture contents to increase and water tables to rise. When plants die, their roots decay, creating passages through which infiltrating precipitation may move rapidly into the subsurface. The loss of root structures through decay can be an important factor in destabilizing slopes. The increased water content in the affected soils and subsurface sediments may destabilize steep slopes, generating slope failures and landslides.

Wildland/Urban Interface

Residential and commercial development has been rapid adjacent to national forest boundaries and on in-holdings. As water supplies become stressed, land managers will be pressured to permit additional municipal drinking-water wells on NFS land. In the future, ground water management is likely to evolve toward total aquifer management. Protection measures such as limitations on activities in recharge areas, reservation of some areas for production of high-quality water, and protection of unique ground water-dependent ecosystems will be incorporated into land management plans. It will no longer be sufficient to manage for operators and users. Managers must recognize that ground water serves diverse functions, some of which are ecological.

In unincorporated areas, residential growth is characterized by the use of individual domestic wells and individual sewage disposal systems (ISDS; also known as septic systems). In the fractured-rock settings typical of much NFS land, proper siting and design of an ISDS is problematic. The traditional ISDS; design is appropriate for installation in areas underlain by sufficient soil thickness and porous media aquifers. The use of these types of ISDS in fractured-rock settings often results in contamination of nearby domestic wells or surface waters. Primary causes include insufficient filtering and treatment by the typically thin soils that overlie fractured bedrock and the difficulty in determining the nature and orientation of ground water flow paths. When properly designed, installed, and operated, some advanced ISDS systems, such as those based on the "mound" concept, have been effective in many areas. These advanced systems, however, can cost substantially more to install and operate.

Monitoring

This section outlines issues that relate to the design and implementation of ground water monitoring programs. These issues were addressed in the report of the Intergovernmental Task Force on Monitoring Water Quality (Franke 1997), and much of this section is taken from that document. Specific details for designing and implementing a ground water monitoring program can also be found in Sanders and others (2000), proper sampling protocols are described in U.S. Geological Survey (1997 to present), and statistical methods for analyzing water-quality data are found in Helsel and Hirsch (1992).

Monitoring Program Objectives

Ground water monitoring is critical for appropriate water-resource management. The hydrological connections between ground water and surface water mandate that monitoring programs for all water resources be closely linked. By acknowledging this close hydrological connection, ground water monitoring can provide critical support to surface water and ground water management programs.

Monitoring of ground water quality is defined as an integrated activity for obtaining and evaluating information on the physical, chemical, and biological characteristics of ground water in relation to human health, aquifer conditions, and designated ground water and surface water uses. With accurate information, the current state of ground water resources on NFS lands can be better assessed, water-resource protection programs can be run more effectively, and long-term trends in ground water quality and the success of land management programs can be evaluated.

Many Forest Service units do not have the capability or sufficient resources to undertake a water-monitoring program in a short timeframe for all aquifers within their jurisdictions. Therefore, it is recommended that the agency combine resources and talents with others to begin a systematic process of sampling aquifers that are the highest priority, such as those that have the largest human water use. Depending on the availability of resources, this approach may extend the amount of time needed to assess all aquifers in a unit's jurisdiction, but the most important ones from a human health perspective will be addressed first. Monitoring ground water in a systematic manner will gradually result in the development of high-quality, comparable data sets that will increase knowledge of the occurrence and distribution of constituents in ground water and environmental settings where different indicators should be included in monitoring programs.

When designing and implementing monitoring programs, it is vital to consider the differences in the spatial and temporal characteristics of ground and surface waters. Ground water has a three-dimensional distribution within a geological framework and is characterized by contrasting aquifer and geological features. In addition, of course, access is limited because ground water must be sampled through a well or spring. Therefore, the design and implementation of a ground water quality monitoring program must be based on a thorough understanding of the unique hydrogeological characteristics of the ground water flow system under investigation and the locations of particular land uses and other contaminant sources likely to affect ground water quality.

An important aspect of any program for monitoring ground water quality is the sharing and using of data from various sources. One such area of exchange is among programs designed to gather background or ambient-monitoring data. Another is among programs designed to gather data about regulatory compliance.

Monitoring programs have the following general objectives:

- Assess background ground water-quality and quantity conditions.
- Comply with statutory and regulatory mandates.
- Determine changes (or lack of change) in ground water quality and quantity over time to define existing and emerging problems, to guide monitoring and management priorities, and to evaluate effectiveness of land and water management practices and programs.

- Improve understanding of the natural and human-induced factors affecting ground water quality and quantity.

Several types of ground water monitoring are conducted by Federal, State, local, and private organizations to accomplish one or more of the objectives stated above.

Background or Baseline Monitoring

Background or baseline monitoring of water resources often is needed when sampling an area for the first time or in advance of the initiation of a new activity that could affect ground water quality. A wide variety of chemical, physical, and biological contaminants may affect ground water resources (Fetter 1999). As a result, background monitoring programs are designed to establish baseline water-quality characteristics and to investigate long-term trends in resource conditions. Parameters are selected to provide data on general ground water conditions or on conditions relevant to a new activity. Baseline concentrations of elements, species, or chemical substances in ground water present are those that occur naturally from geological, biological or atmospheric sources, or from existing human activities. Established water-quality limits may be exceeded by natural or anthropogenic processes for various elements.

Monitoring for Specific Land-use Impacts

These monitoring programs typically focus on assessing the impact from stressses or contaminant sources that are related to specific land uses. For these monitoring efforts, parameters are identified on the basis of a thorough understanding of the resource to be evaluated and the sources of stress or contamination.

CASE STUDY: GROUND WATER MONITORING IN THE TURKEY CREEK WATERSHED, JEFFERSON COUNTY, CO

The Turkey Creek watershed in Jefferson County, southwest of Denver, CO, is an area of rapidly developing communities in the foothills of the Colorado Rocky Mountains. About 5,000 households in the watershed depend on domestic wells for their water needs, and individual septic-disposal systems are used for wastewater. County government agencies are concerned about the impacts of the development on water quality and water quantity.

To understand the hydrological conditions in the Turkey Creek watershed, the USGS and Jefferson County undertook a cooperative study to evaluate the water resources of the watershed from 1998–2001. A critical component of this study was the establishment of monitoring networks for both ground water levels and ground water quality. These networks provided baseline hydrologic information as well as data necessary for the construction and calibration of a precipitation-runoff model of the study area.

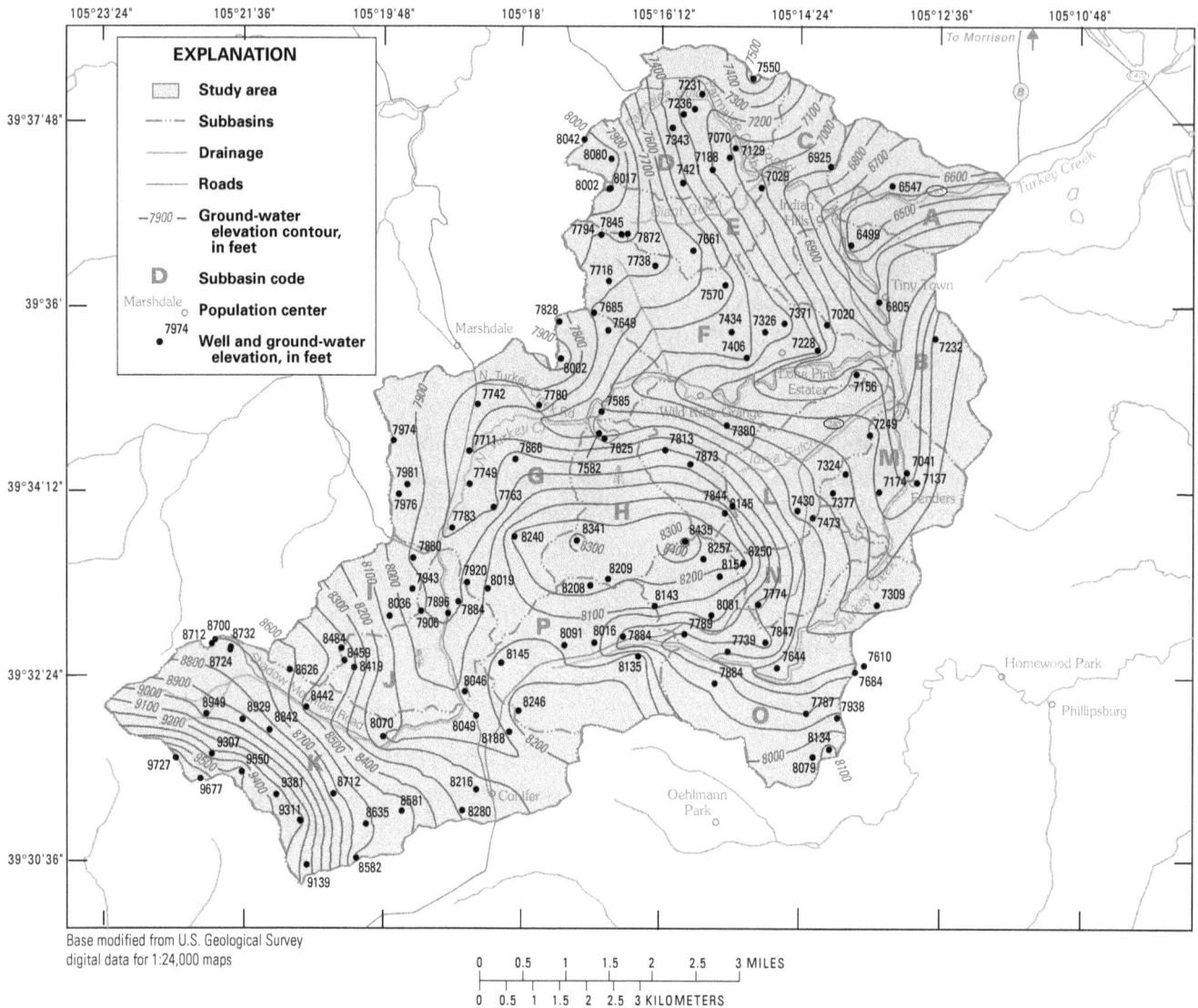

Figure 28. Ground water table in the Turkey Creek watershed, September 2001.

Ground water levels were monitored monthly at 15 monitoring wells, beginning in 1999. These wells are no longer used by homeowners, and are considered reliable indicators of static water levels. Three of the wells are shallow, hand-dug wells, and the remaining wells are completed at depths ranging from 70 to 505 feet. Water levels also were measured in 131 domestic wells from September 24 to October 4, 2001 to complete a water-table map for that time period. The resulting map (fig. 28) can be used to indicate areas of ground water recharge and discharge and directions of ground water flow. The water table generally mimics the topography, with ground water flowing from higher recharge areas to lower discharge areas near streams.

Water-quality data were obtained quarterly from 22 surface water sites and 110 wells and springs from October 1998 to September 1999. A few miscellaneous samples were obtained during 2000 and 2001 to fill data gaps. Samples were analyzed for temperature, specific conductance, major ions, nutrients, bacteria, and minor elements from 1998 to 1999 (fig. 29). During the 2000–01

80

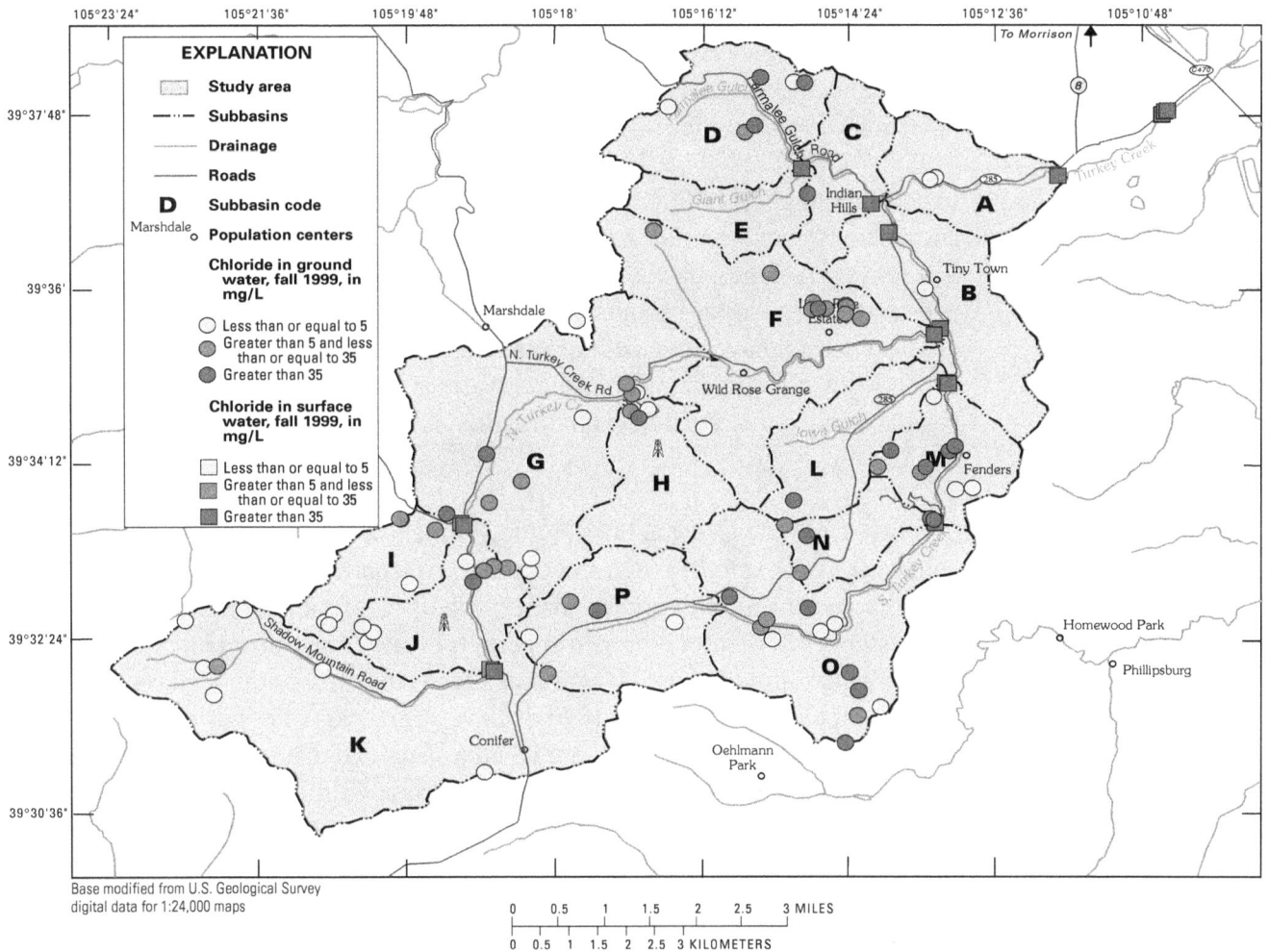

Figure 29. Chloride concentrations in the Turkey Creek watershed, fall 1999.

sampling, water was analyzed for bromide, selected inorganic ions, wastewater compounds indicative of septic-tank effluent, and tritium, which is indicative of modern recharge. The water-quality data indicate that some wastewater compounds are present in the ground water and surface water, and that most of the ground water has been recharged within the past 50 years.

For additional information see Bossong and others (2003). Also see http://water.usgs.gov/pubs/wri/wri03-4034/.

Monitoring for Facility-based Compliance

Monitoring is often needed to be certain that specific facilities are complying with specific regulatory requirements or permit conditions. These efforts may be designed to comply with various laws such as RCRA, or to support remedial activities such as those under CERCLA (Superfund).

Monitoring of Ground Water-dependent Ecosystems

Monitoring is required to inform management and to help develop an understanding of ecological processes in ground water-dependent ecosystems. It should address the environmental condition of ground water-dependent ecosystems at particular points in time and the trend in condition over time. Subject to resource availability, such monitoring could address key ecological processes and any changes in vulnerability to processes or events that threaten the integrity of the ecosystem. Monitoring of important processes such as water regime and allocation, water quality, and usage of ground water will enable detection of changes detrimental to the health of the ecosystem.

In addition to contributing flow to surface waters, ground water directly sustains wetlands, riparian zones, meadows, marshes, some forest tree stands and some grasslands, as well as aquatic species in lakes, streams, cave systems, and springs. The loss of ground water flow to these ground water-dependent ecosystems can have adverse impacts on the flora and fauna of the NFS. The dependency of ecosystems on ground water is based on at least one of three basic ground water attributes: quality, flux, and level (Sinclair Knight Merz Pty Ltd. 2001).

Quality. Ground water quality is typically measured in terms of electrical conductivity (or salinity), nutrient content, concentrations of major ions, and/or concentrations of contaminants such as metals and organic chemicals. Ecosystems and their component species typically function adequately over certain ranges in water quality. Outside these ranges, the composition and health of the ecosystem is likely to decline. A ground water attribute can become important to an ecosystem when a sustained change in quality or trend away from the natural water-quality state occurs. Salinity is typically key inorganic indicator of ground water quality for such ecosystems. Terrestrial ecosystems may also be sensitive to ground water contamination by nutrients, pesticides, or metals, but little is known about most ecosystem responses. Phytotoxicity of metals, however, has been established for some plant species (Kabata-Pendias and Pendias 2000).

Flux. Ground water flux or flow is the rate of surface or subsurface discharge of an aquifer. It is relevant to the provision of an adequate quantity of water to sustain an ecosystem or of a sufficient quantity to dilute more saline water (in wetland systems) to allow an ecosystem to function. Quantity of water is critical to ecosystems that occupy discharged ground water, such as cave systems, aquatic ecosystems in baseflow-dependent streams and many ground water-fed wetlands, or ecosystems whose sole or principal source of water is ground water. For terrestrial vegetation, the ground water flux needs to be sufficient to sustain a level of uptake by vegetation that at least partly satisfies evaporative demand.

Level. Ground water level is the depth of the water table. It is relevant to a broad range of ecosystems, including wetlands fed by unconfined aquifers, many coastal lacustrine and estuarine ecosystems, some cave and aquifer ecosystems, and baseflow-dependent ecosystems. The ecosystem's location or usage of ground water depends on the level of the water table remaining within a certain range. Aquifer pressure has a similar role in ecosystems fed by confined aquifers to that of level in systems fed by unconfined aquifers. It determines discharge rates from springs and from fractured bedrock aquifers.

The response of ecosystems to change in these attributes is variable. There may be a threshold response in some cases, whereby an ecosystem collapses completely if a certain required attribute value is not met. Examples might be springs or fens supported by ground water discharge. These would cease to exist if pressures in the supporting aquifer fell to the point where no further surface discharge occurred. In other cases, a more gradual change in the health, composition, or ecological function of communities is expected. For example, an ecosystem may change slowly in response to gradually increasing ground water salinity.

An assessment of ecosystem dependency on ground water can be performed by identifying ecosystem traits that imply such dependency (Sinclair Knight Merz Pty Ltd. 2001). A high level of dependency on ground water makes an ecosystem vulnerable to change in water regime. Many of such ecosystems have relatively high levels of endemism. The following checklist can be used to help determine ground water dependency:

- Is the ecosystem identical or similar to another that is known to be ground water dependent?
- Is the distribution of the ecosystem associated with surface water bodies that are or are likely to be ground water dependent? Examples are permanent wetlands and streams with consistent or increasing flow along the flow path during extended dry periods.
- Is the distribution of the ecosystem consistently associated with known areas of ground water discharge from springs or seeps?
- Is the distribution of the ecosystem typically confined to locations where ground water is known or expected to be shallow, such as topographically low areas and major breaks of topographic slope?
- Does the ecosystem withstand prolonged dry conditions without obvious signs of water stress?
- Is the vegetation community known to function as a refuge for mobile fauna during times of drought?
- Does the vegetation in a particular community support a greater leaf area index and more diverse structure than those in nearby areas in somewhat different positions in the landscape?
- Does expert opinion indicate that the ecosystem is ground water dependent?

CASE STUDY: MONITORING OF GROUND WATER-DEPENDENT ECOSYSTEMS, ARROWHEAD TUNNELS, SAN BERNARDINO MOUNTAINS, CA

Underground tunnel construction can disrupt ground water flow systems and cause dewatering of overlying springs and riparian areas that provide valuable habitat for flora and fauna. The surface resources monitoring and mitigation plan for the Arrowhead East and West Tunnels Project is an example in which the potential for ground water disruption was recognized and addressed. The Forest Service objective is to maintain ecosystem health at each of the surface water features being monitored.

The Arrowhead East and West tunnels are part of the Metropolitan Water District (MWD) of Southern California's Inland Feeder System, which connects the California and Colorado Aqueducts in the southern part of California. The project consists of two 16-foot diameter tunnel segments that pass under the San Bernardino National Forest (fig. 30). The combined length of the tunnels will be more than 8 miles and depths will reach 2,040 feet under the San Bernardino Mountains. The tunnels cross active splays of the San Andreas Fault in three locations. Ground water levels are as high as 1,100 feet above the tunnels, and significant ground water inflows have been encountered during the tunneling operations.

A key requirement of the Arrowhead Tunnels project is to protect the water resources in the San Bernardino National Forest. Limits have been placed on ground water inflows into the tunnels by the Forest Service under a special-use permit issued to the MWD. In 1993, the MWD and the Forest Service recognized the potential for construction of the Arrowhead Tunnels to affect local surface water resources in the San Bernardino Mountains, and they adopted a water-resources monitoring and mitigation plan that will provide data that can be used to identify construction-related effects. The plan targets selected ground water-dependent surface water and biological resources in the vicinity of the tunnel segments.

A total of 126 spring, stream, rain gage, and well sites were identified for monitoring during the preconstruction, construction, and postconstruction periods (fig. 30). The monitoring effort has provided an unprecedented amount of data about the hydrological characteristics of the ground water regime in the project area. These data have proven to be quite valuable for assessing the hydrological trends across the mountains and for identifying the variables that most significantly influence baseflow at the spring and stream monitoring points.

In addition, biological monitoring and several focused biological surveys have been completed for mollusks (fig. 31), amphibians (fig. 32), birds, and riparian vegetation. These have provided detailed information about the plant and animal species in the project area that depend on ground water discharge and their responses to normal fluctuations in rainfall, temperature, and wildfires. Spring snails (fig. 31) in the Transverse Range of Southern California, are particularly good indicators of spring ecosystem health because they occupy only springs that are minimally disturbed and have persisted for thousands of years. They do not occupy habitats that are scoured by floods, or that periodically dry, and they are susceptible to establishment of nonnative species and cultural activities affecting the quality of spring-fed aquatic habitats.

Figure 30. Aerial photograph of showing the East Tunnel alignment and monitoring points on the San Bernadino National Forest.

Regression models have been developed to help estimate what the seasonally adjusted baseflow should be at each surface-monitoring site within 2,500 feet of the tunnels. In addition to surface-related variables, ground water levels in wells, tunnel-heading inflow, probe-hole flow, and portal discharge are regularly measured. These data will provide early warning of the potential for a surface-related impact from tunnel construction. They also will be used to corroborate the occurrence and to assess the magnitude and extent of tunnel-related surface impacts.

The flow chart in figure 33 shows the mitigation triggers that would occur if a tunnel-related hydrologic effect were suspected. During the supplemental biological monitoring phase, plant-water potential, general observations of plant health, soil moisture, and animal condition and habitat will be evaluated and compared with reference sites. Indicators that provide important information for evaluations include (1) willow or sycamore trees that begin dropping leaves early in the season, (2) the herbaceous understory that begins to desiccate early in the season, (3) soil moisture readings that indicate unusual drying of soils, and (4) water potential measurements at predawn that are increasingly elevated. For more information on spring snails see Sada (2002).

85

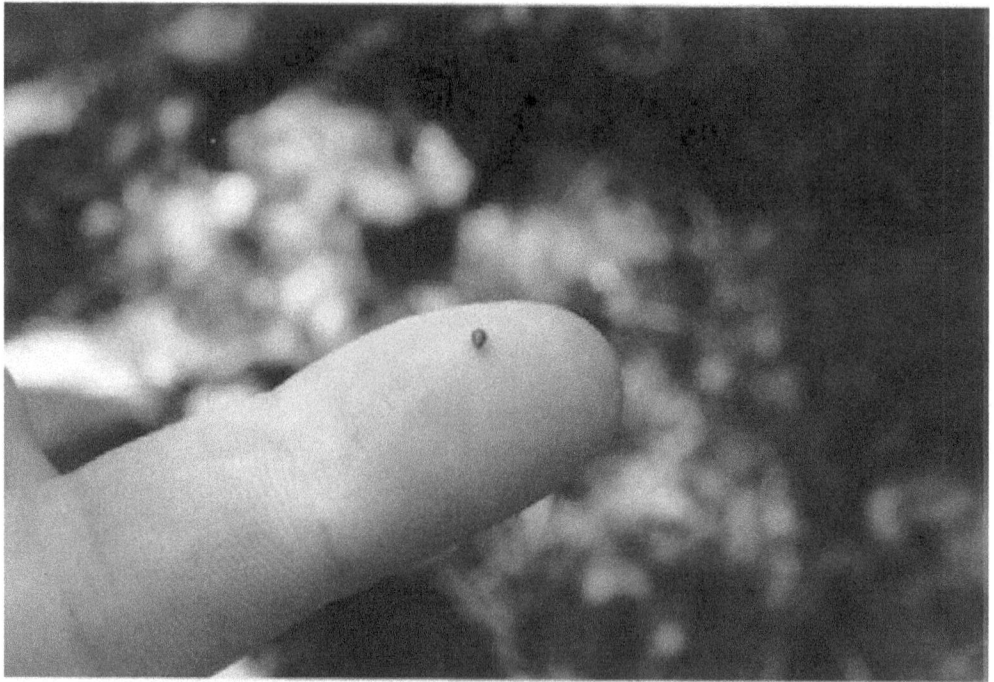

Figure 31. Spring snail, **Pyrgulopsis californiensis.**

Figure 32. Western spadefoot toad, **Spea hammondii.**

86

Monitoring for Environmental Change

Geoindicators have been developed to assist in assessments of natural environments and ecosystems. They provide an approach for identifying rapid changes in the natural environment (Berger and Iams 1996). An international working group of the International Union of Geological Sciences developed geoindicators to assess common geological processes occurring at or near the Earth's surface that may undergo significant change in magnitude, frequency, trend, or rate over periods of 100 years or less. Geoindicators measure both catastrophic events and those that are more gradual but evident within a human lifespan. Geoindicators that focus on environmental changes in ground water systems are described here. For the purpose of this technical guide, they are called hydrogeoindicators. As descriptors of hydrogeological processes that operate in many settings, hydrogeoindicators can be used by the Forest Service to monitor natural as well as human-induced changes in ground water systems and in the ecosystems they sustain.

The most effective use of hydrogeoindicators is in environmental monitoring programs. They are designed for use on local or national scales. They can help to answer four basic questions:

1. What is happening in the environment (conditions and trends)?
2. Why is it happening (causes, links between human influences and natural processes)?
3. Why is it significant (ecological, economic, and health effects)?
4. What are we doing about it (implications for planning and policy)?

The use of hydrogeoindicators presents several specific challenges. One is to define more closely the thresholds or critical levels involved, so that it is possible to specifically express the relative stability of a particular environment to management. For each indicator, target, trend, or threshold values will need to be set. If a threshold is reached, action of some type should be required. Eleven important hydrogeoindicators are presented in table 2. Appropriate indicators can be selected from this list depending on the terrain and the environmental issues under consideration.

Edmunds (1996) proposed a monitoring scheme for ground water designed to detect changing conditions using a set of parameters that have global or regional significance and undergo changes over a time scale of 50 to 100 years. The primary and secondary indicators shown in table 3 monitor both natural changes in ground water chemistry and effects from human influences.

These indicators have been developed from standard approaches used in geology, geochemistry, geophysics, geomorphology, hydrology, and other earth sciences. For the most part, the expertise and technology already exist to monitor and analyze the resulting data and most indicators are relatively simple and inexpensive to apply.

Water Resource Mitigation Flow Chart

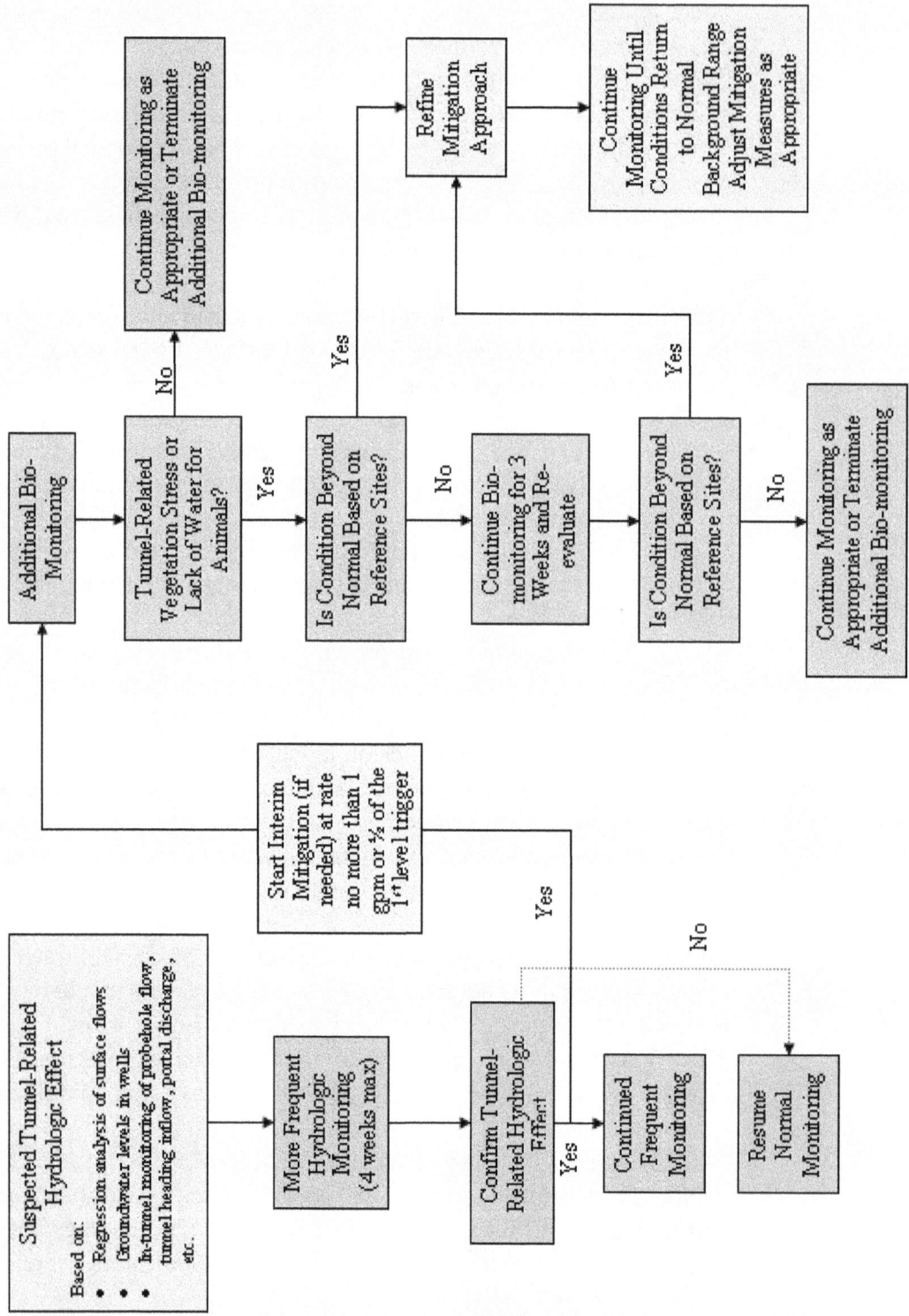

Additional Bio-Monitoring

Tunnel-Related Vegetation Stress or Lack of Water for Animals? — No → Continue Monitoring as Appropriate or Terminate Additional Bio-monitoring

Yes ↓

Is Condition Beyond Normal Based on Reference Sites? — Yes → Refine Mitigation Approach → Continue Monitoring Until Conditions Return to Normal Background Range Adjust Mitigation Measures as Appropriate

No ↓

Continue Bio-monitoring for 3 Weeks and Re-evaluate

Is Condition Beyond Normal Based on Reference Sites? — Yes → (to Refine Mitigation Approach)

No →

Continue Monitoring as Appropriate or Terminate Additional Bio-monitoring

Start Interim Mitigation (if needed) at rate no more than 1 gpm or ½ of the 1st level trigger

Suspected Tunnel-Related Hydrologic Effect

Based on:
- Regression analysis of surface flows
- Groundwater levels in wells
- In-tunnel monitoring of probehole flow, tunnel heading inflow, portal discharge, etc.

More Frequent Hydrologic Monitoring (4 weeks max)

Confirm Tunnel-Related Hydrologic Effect — Yes → Continued Frequent Monitoring

No → Resume Normal Monitoring

Yes (up to Start Interim Mitigation)

88

Figure 33. Mitigation flow chart for the Arrowhead Tunnels Project.

Ecosystem management, reporting, and planning generally focus on biological issues such as biodiversity, threatened and endangered species, exotic species, and biological and chemical parameters that describe air and water quality. Much less attention is paid to the physical processes that shape the landscape—the natural, changing foundation on which humans and all other organisms live and function.

Hydrogeoindicators can help answer Forest Service resource management questions about what is happening to the hydrological environment, why it is happening, and whether it is significant. They can establish baseline conditions and trends, so that human-induced changes can be identified. Applying this approach will provide science-based information to support resource management decisions and planning. Hydrogeoindicators help non-geoscientists focus on key geological issues. They can help forests managers to anticipate changes that might occur in the future, and to identify potential management concerns from a hydrogeological perspective.

Hydrogeological processes are integral to forest management and planning. When measures of natural change are omitted from monitoring and planning, the assumption that natural systems are stable, fixed, and in equilibrium is perpetuated. Natural systems are dynamic, and some may be chaotic; change is the rule, not the exception. Using hydrogeoindicators shifts management actions from response (crisis mode) to long-range planning, so issues can be recognized before they become serious concerns.

Hydrogeoindicator	Significance	Types of monitoring sites	Method of measurement	Frequency of measurement	Thresholds
Ground water quality	The chemical composition of ground water can be used as a measure of its suitability for human and animal consumption, irrigation, and for industrial and other purposes. It also influences ecosystem health and function, so it is important to detect change and early warnings of change, both natural and resulting from human activity.	Wells, springs, wetlands, adit discharges, lakes, stream baseflow. Focus on major aquifers providing water supplies or substantial discharge to important surface waters. Monitor downgradient of potential problem areas such as mines, urban areas, waste disposal sites, burned areas, and timbered areas. Relate individual pollutants to their sources; include sampling of potential sources when they can be identified and accessed. Wherever possible, integrate monitoring with other national, State, or local water-quality networks.	Standard sampling and laboratory techniques and equipment. Some monitoring can be conducted remotely using dataloggers and sensors placed in wells or at points of ground water discharge. In many cases, it will be important to collect data over a time period sufficient to be able identify normal seasonal variability. Statistical analysis of temporal and synoptic data may be appropriate. [Specialized sampling and analysis for isotopes]	Usually on a seasonal or annual basis; maximum frequency of 4 times a year is suggested to detect changes in shallow ground water sources, but annual measurements are often sufficient for deeper sources.	Water-quality standards for applicable beneficial uses, established trends, statistically significant changes in concentration.
Ground water level (elevation)	The availability of water is of fundamental importance to the sustainability of life. Ground water levels are essential to be able to identify the extent of the resource and to determine the recharge. Regularly measure water levels in wells and boreholes and/or ground water-fed surface waters. Results are the simplest indicator of changes in ground water resources. The level of ground- water is an essential parameter to understand the ground water system.	Boreholes, wells, and ground water discharge areas associated with springs, streams, lakes, and wetlands representative of the particular aquifer.	Depth to the water table can be measured manually, with automatic water-level recorders, or with pressure transducers. Standard hydrogeological methods are used to calculate a water balance for the ground water system of interest.	At least monthly to reflect seasonal as well as annual changes. The state of fossil aquifers should be assessed at about 5-year intervals. Water levels can be measured both seasonally and annually over decades to determine overall trends.	A threshold is crossed when the rate of extraction exceeds the rate of recharge, and a sustainable, renewable resource becomes a non-renewable, mined one. Drying up of wells, springs, wetlands, and so on.

Table 2—Cont.

Hydrogeoindicator	Significance	Types of monitoring sites	Method of measurement	Frequency of measurement	Thresholds
Vadose zone chemistry	Changes in recharge rates have a direct relationship to water resource availability. The unsaturated zone may store and transmit contaminants, the release of which may have a gradual or sudden adverse impact on ground water quality.	Unconsolidated sediments or consolidated porous media (sand, till, sandstone, chalk, calcarenite, volcanic ash) on relatively level terrain that has negligible surface runoff. The best records are obtained where the unsaturated zone is 10–30m thick, and where sediments and flow are relatively homogeneous.	Sampling from either material samples from borings or from lysimeters. Samples can be collected from borings/wells placed by hollow stem auger, percussion, air-flush rotary, or dual tube drilling. Pore water is extracted from sediments by high-speed centrifuge (drainage or immiscible liquid displacement) or, for nonreactive components such as Cl and NO_3, by elution with de-ionized water. For isotopic samples (^3H, $\delta^{18}O$, δ^2H), vacuum distillation may be used. Lysimeters are monitoring devices designed to collect soil water, either as drainage (basins) or from within the matrix (suction).	For collected soil samples, 5- to 10-year intervals to confirm movement of solutes toward the water table. For soil water, sampling quarterly or more frequently, to ensure proper operation of the equipment, until conditions are identified. May be repeated with new equipment installations on a similar 5- to 10-year basis.	Vadose zone begins to measurably affect saturated zone water quality.
Stream and spring baseflow	Ground water discharge to streams and springs is vital for the regulation and maintenance of aquatic health and biodiversity. Human-induced depletion of baseflow has major implications for the health of riparian ecosystems.	Stream channels and springs.	Standard techniques for measuring streamflow, hydrograph separation. Synoptic sampling—tracer injection studies, streambank piezometers.	Continuous to periodic.	Established critical flow for sustaining healthy aquatic ecosystems.
Surface water quality	The quality of surface water in rivers, streams, lakes, ponds, and wetlands is influenced by interactions with ground water. The bulk of the solutes in surface water are often derived from ground water baseflow where the influences of water-rock interactions are important.	Stream channels, lakes, wetlands, and springs.	Standard sampling and laboratory techniques and equipment. Statistical analysis of data. [Specialized sampling and analysis is needed for isotopes and trace metals.] Bioindicators for inferring past lake- water chemistry.	4–6 times yearly for major ions, twice yearly for radionuclides and organic chemicals. Continuous, real-time monitoring systems provide the most complete information for field parameters.	Water-quality standards and trends. Human and aquatic health criteria.

Table 2 – Cont.

Hydrogeoindicator	Significance	Types of monitoring sites	Method of measurement	Frequency of measurement	Thresholds
Lake levels	Lakes are sensitive to local climate and to land-use changes in the surrounding landscape. Lakes can also be valuable indicators of near-surface ground water conditions. Where not directly affected by human activities, lake level fluctuations are excellent indicators of drought conditions. Lake level fluctuations vary with the water balance of the lake and its catchment, and may reflect changes in shallow ground water resources.	Lakes with ground water exchange.	Shoreline gauges. Areal extent from successive aerial photos, satellite images, and geomorphology.	Lake level monthly to annually. Areal extent every 5 years.	Critical inflow-outflow-storage water balance perturbation.
Subsurface temperature	The thermal regime of soils and bedrock strongly influences the soil ecosystem, near-surface chemical reactions involving ground water, and the ability of these materials to sequester or release greenhouse gases. It may affect the type, productivity, and decay of plants, the availability and retention of water, the rate of nutrient cycling, and the activities of soil microfauna.	Sites remote sites from human disturbances, bodies of surface water, or areas of high geothermal flow where the ground cover is left undisturbed. The best results are obtained from measurements in relatively impermeable bedrock or where there has been minimal ground water movement.	Data loggers, thermocouples, thermistors.	Once every 5 years for deep boreholes, more frequently (as often as twice daily) for near-surface temperatures.	Operationally defined change in temperature.
Wetlands	Wetlands have high biological productivity and diversity. They are important for wildlife habitat, water storage, and human recreation. Wetlands can affect local hydrology by acting as filters, sequestering and storing heavy metals and other pollutants, and serving as flood buffers.	Ground water supported fens and marshes.	Areal extent and distribution. Permanent transects and plots can be set up for ease of data comparison and establishing temporal trends in vegetation distribution, surface morphology, accumulation rates, hydroperiods, water levels, and hydrochemistry. Piezometers, wells, and weirs can be used. Variations in the chemistry of water inflows and outflows, changes in water levels and in seasonality of flow patterns.	Every 5–10 years for distribution, extent, and structure; for water levels, and hydrochemistry, initial measurements should be weekly to monthly (more frequently in times of rapid change such as spring thaw) until important times and parameters have been identified, then less frequently.	Change in areal extent, vegetation community type, water-flow regime.

Table 2 – Cont.

Hydrogeoindicator	Significance	Types of monitoring sites	Method of measurement	Frequency of measurement	Thresholds
Karst features	Karst landscapes occupy up to 10% of the Earth's land surface, and as much as a quarter of the world's population is supplied by karst water. The karst system is sensitive to many environmental factors. Instability of karst surfaces leads annually to millions of dollars of damage to roads, buildings, and other structures. Radon levels in karst ground water tend to be high in some regions.	Caves allow direct observation and mapping of underground features and their relation to the surface and to ground water flow. Wells, borings, and quarries may be less useful as monitoring sites because they may provide only discontinuous points of information.	Pumping tests on wells, dye tracing. Hydrological and geochemical measurements of springs, sinking streams, drip waters into caves, and cave streams provide records of short-term changes in water quality and chemical processes. In built-up areas, locate buried cavities and monitor their potential for collapse, using a combination of geophysical surveys, exploratory drilling, and repeated leveling.	Continuous measurements are needed to interpret the karst system. Surface features, ground water chemistry, and contamination in karst terrains are notoriously unstable and can change rapidly.	Operationally defined change between dissolution and precipitation of calcite, critical change in water flow or water quality.
Slope failure	Slope failure is one of the most widespread causes of land disturbance, so the initiation and development of landslides should be closely monitored. Ground water is a major factor in causing landslides.	The highest part (crown) of landslides and other potential slope failures is generally the most important place for monitoring cracks, subsidence, and sagging. Upheaval or buckling generally begins in the toe area. As failure progresses and the slide or flow develops, cracks and ground subsidence may form at any point, including the toe.	Satellite images, air photos, repeated conventional surveying, installation of various instruments to measure movements directly, and inclinometers to record changes in slope inclination near cracks and areas of greatest vertical movement. Subsurface methods include installation of inclinometers and rock noise instruments, and geophysical techniques for locating shear surfaces.	Dictated by changes in rate of crack propagation and ground deformation and by the degree of potential damage if the slope fails. If little activity has taken place in a particular area, re-assessment can be delayed for several years or more. Critical periods for monitoring are during and immediately after intense rains and rapid snowmelt.	Slope instability affects or threatens structures such as roads or buildings.
Subsidence/ Surface displacement	Subsidence from extraction of ground water or oil and gas, and mining activities can damage buildings, foundations, and other structures. Displacements of the ground surface can be used to assess and warn of environmental problems, especially in areas liable to subsidence from bedrock solution, mining and fluid extraction.	Areas extracting ground water, oil or gas, underground mines, active fault zones, and reservoirs.	Repeated precise leveling and ground surveys. Standard geodetic techniques, especially using Global Positioning System and laser range finders.	Depends on the nature of the movement taking place.	A threshold is crossed when the rate of movement crosses an operationally defined value or range, affects or threatens structures.

93

Table 3. Recommended indicators in the ground water environment (after Edmunds 1996).

Priority	Issue	Primary indicators								Secondary indicators	Frequency of measurement (yrs)
		Water level	HCO_3		pH	DOC[2]		Cl	SO_4		
*	Changing water table	x								Spring discharge	0.25
*	Total ground water reserves	x								Water-quality indices; storage changes	5
*	Acid neutralization		x		x					Al, Ca	0.5
	Salinity							x		Mg/Cl, Br, $\delta^{18}O$, δ^2H, total dissolved solids, SpC[3]	0.5
	Agricultural impact					x	x			K, Na, PO_4, pesticides	0.5
	Urban industrial impact		x			x		x		B, PO_4, solvents, metals	0.5
	Radioactive contamination			x						^3H, ^{36}Cl, ^{85}Kr	2
	Aquifer redox status						x			Eh, Fe^{2+}, HS	0.5
*	Land-use/forestry change						x	x			2
	Depletion of paleowater	x				s				$\delta^{18}O$, δ^2H, ^{14}C, CFC's	2
*	Changing recharge and climatic influence							x		$\delta^{18}O$, δ^2H,	2
*	Mining impact				x				x	Metals	0.5

[1]DO = dissolved oxygen
[2]DOC = dissolved organic carbon
[3]SpC = specific conductance (electrical conductivity at 25°C)

94

Staffing and Resource Needs

In the United States, management of the development and use of ground water resources is primarily the responsibility of State and local governments. No Federal law or regulation applies across the entire country. Increasingly, the Forest Service is involved in ground water issues to ensure that ground water users or ecological resources on NFS lands will not be impacted and that development will not impair ground water quality. Restoration or remediation of contaminated ground water is typically achieved under the authorities established in Federal laws; for example, CERCLA, RCRA, SDWA, and CWA. Remediation projects are typically directed and overseen by appropriate State or Federal agencies. The following staff and resource requirements are considered essential for effective management of ground water resources on NFS lands.

Expertise

A staff with the pertinent expertise is critical for any ground water management organizational unit or program. The study of ground water is interdisciplinary. It requires knowledge of many of the basic principles of geology, physics, chemistry, and mathematics. The most appropriate disciplines are hydrogeology, hydrology, and geochemistry. Ground water occurs, flows, and obtains its chemical signature in the geological environment; therefore, it is critical to be able to characterize and understand the geological environment and its control on the movement and chemistry of ground water. Knowledge in the fields of biology, geophysics, soil science, and statistics, and geographic information systems (GIS) provides the hydrogeologist with additional tools needed to describe ground water flow systems and to assess human impact on these systems.

Many universities offer specific degrees in these areas. It is not appropriate to staff a ground water program with people who have training and university degrees that do not include these areas. Unfortunately, it has been very common to staff ground water programs in other organizations with people who have some experience in water-related issues but no formal training as ground water scientists.

Hardware/ Software

Ground water scientists rely on data analysis, mapping and analytical and numerical models to help develop and evolve conceptual understandings related to ground water flow, chemistry, and interaction with surface water. Sound conceptual understandings are essential for wise management of ground water resources in a given aquifer or area.

Computer models. Many sophisticated models have been developed for simulating ground water flow and contaminant transport. Purchasing and using these models can be expensive and time consuming. Public domain ground water models, such as MT3D and the USGS code MODFLOW, have been extensively used and improved and are available at nominal cost. The same is true for public domain geochemical speciation and mixing models such as EPA's MINTEQA2 or USGS' PHREEQC. Other proprietary software for modeling or developing model inputs is also available.

95

GIS. In the past 10 years the development and use of GIS software has increased dramatically. GIS software can be used to perform spatial analyses and develop spatially-based input files for other programs. GIS technology is essential for performing ground water inventories.

Other applicable software. Geological software packages such as ROCKWORKS include a number of analytical and semi-analytical programs, data plotting programs, and cross-section programs. Statistical analysis of water-quality data can be performed using WQStat Plus. Geochemical analysis of water chemistry data can be evaluated with Aquachem.

Field and Laboratory Requirements

A ground water program requires field staff and access to facilities for water-quality analysis. It is common for many State and Federal water-quality programs to require responsible parties to bear the analytical costs, but this requirement may not be realistic when a ground water inventory is needed. There are hundreds of EPA-certified water-quality laboratories across the United States. EPA certification ensures the use of consistent sample management protocols and analytical and reporting methods. All analytical water-quality testing in support of Forest Service or special-use activities on NFS lands should be conducted by EPA- or State- certified laboratories.

Some standard water-quality measurements should be conducted in the field at the time of sampling. Standard field equipment for the hydrogeologist includes a pH meter, specific conductance meter, temperature meter, water-level indicator, bailer, sampling pump, and water-sampling equipment.

Training

It is imperative that any ground water management staff have access to continuing education. The fields of hydrogeology, hydrology, and geochemistry are dynamic. New things are learned and new tools and techniques are being developed regularly. Budgets for ground water programs should include adequate funds for training. A number of national and international professional organizations specialize in ground water and provide training opportunities. These include the National Ground Water Association, the International Association of Hydrogeologists, the Geological Society of America, the American Geophysical Union, and the American Institute of Hydrology. Many States have active ground water or water-resources associations. In addition, certifications for ground water professionals can be acquired through the National Ground Water Association, the American Institute of Hydrology, and many States.

Part 3. Hydrogeologic Principles and Methods of Investigation

Basic Hydrogeologic Principles

Hierarchical Classification of Aquifers

The following discussion is abstracted from Heath (1984), to which the reader is referred for additional details of the ground water characteristics of individual regions. Additionally, an updated (2001) "Ground Water Atlas of the United States" is available on line from the USGS at http://capp.water.usgs.gov/gwa/gwa.html. The atlas consists of 13 chapters that describe the ground water resources of regions that collectively cover the 50 States, Puerto Rico, and the U.S. Virgin Islands. Definitions of common hydrogeological terms and concepts are presented in appendix II of this technical guide.

Ground Water Regions of the United States

To divide the country into ground water regions, a classification was developed that identifies features of ground water systems that affect the occurrence and availability of ground water. The five features of this classification are as follows:

(1) the components of the system and their arrangement (confined and unconfined aquifers, confining units),
(2) the nature of the water-bearing openings of the dominant aquifers (primary vs. secondary porosity),
(3) the mineral composition of the rock matrix of the dominant aquifers (soluble vs. insoluble),
(4) the water storage and transmission characteristics of the dominant aquifers, and
(5) the nature and location of recharge and discharge areas.

The first two features are primary criteria used in all delineations of ground water regions. The remaining three are secondary criteria that are useful in subdividing regions into more homogeneous areas. On the basis of the these criteria, the United States, Puerto Rico, and the U.S. Virgin Islands are divided into 15 ground water regions plus alluvial valley aquifers (fig. 34).

The nature and extent of the dominant aquifers (fig. 35) and their relation to other units of the ground water system are the primary criteria used in delineating the regions. Consequently, the boundaries of the regions generally coincide with major geological boundaries, rather than with drainage divides.

A

ALASKA

HAWAII

PUERTO RICO AND
VIRGIN ISLANDS

B

ALASKA

HAWAII

PUERTO RICO AND
VIRGIN ISLANDS

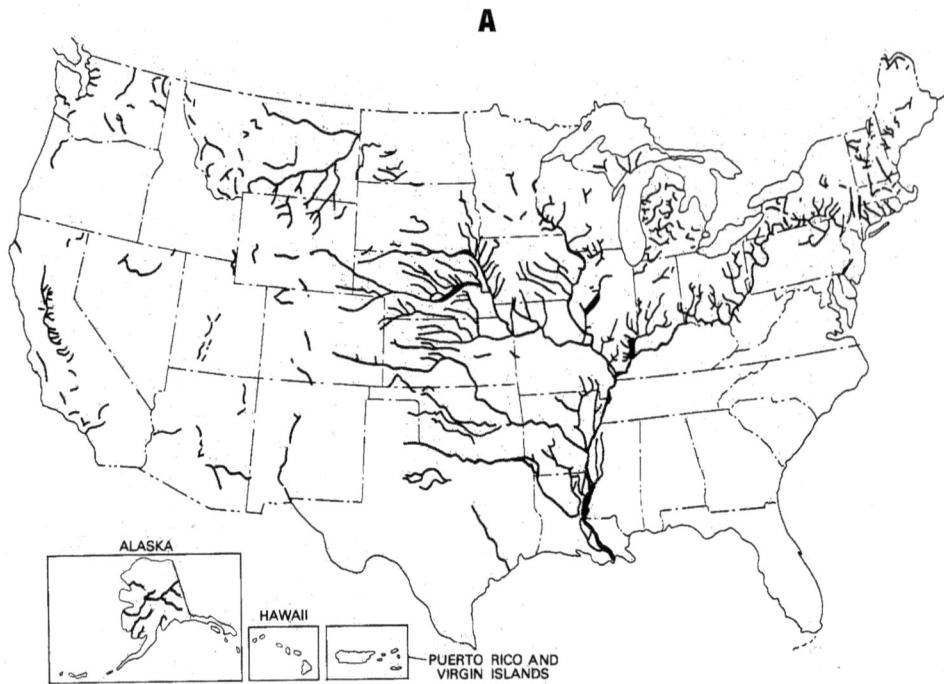

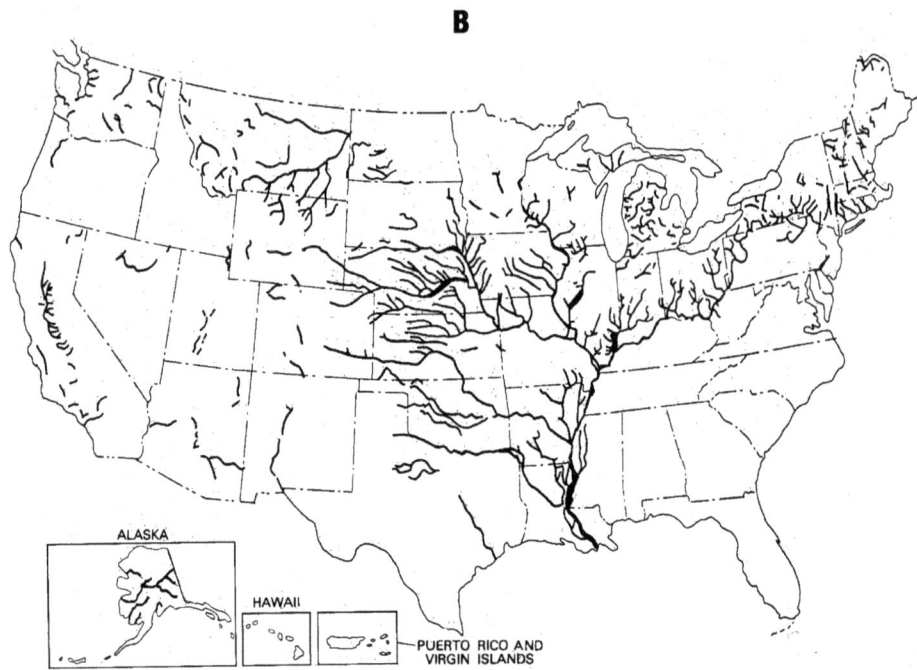

Figure 34. (A) Ground water regions of the United States, and (B) alluvial valley aquifers (Heath 1984).

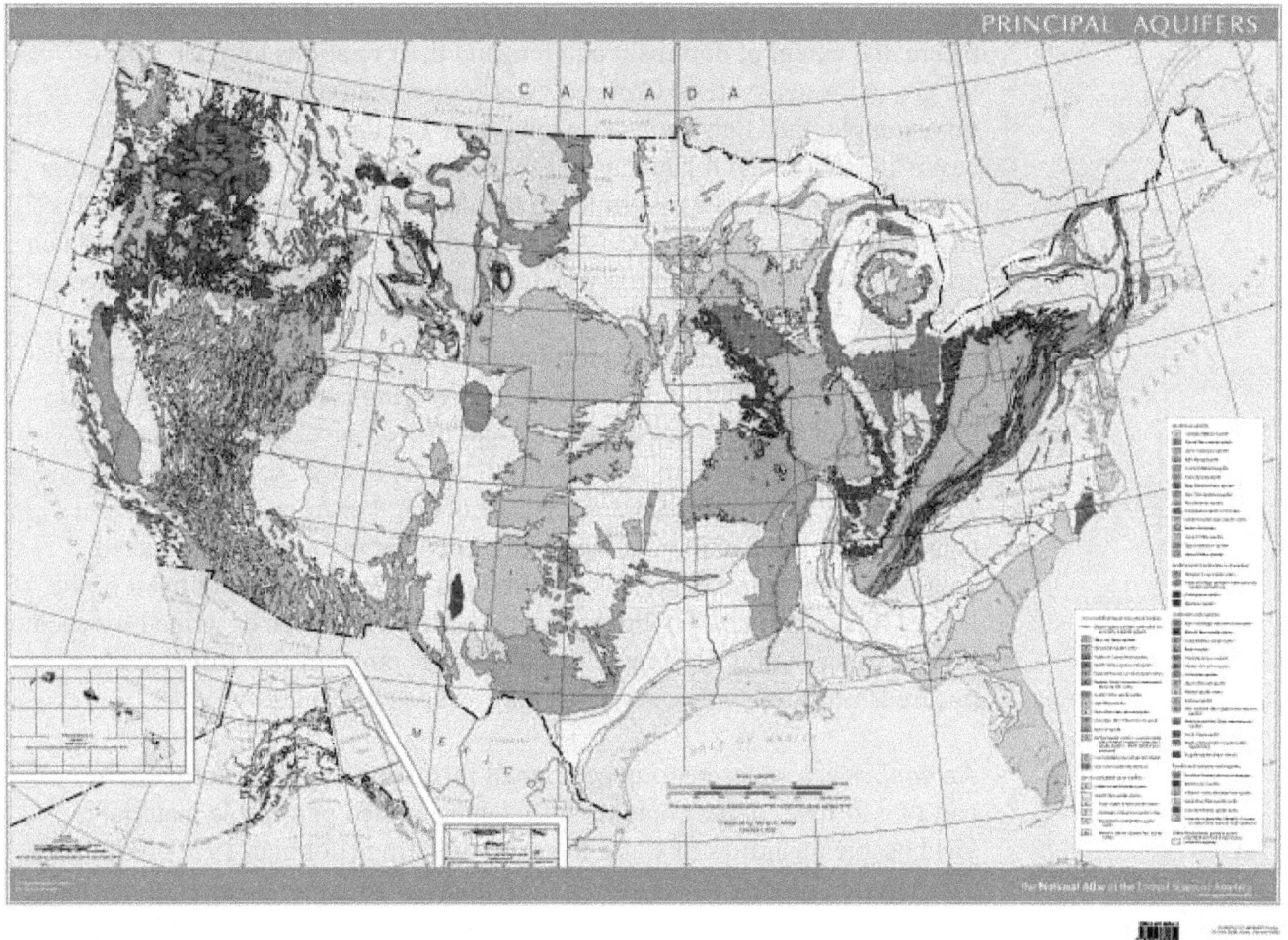

Figure 35. Principal aquifers of the United States (Miller 1998).

A Classification Framework for Ground Water

The first step in an inventory is to identify and map the areal extent of aquifers. The classification framework for surface water employs a hierarchy of units for resource characterization and management purposes (watersheds, basins, hydrologic units, and so on). A similar framework for ground water can be useful for understanding, classifying and mapping ground water resources (Maxwell and others 1995). The hierarchical classification presented here is based on mappable features that control ground water occurrence, flow and quality. In order of descending scale, the following is the hierarchy of units:

> Ground water regions
>> Hydrogeological settings
>>> Aquifers
>>>> Aquifer zones
>>>>> Aquifer sites

Ground water regions are geographic areas where the composition, arrangement, and structure of rock units that affect the occurrence and availability of ground water are similar. Heath (1984, 1988) built on the work of Meinzer (1923) and Thomas (1952) to map 15 ground water regions in the United States (fig. 15). Ground water regions coincide closely with

99

physiographic provinces (Fenneman 1938) and their boundaries reflect the nature and extent of dominant aquifers and their relations to other units of the ground water system. Ground Water regions can underlie large areas. For example, the High Plains Ground Water Region underlies eight river basins in Nebraska, Oklahoma, and the Texas Panhandle. Most ground water regions underlie tens of thousands to hundreds of thousands of square miles. Exceptions are segments of the Alluvial Valleys Ground Water Region, which are so narrow that they typically underlie tens to hundreds of square miles.

Ground water regions have been subdivided into hydrogeological settings (Aller and others 1987). A *hydrogeological setting* is defined as a composite description of all the major geological and hydrological factors that affect and control ground water movement into, through, and out of an area. It is a mappable unit with common hydrogeological characteristics, and as a consequence, has common vulnerability to contamination by introduced pollutants (Aller and others 1987). Although not yet mapped for most of the United States, a suite of hydrogeological settings has been described for each ground water region. Hydrogeological settings range in size from tens to hundreds of square miles. A typical map scale is 1:250,000.

An *aquifer* is a water-bearing geological formation, group of formations, or part of a formation that contains sufficient saturated permeable material to yield usable quantities of water to a well or spring (Lohman 1972). Within each aquifer, ground water moves from areas of recharge to areas of discharge. Flow direction, velocity, and discharge rates are controlled by aquifer porosity, hydraulic conductivity, and hydraulic gradient. Aquifers range in area from a few to hundreds of square miles. The recommended mapping scale is in a range from 1:24,000 to 1:63,000.

Aquifer zones are subdivisions of aquifers with differing hydrological conditions. Aquifer zones include recharge and discharge areas as well as confined and unconfined areas. Locally important hydraulic connection to surface-water systems that may be obscured at coarser hierarchical levels are identified at this level. Recharge may occur through direct precipitation, losing streams and lakes, or leakage from other aquifers. Discharge may occur to springs, seeps, gaining streams, lakes, and wetlands, by evapotranspiration, or by seepage into adjacent aquifers. Recharge zones are usually greater in area than discharge zones. Regionally significant recharge and discharge zones can occur in discrete localized areas. Recharge can be through fault zones or sinkholes; discharge can be through springs, and so on. Any one stream, lake, or wetland may have both gaining and losing portions, but in certain locations, either recharge or discharge may dominate.

It is not uncommon for a single aquifer to include areas where the ground water is confined as well as areas where it is not confined. This condition occurs because many aquifers outcrop or subcrop along part of their areal extent and are buried beneath other geologic units along other portions of their areal extent. These areas should be considered distinct aquifer zones.

Recharge and discharge can also occur at *aquifer sites*, which are specific features such as sinks and springs. Sinks and springs may be single points, clusters of points, or linear features along streams. They are most common in karst areas. A spring is ground water that naturally discharges from a geologic unit or aquifer onto the land surface or into surface waters. Sinks are commonly formed by the dissolution of soluble bedrock or semiconsolidated sediments (e.g., calcite-, dolomite-, and gypsum-bearing materials).

Geology and Ground Water

Ground water occurs in openings in the rocks that form the Earth's crust. The volume of the openings and the other water-bearing characteristics of the rocks depend on the mineral composition, age, and structure of the rocks. Therefore, to understand the occurrence of ground water in an area, it is necessary to have an understanding of the geology of that area. Below is a summary of a detailed discussion of geology and ground water by Heath (1984).

The United States is underlain by many different rock types. The nature of the water-bearing openings (porosity) in these rocks depends to a large extent on the geological age of the rocks as well as the processes that formed and may have subsequently modified the rocks. The youngest rocks are unconsolidated sedimentary deposits such as sand, gravel, clay, and glacial till, as well as volcanic rocks. The openings in sedimentary rocks generally are pores between the mineral grains (fig. 36). The openings in volcanic rocks include cooling fractures, pores in ash deposits, and lava tubes. Both of these geologically young rocks tend to be able to store and transmit more water than do older rocks of the same type.

At the time of their formation, crystalline rocks, such as granite, do not contain any appreciable porosity. Over the course of geological time, various tectonic forces and release of confining pressure cause the rocks to break along horizontal and vertical sets of fractures, which can then serve as water-bearing openings (fig. 36). Similar fractures can also form in sedimentary rocks that have been deeply buried and then are exposed by erosion of overlying rocks.

Carbonate rocks (limestone and dolomite) are soluble in weak acidic solutions, such as rainwater that percolates through the soil. As the rocks dissolve, often along existing fractures or bedding planes, these openings can enlarge to form large passages, sinkholes, and caverns (fig. 36). Areas in which these processes have occurred are called karst areas. Such enlarged solution openings can contain and transmit huge quantities of ground water, and these areas can have true "underground rivers." Karst areas in which sinkholes are common are particularly vulnerable to contamination from the surface because the contaminants can travel rapidly to the water table.

Although nearly all rock types can contain ground water, the earth materials that are most important as sources of ground water include sand and gravel, limestone/dolostone, sandstone, and extrusive volcanics, such as basalt and rhyolite. Earth materials with limited fractures and few or extremely small

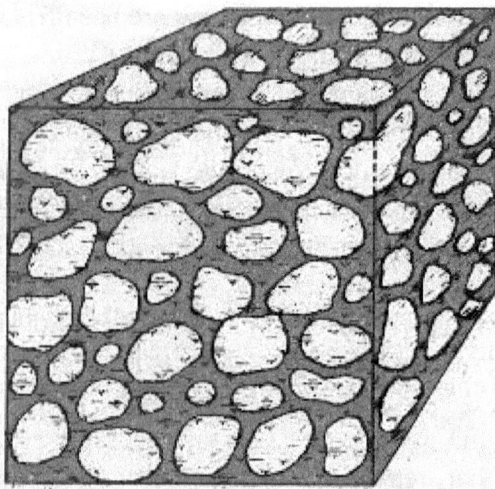

A. Pores in unconsolidated sedimentary deposit

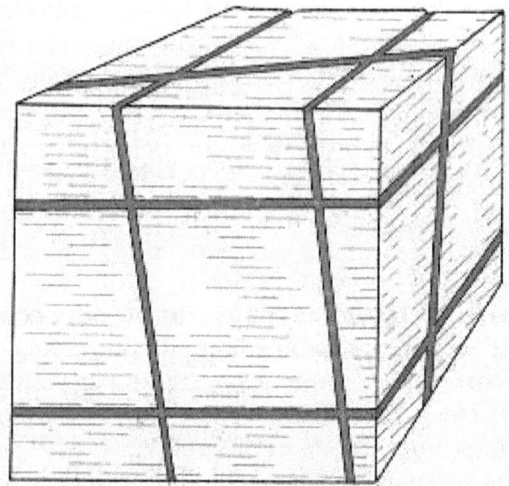

B. Fractures in intrusive igneous rocks

C. Caverns in limestone and dolomite

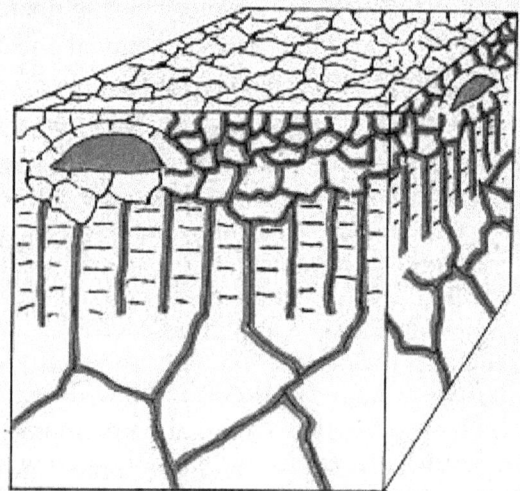

D. Lava tubes and cooling fractures in extrusive igneous rocks

Figure 36. Types of openings in selected water-bearing rocks. The size of the blocks can range from a few millimeters (A) to tens of meters (C and D) (Heath 1983).

intergranular openings generally do not readily yield water to wells and act to impede ground water flow. Earth materials that primarily act as barriers to ground water flow include silt, clay, shale, glacial till, and unfractured crystalline rocks. Clay deposits are composed of microscopic, flat particles that form an irregular (but very open) structure laced with very small pores. The pores are so small that most of these openings are occupied by water that is bound to the surface of the clay particles. Only minute amounts of water within these deposits are free to move. Although typically composed of up to 50 percent water, saturated clays may release less than 1 percent of that water when allowed to drain freely by gravity.

Of primary interest in hydrogeology is the capability of the various rock units to store and transmit water. Aquifers are identified on that basis. Geological units can be categorized as potential aquifers by describing the rock unit, interpreting the environment in which the rock unit was deposited, and interpreting the post-depositional conditions experienced by the unit. For purposes of assessing ground water potential, rock units generally have the following characteristics:

Massive Shale/Clay/Silt/Glacial Till. Thick-bedded shale, claystone, siltstone, glacial till, or clay typically yields only small quantities of water from fractures. Wells drilled into these units are often dry. These units often serve as confining layers in sedimentary sequences, producing artesian aquifers where an aquifer exists below the unit and that aquifer is connected to a recharge area. These units are distributed throughout the country, and are often found in conjunction with water-bearing units resulting in complex ground water flow systems.

Unweathered Metamorphic/Intrusive Igneous Rock. Consolidated bedrock of metamorphic or igneous origin contains very little or no primary porosity and yields water only from fractures or joints within the rock (secondary porosity). Typically, well yields are very low; dry holes very often occur, or wells go dry after producing only for a short time. Very low yields are sometimes obtained from fractures. In general, however, these units neither store nor transmit much water and are of only minor importance as aquifers.

Weathered Metamorphic/Igneous Rock. Unconsolidated material, commonly termed regolith or saprolite, is derived by weathering of the underlying consolidated bedrock, and contains only primary porosity. Water generally moves readily in this rock, but well yields are commonly low because the available thickness often is insufficient to adequately supply a well.

Bedded Sandstone/Limestone/Dolostone. Typically, thin-bedded sequences of consolidated sedimentary rock contain substantial porosity. The primary porosity in sandstones is generally substantial and minor in limestones and dolostones, while the secondary porosity in limestones and dolostones may be considerable.

Bedded Shale/Clay/Silt. Thin-bedded shale typically contains some secondary porosity as fractures and minor primary porosity along bedding planes. Thin-bedded clay and silt contains substantial primary porosity, but generally does not yield adequate water to supply a well.

Massive Sandstone. Consolidated sandstone bedrock contains both primary and secondary porosity and is typified by thicker deposits than the bedded sandstone deposits, which are distinguished by several identifiable beds over a distance of a few tens of feet. Sandstone is most important as a source of ground water where the cementing minerals have been deposited only around the points of contact of the sand particles, resulting in appreciable intergranular porosity. Bedding plane openings and other fractures in sandstone may also yield substantial amounts of water to wells. Sandstone is an important source of ground water in the north-central part of the country, in Texas, and in a narrow zone west of the Appalachian Mountains.

Massive Limestone/Dolostone. Consolidated limestones and dolostones are generally characterized by substantial secondary porosity, usually from fractures, but they can also have significant porosity developed from solution cavities that form along fractures. Limestones and dolostones are the sources of some of the largest well and spring yields in the United States. Yields of thousands of gallons per minute are common from springs and wells that are developed in carbonate rocks. These rocks underlie large areas in the Southeastern and Central United States.

Sand and Gravel. Unconsolidated mixtures of sand- to gravel-sized particles contain varying amounts of fine materials. The fine materials can be clays and silts and limit the interconnectivity of the porosity, or they can be sands and not substantially limit that connectivity. In the latter case, these materials are capable of being very productive aquifers if a sufficient thickness of material is present. Sands and gravels that contain only small amounts of fine materials are termed "clean." Their ability to move ground water can be high, and wells in them can be highly productive. Sand and gravel deposits from glacial activity, stream deposits, or mass movements, such as landslides and debris flows, are the sources of much of the ground water pumped from wells in the United States. These deposits occur throughout most of the country. The importance of sand and gravel deposits as a source of ground water is a result of both their widespread distribution and their capacity to yield water to wells at large rates.

Volcanic Rock. Consolidated extrusive igneous rock generally contains secondary porosity along fractures, interflow zones, and in vesicles. When well fractured, it often has high well yields. Basalts, rhyolites and other volcanic rocks are also among the most productive water-bearing formations. Basalt, which may be composed of thick layers that represent individual lava flows, is common in the northwestern United States. Large amounts of ground water can be pumped from both fractures within the flow units and from coarse-grained sediments that may be present between the individual lava flows. Lava tubes, common in Hawaii's volcanic rocks, act as channels for ground water flow.

Karst/Fractured Limestone/Dolostone. Consolidated limestone/dolostone that has been dissolved to the point in which large, open interconnected cavities are present is known as karst. Both karstic and fractured limestone/dolostone are capable of very large well yields, but water quality may be more like surface water than most ground water.

Ground Water Flow Systems

Recharge and Discharge

The addition of water to an aquifer is called recharge. It often occurs through infiltration of rainwater or snowmelt through the surface soil, followed by downward percolation through the unsaturated zone. The portion of infiltrating water that percolates to the water table is termed recharge. The amount of recharge by precipitation depends on factors such as the amount of rainfall, soil type, subsurface geology, slope, aspect, depth to the water table, and vegetation cover. Rates of recharge can range from less than an inch per year in the desert areas of the Southwest to more than 30 inches per year in karst areas of the Southeast. Other mechanisms of naturally occurring recharge include infiltration from streams and lakes and ground water flow from adjacent aquifers. Recharge can also be artificially created through establishment of infiltration ponds and galleries and by injection of water through wells.

Ground water leaves an aquifer (known as discharge) by several mechanisms. In areas where the water table is relatively shallow, transpiration by plants or direct evaporation from the water table is a common discharge mechanism. A large percentage of the baseflow to streams can be made up of discharged ground water (fig. 37). Ground water also discharges to ponds and lakes, wetlands, and the sea, as well as to adjacent aquifers. In addition, water withdrawn from wells accounts for the discharge of millions of gallons of ground water each day.

Under natural conditions, the ground water system develops a quasi equilibrium (or "steady state") with its recharge and discharge. That is, averaged over some period of time, the amount of water entering the system is about equal to the amount of water leaving the system. Because the system is in equilibrium, the amount of water stored in the system is constant or varies about some average condition in response to annual or climatic variations. As humans develop the ground water resources of an area, the natural system equilibrium begins to change (becomes "transient"), and the result can be an increase in recharge, a decrease in discharge, removal of ground water from storage in the aquifer, or a combination of all three.

An example of a recharge/discharge system in the Basin and Range hydrogeologic setting is shown in figure 38. Most recharge to basin aquifers occurs from precipitation falling on bedrock highlands. This water makes its way to the ground water reservoir along the basin margins, and from losing reaches of the larger intrabasin streams.

At present, our ability to quantify recharge and discharge is limited, and no uniformly acceptable methods exist for measuring recharge and discharge fluxes (National Research Council 2004, De Vries and Simmers 2002, Halford and Meyer 2000). Methods that have been used successfully in specific situations include measurements in surface water using channel water budget, baseflow discharge, seepage meters, heat tracers, isotopic tracers, solute mass-balance, and watershed modeling. Measurements in ground water using age dating, environmental tracers (CFCs, $^3H/^3He$, ^{14}C), Darcy's Law, and numerical modeling also have been successful. In general, the interconnected nature of the hydrologic system necessitates that some combination of information from both surface and ground waters be used to in order to develop a comprehensive view of an aquifer's water budget.

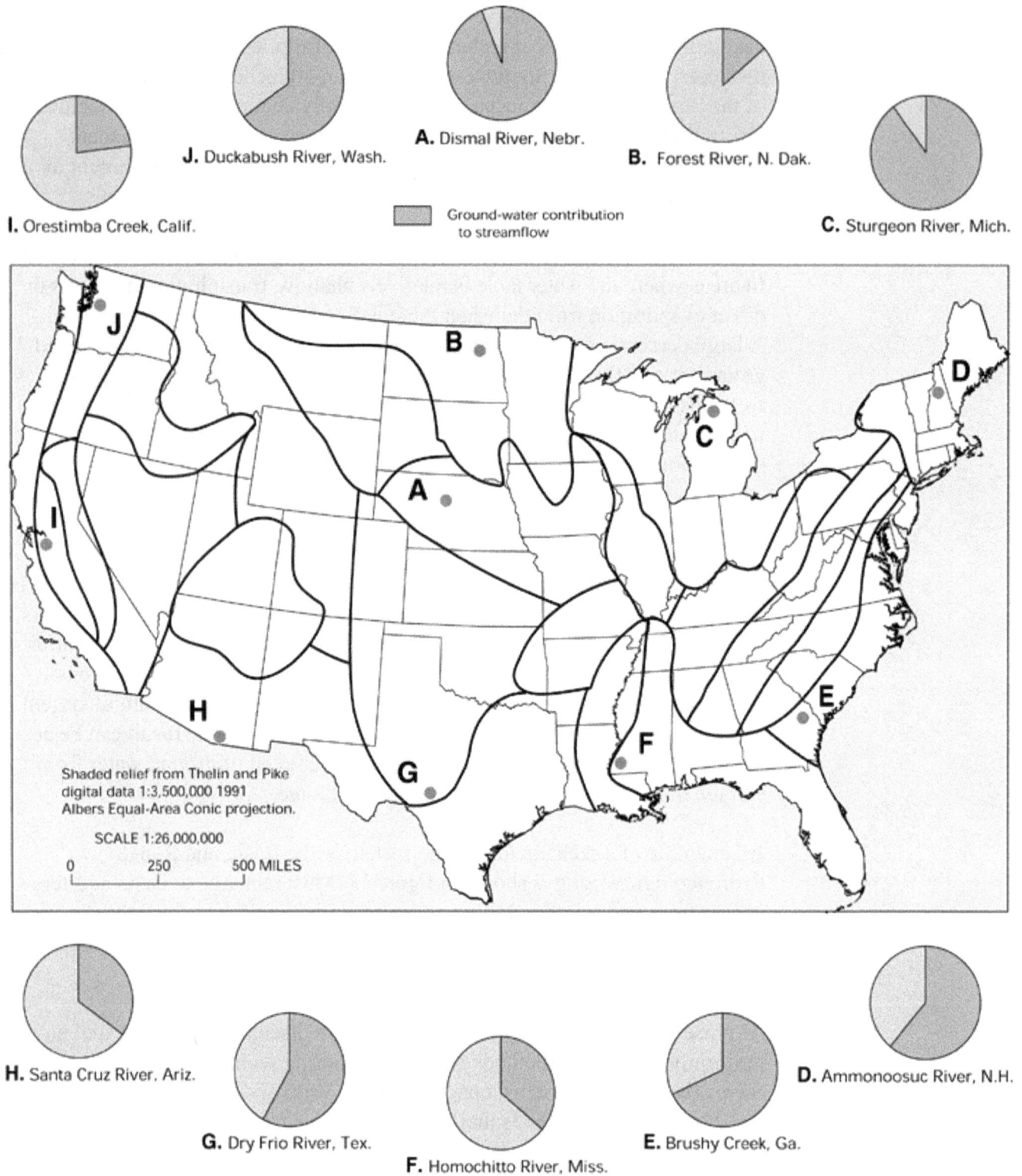

Figure 37. In the conterminous United States, 24 regions were delineated by the USGS where the interactions of ground water and surface water are considered to have similar characteristics. The estimated ground water contribution to streamflow is shown for specific streams in 10 of the regions (Winter and others 1998).

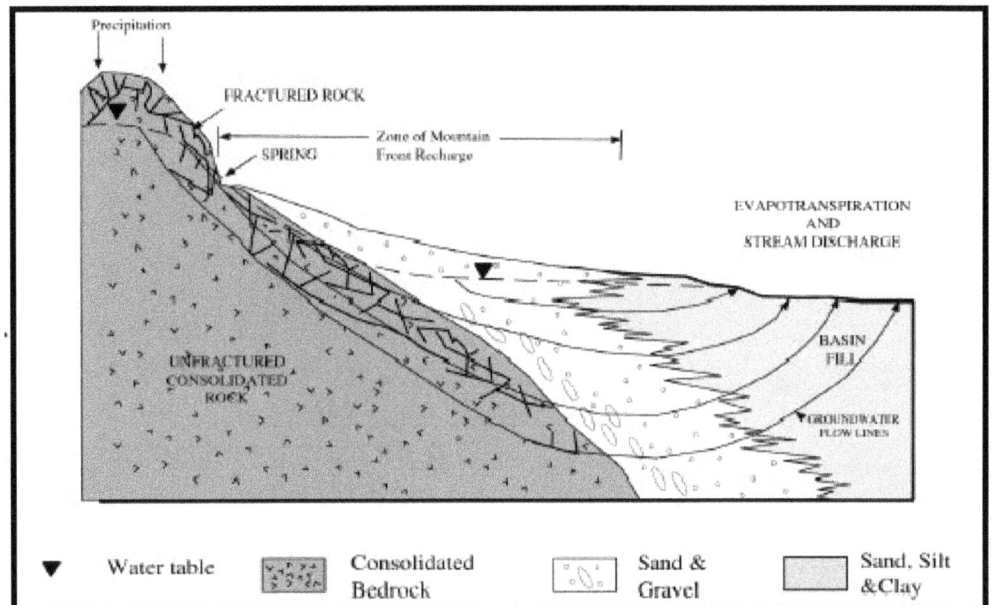

Figure 38. Two-dimensonal conceptual model of a ground water recharge system in a Basin and Range hydrogeological setting (Mifflin 1988).

Movement of Ground Water

Ground water moves from areas of high hydraulic head (usually upland areas) to areas of low hydraulic head (such as lowland areas, marshes, springs, and rivers). This allows a hydrogeologist to make use of water-level data obtained from wells, springs, and surface water features to determine the direction of ground water movement, both horizontally and vertically, as well as to estimate the quantity of ground water flow.

A potentiometric surface is an imaginary surface that represents the total head in an aquifer. It represents the height above a datum plane at which the water level stands in tightly cased wells that penetrate the aquifer in multiple locations. The water table is a special type of potentiometric surface. Potentiometric-surface maps can be constructed from water-level data by plotting these data on a map and contouring the interpreted surface based on these data. The potentiometric contours are also called equipotential lines. In most cases, the direction of ground water flow is perpendicular to the potentiometric contour. Figure 39 shows an example of a potentiometric-surface map and the inferred directions of ground water flow. Figure 40 shows an example of a cross section used to infer vertical ground water flow directions.

Regional and Local Flow Systems

The areal extent of ground water flow systems varies from a few square miles or less to tens of thousands of square miles. The length of ground water flow paths ranges from a few feet to tens, and sometimes hundreds, of miles. A deep ground water flow system with long flow paths between areas of recharge and discharge may be overlain by, and in hydraulic connection with, several shallow, more local, flow systems (fig. 41). Thus, the definition of a ground water flow system is to some extent subjective and depends in part on the scale of interest in a given study.

107

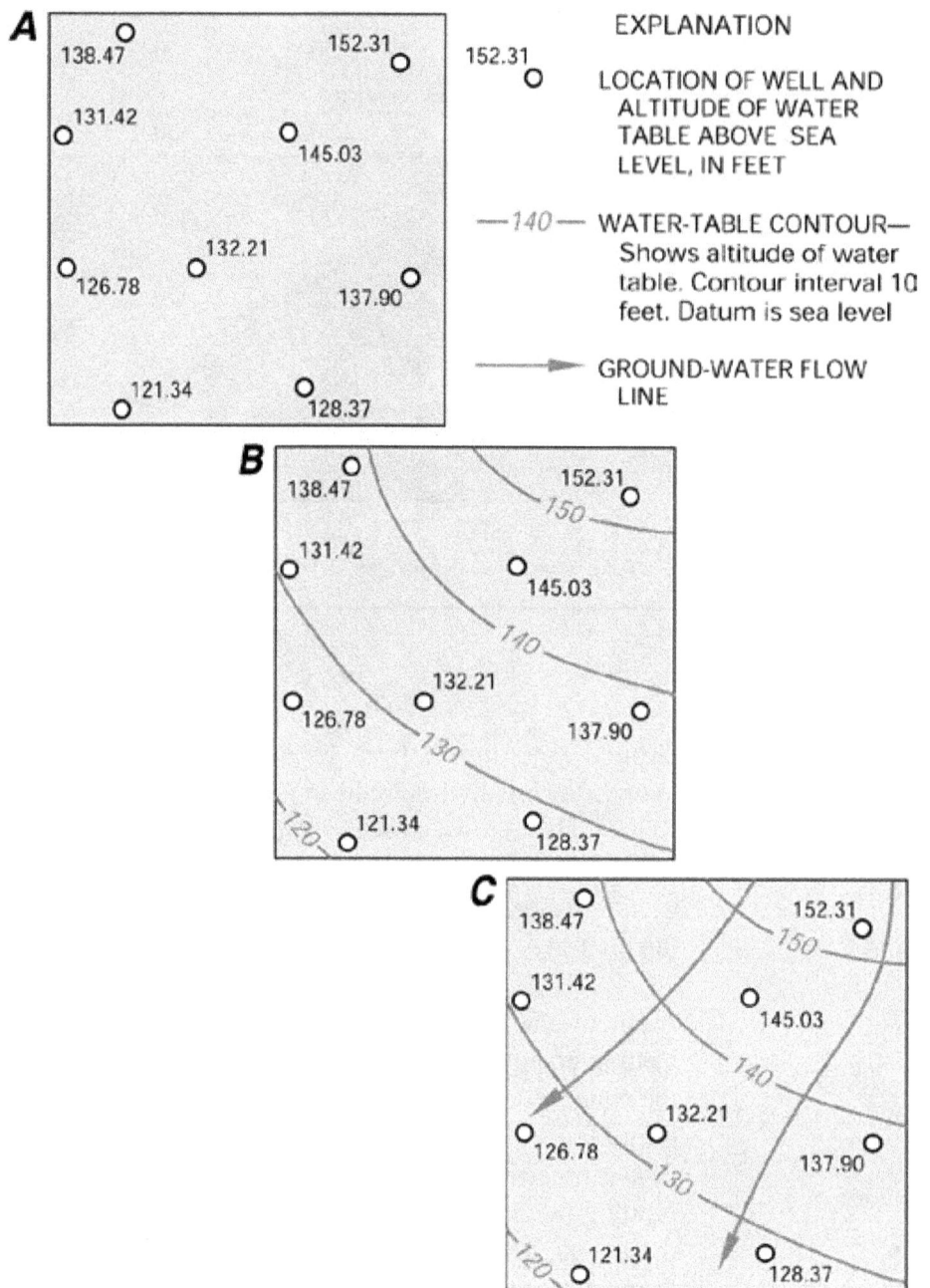

Figure 39. *Using known altitudes of the water table at individual wells (A), contour maps of the water-table surface can be drawn (B), and directions of ground water flow (C) can be determined (Winter and others 1998).*

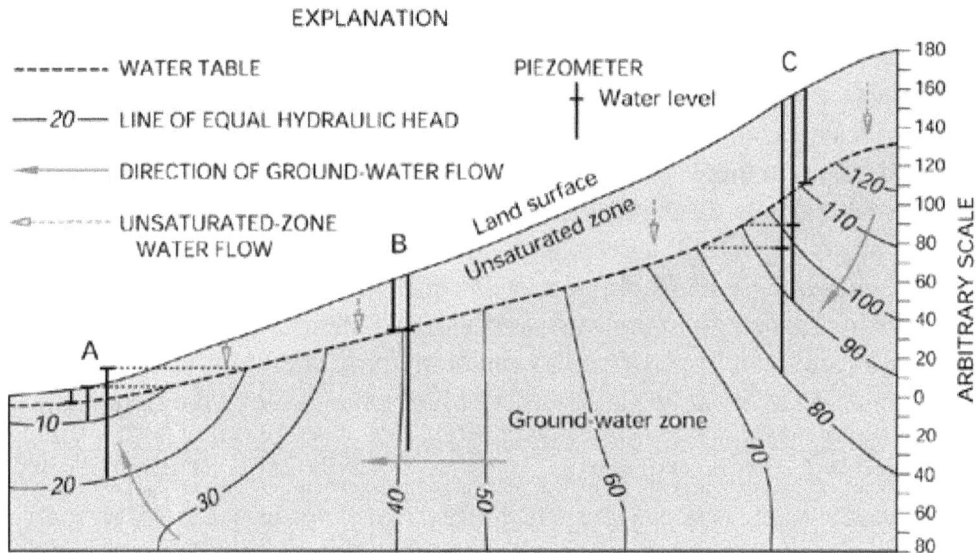

EXPLANATION

- - - - - - - WATER TABLE

— 20 — LINE OF EQUAL HYDRAULIC HEAD

←———— DIRECTION OF GROUND-WATER FLOW

←········· UNSATURATED-ZONE
          WATER FLOW

PIEZOMETER
Water level

Figure 40. If the vertical distribution of hydraulic head in a vertical section is known from nested piezometers (wells completed at discrete intervals below land surface), vertical patterns of ground water flow can be determined (Winter and others 1998).

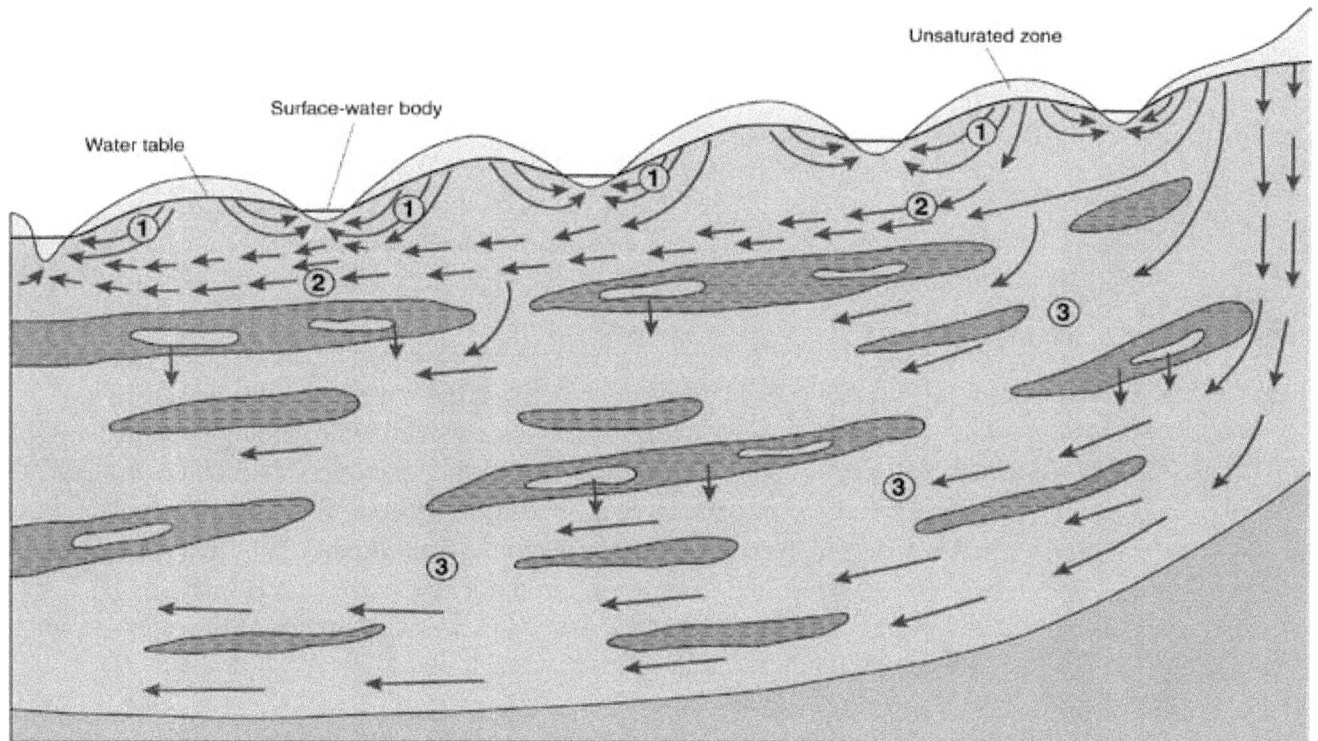

EXPLANATION

High hydraulic-conductivity aquifer

Low hydraulic-conductivity confining unit

Very low hydraulic-conductivity bedrock

← Direction of ground-water flow

① Local ground-water subsystem

② Subregional ground-water subsystem

③ Regional ground-water subsystem

Figure 41. A regional ground water flow system entails substyems of at different scales and a complex hydrogeological framework (after Sun 1986).

Significant features of the flow system depicted in figure 22 include (1) local ground water subsystems in the upper water-table aquifer that discharge to the nearest surface water bodies (lakes or streams) and are separated by ground water divides beneath topographically high areas; (2) a subregional ground water subsystem in the water-table aquifer in which flow paths originating at the water table do not discharge into the nearest surface water body but into a more distant one; and (3) a deep, regional ground water flow subsystem that lies beneath the water-table subsystems and is hydraulically connected to them. The hydrogeologic framework of the flow system exhibits a complicated spatial arrangement of high hydraulic-conductivity aquifer units and low hydraulic-conductivity confining units. The horizontal scale of the figure could range from tens to hundreds of miles.

Ground water study areas can range from less than 1 square mile to several square miles for local-scale studies, to hundreds of square miles for regional studies. Examples of local studies include those associated with problems involving drainage from individual mines, leaking underground storage tanks, accidental spills of hazardous materials, and parts of hydrogeological units that are near heavily pumped public-water-supply wells in which contaminated ground water is present. Regional studies include those covering an entire basin or region (Brahana and Mesko 1988). Basin-wide studies can include proposed mine or well-field development, ground water/surface water interaction problems, or mine reclamation. Regional-scale studies are generally associated with resource inventories that cover more than one drainage basin. The general types of information needed for studies of each of these scales are similar, but the amount of detail needed for each can be very different.

Attributes of local-scale studies of ground water are listed in table 4. Because of the range in possible project objectives, these attributes are quite general. Many local-scale studies, particularly on surficial hydrogeological units, are based on newly constructed project wells, the locations of which are guided by patterns of flow in the local ground water system. Examples of local-scale studies include (1) local-scale aquifer assessments, (2) early warning monitoring studies, (3) monitoring of point sources of contamination, (4) flow-path water-quality studies, and (5) local-scale studies of the interactions between ground water and surface water.

Local-scale assessments typically are used for areas and volumes of hydrogeological units in which potentially high concentrations of contaminants or locally high variability in water quality or quantity are expected. Early warning monitoring studies are conducted in areas where important ground water bodies are vulnerable to gradual inflow of contaminated ground water or to changes in water budget. Frequent sampling of monitoring wells is characteristic of local-scale studies.

Table 4. General attributes of local-scale assessments of ground water systems.

Attribute	Explanation
General objectives	Objectives of local-scale ground water assessments and research studies range from a survey of local-scale occurrence and distribution of water-quality and aquifer characteristics, to research studies on the transport and degradation of selected analytes, particularly in surficial hydrogeological units. The focus of many local-scale water-quality studies is to relate water quality explicitly to the ground water flow system.
Volume of earth material targeted for sampling	Most frequently, a small part of a hydrogeological unit.
Existing wells or new wells	Likely new wells, possibly supplemented by existing wells.
Number of wells to be sampled	Variable, depending on objectives and project design.
Well-selection strategy	Depending on study objectives, locations for new wells may be selected randomly or nonrandomly. Nonrandom locations may, for example, be in relation to the local ground water flow system and additional physical and cultural features, such as surface-water bodies, potential sources of contamination, and discharge locations, including water wells.
Temporal sampling strategy	Depends on study objectives; the objectives of many types of local-scale studies would require multiple samples from at least some of the wells.
Selection of target analytes	Analytes are targeted to meet study objectives.

For local-scale problems, information from a just a few wells may be sufficient, but the information needed for each of the wells may be extensive. Aquifer characteristics and water-quality samples might need to be defined for several intervals within the well, and the lithology may need to be described at intervals of less than a foot vertically within the well. Characterization of the physical and hydrologic properties of many individual fractures within the bedrock may also be required. Stream discharge may also need to be defined throughout many small segments of the stream for a local study. Topographic information, typically available from USGS quadrangle-scale (1:24,000) topographic maps, may not be of sufficient detail for a local study.

Regional-scale assessments or occurrence and distribution surveys of hydrogeological units are characterized by a wide spatial coverage and a broad array of analytes. The principal purposes of these broad surveys are (1) to provide evidence for naturally occurring constituents, including natural and anthropogenic contaminants that are present in water samples derived from a hydrogeological unit; (2) to provide an indication of contaminant concentrations by geographic location; and (3) to define general aquifer characteristics. General attributes of regional-scale studies of ground water are listed in table 5.

Table 5. Attributes of a regional-scale assessment of a hydrogeologic unit or group of units (occurrence and distribution survey).

Attribute	Explanation
General objective	To supplement existing data by providing a broad overview of ground water in a targeted hydrogeological unit or group of units—an occurrence survey and the beginning of a study of spatial distribution of water-quality constituents and aquifer characteristics in the hydrogeological unit(s).
Volume of earth material targeted for sampling	Generally, an entire hydrogeological unit or group of units; in thick hydrogeological units in which significant changes in water quality with depth are known or anticipated, dividing the hydrogeological unit into two or more parts based on lithology, depth, or both may be advisable. Sampling of these parts would then be structured separately.
Existing wells or new wells	Generally, existing wells are used exclusively.
Number of wells to be sampled	The number depends, in part, on the quality and breadth of existing water-quality data and on the known or anticipated spatial variability in water quality and aquifer heterogeneity; for example, in some surficial hydrogeological units, a considerably larger number of wells may be needed for a reasonable occurrence survey compared to some deeper confined hydrogeological units.
Well-selection strategy	A random component in well selection is usually highly desirable; sampling as few different types of wells as possible is advisable as long as the desired spatial coverage is achieved.
Temporal sampling strategy	Most wells are sampled once or twice unless (1) the entire assessment survey is repeated at some later time (generally 10 years or more).or (2) a well is selected to be part of a long-term monitoring (trend) network that is sampled at a fixed time interval.
Selection of analytes	Broad array of analytes, encompassing all project and monitoring-program objectives.

For a regional-scale study, the information needs generally are the same as those in a local study, but the level of detail required is usually less. Well information is still needed, but data may have to be collected from many wells over the entire study area. At each well, aquifer characteristics and water-quality samples that are representative of the area surrounding each well may suffice, and lithology may be described in very general terms. Characterization of only major fracture sets may suffice, as will streamflow information for only major stream reaches and tributaries. Topographic information from regional-scale (1:100,000 or 1:250,000) maps may suffice.

Shallow, Intermediate, and Deep Aquifers

Shallow aquifers generally are the focus of local-scale inventories, but the term "shallow" is relative. In parts of New England or southern Florida, shallow aquifers generally occur within the first 30 feet or so of the surface. In parts of the arid Southwest, where the depth to the water table is much greater, shallow aquifers may be at depths of hundreds of feet. Shallow aquifers tend to be recharged relatively quickly through infiltration of precipitation, and ground water is generally young (less than 50 years old). These aquifers also tend to be drained by small streams, or, in agricultural areas, by drainage tiles or ditches. They may respond rapidly to local stresses. Understanding of ground water/ surface water relationships is particularly critical in assessments of shallow aquifers, and these aquifers can be very susceptible to contamination from surface sources.

The amount of ground water in shallow aquifers can be highly variable, depending on the season or climatic conditions. During winter and spring, when evapotranspiration is low and precipitation is high, water levels may rise significantly, adding a large volume of water into storage. During summer and fall, when evapotranspiration is high and precipitation may be low, water levels tend to fall, draining ground water from storage. Ground water flow rates and directions, therefore, may be variable during the year or over a period of several years. Because of this temporal variability, data must be collected over relatively short intervals—from hourly or daily to weekly or monthly, depending on the objectives of the study. Shallow aquifers are most susceptible to periods of drought, and monitoring of ground water conditions is critical during those times.

Deep aquifers are often the focus of basin-wide or regional-scale inventories. As with shallow aquifers, the term "deep" is relative, depending on hydrogeologic and climatic conditions. Deep aquifers in the Eastern United States may be at depths of hundreds of feet, but in the western part of the country, deep aquifers may be several thousand feet deep. Deep aquifers generally receive less recharge than shallow aquifers, and recharge mechanisms are variable. A deep aquifer may receive recharge from precipitation at outcrop areas that can be hundreds of miles away from the area of study, or recharge may occur by leakage from overlying or underlying aquifers. Ground water ages of deep aquifers generally are much greater than those of shallow aquifers. The ground water age of a deep aquifer can be on the order of thousands of years. Discharge of water from a deep aquifer tends to occur only to large, regional rivers or to fracture- or fault-controlled springs that are connected to the aquifer. Deep aquifers are often confined and hydraulically isolated from overlying shallow aquifers, and ground water flow direction can differ from that of the overlying aquifers. The chemical quality of deep-aquifer water is often different from the quality of water in shallow aquifers. Deep aquifers may have high dissolved-solids concentrations because of dissolution of minerals along the long flow paths in the aquifer. Because of their hydraulic isolation, deep aquifers tend to be less susceptible to anthropogenic contamination than shallow aquifers. Deep aquifers tend to be less affected by short-term drought conditions, and respond very slowly to changing climatic conditions. Data collection frequency, therefore, generally usually can be less (quarterly or annually) than that needed for shallow aquifers. Intermediate aquifers are transitional between shallow and deep aquifers, and have characteristics of both types of aquifers.

Ground Water Development and Sustainability

A ground water system consists of a mass of water flowing through the pores or cracks below the Earth's surface. This mass of water is in motion. Water is constantly added to the system by recharge from precipitation, and water is constantly leaving the system as discharge to surface water and as evapotranspiration. Each ground water system is unique in that the source and amount of water flowing through the system is dependent on external factors such as rate of precipitation, location of streams and other surface water bodies, and rate of evapotranspiration. The one common factor for all ground water systems, however, is that the total amount of water entering, leaving, and being stored in the system must be conserved. An accounting of all the inflows, outflows, and changes in storage is called a water budget (Alley and others 1999).

Human activities, such as ground water withdrawals and irrigation, change the natural flow patterns, and these changes must be accounted for in the calculation of the water budget. Because any water that is used must come from somewhere, human activities affect the amount and rate of movement of water in the system, entering the system, and leaving the system.

Some hydrologists believe that a predevelopment water budget for a ground water system (that is, a water budget for the natural conditions before humans used the water) can be used to calculate the amount of water available for consumption (or the safe yield). In this approach, the development of a ground water system is considered to be "safe" if the rate of ground water withdrawal does not exceed the rate of natural recharge. This concept has been referred to as the "Water-Budget Myth" (Bredehoeft and others 1982). It is a myth because it is an oversimplification of the information that is needed to understand the effects of developing a ground water system. As human activities change the system, the components of the water budget (inflows, outflows, and changes in storage) also will change and must be accounted for in any management decision. Understanding water budgets and how they change in response to human activities is an important aspect of ground water hydrology; however, a predevelopment water budget by itself is of limited value in determining the amount of ground water that can be withdrawn on a sustained basis.

Under predevelopment conditions, the ground water system is generally in long-term equilibrium. That is, averaged over some period of time, the amount of water entering or recharging the system is approximately equal to the amount of water leaving or discharging from the system. Because the system is in equilibrium, the quantity of water stored in the system is constant or varies about some average condition in response to annual or longer term climatic variations. This predevelopment water budget is shown schematically in figure 42(A).

We also can write an equation that describes the water budget of the predevelopment system as:

Recharge (water entering) = Discharge (water leaving).

A

B

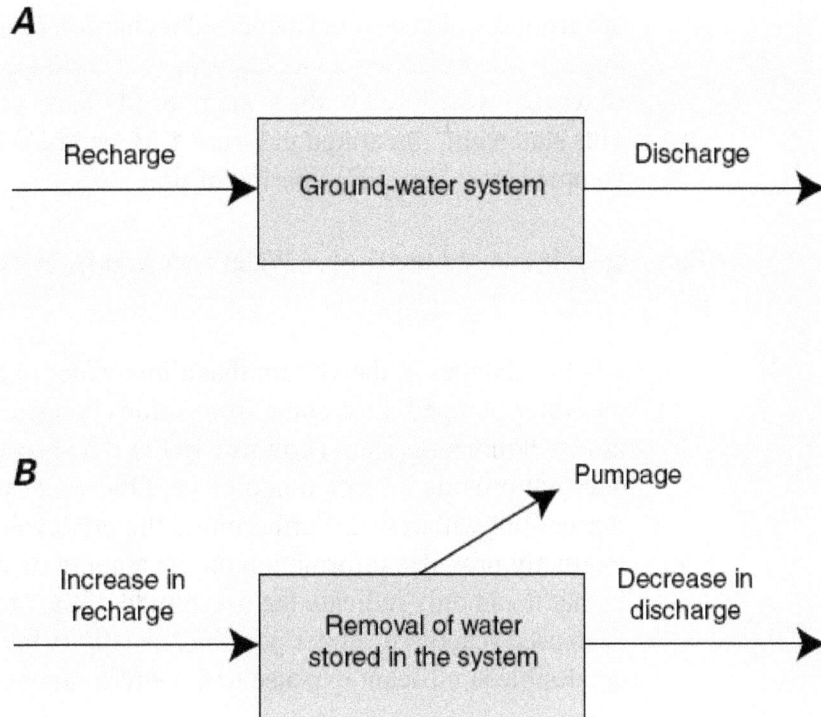

Figure 42. Diagrams illustrating water budgets for a ground water system for predevelopment and development conditions. (A) Predevelopment water-budget diagram illustrating that inflow equals outflow. (B) Water-budget diagram showing changes in flow for a ground water system being pumped. The sources of water for the pumpage are changes in recharge, discharge, and the amount of water stored. The initial predevelopment values do not directly enter the budget calculation (Alley and others 1999).

The water leaving often is discharged to streams and rivers and is called baseflow. The possible inflows (recharge) and outflows (discharge) of a shallow ground water system under natural (equilibrium) conditions are listed in table 6.

Humans change the natural or predevelopment flow system by withdrawing (pumping) water for use, changing recharge patterns by irrigation and urban development, changing the type of vegetation, and other activities. Focusing our attention on the effects of withdrawing ground water, we can conclude that

Table 6. Possible sources of water entering and leaving a shallow ground water system under natural conditions.

Inflow (recharge)		Outflow (discharge)	
1.	Areal recharge from precipitation that percolates through the unsaturated zone to the water table.	1.	Discharge to streams, lakes, wetlands, saltwater bodies (bays, estuaries, or oceans), and springs.
2.	Recharge from losing streams, lakes, and wetlands.	2.	Ground water evapotranspiration.
3.	Flow from an adjacent aquifer.	3.	Flow to an adjacent aquifer.

115

the source of water for pumpage must be supplied by (1) more water entering the ground water system (increased recharge), (2) less water leaving the system through other discharge mechanisms (decreased discharge), (3) removal of water that was stored in the system, or (4) some combination of these three. This statement, illustrated in figure 42B, can be written in terms of rates (or volumes over a specified period of time) as:

Pumpage = Increased recharge + Water removed from storage + Decreased discharge.

It is the changes in the system that allow water to be withdrawn. That is, the water pumped must come from some change of flows and storage in the predevelopment system (Lohman 1972). The predevelopment water budget does not provide information on where the water will come from to supply the amount withdrawn. Furthermore, the predevelopment water budget only indirectly provides information on the amount of water perennially available, in that it can only indicate the magnitude of the original discharge that can be decreased (captured) under possible, usually extreme, development alternatives at possible significant expense to the environment.

Regardless of the amount of water withdrawn, the system will undergo some drawdown in water levels in pumping wells to induce the flow of water to these wells, which means that some water initially is removed from storage. Thus, the ground water system serves as both a water reservoir and a water-distribution system. For most ground water systems, the change in storage in response to pumping is a transient phenomenon that occurs as the system readjusts to the pumping stress. The relative contributions of changes in storage, changes in recharge, and changes in discharge evolve with time. The initial response to withdrawal of water is changes in storage. If the system can come to a new equilibrium, the changes in storage will stop and inflows will again balance outflows and can be written as follows:

Pumpage = Increased recharge + Decreased discharge.

Thus, the long-term source of water to discharging wells is typically a change in the amount of water entering and/or leaving the system. How much ground water is available for use depends on how these changes in inflow and outflow affect the surrounding environment and what the public defines as undesirable effects on the environment.

In determining the effects of pumping and the amount of water available for use, it is critical to recognize that not all the water pumped is necessarily consumed. For example, not all the water pumped for irrigation is consumed by evapotranspiration. Some of the water returns to the shallow ground water system as infiltration (irrigation return flow). Most other uses of ground water are similar in that some of the water pumped is not consumed but is returned to the system. However, depending on the source of the water pumped and the nature of the ground water system, the portion not consumed may be lost from the part of the system from which it was withdrawn. Thus, it is important to differentiate between the amount of water pumped and the amount of water consumed and to understand the source of the water pumped and the recharge location of the water not consumed when estimating water availability and developing sustainable management strategies.

The possibilities of severe, long-term droughts and climate change also should be considered. Long-term droughts, which virtually always result in reduced ground water recharge, may be viewed as a natural stress on a ground water system that in many ways has effects similar to ground water withdrawals through reductions in ground water storage and accompanying reductions in ground water discharge to streams and other surface water bodies. Because a climate stress on the hydrologic system is added to the existing or projected human-derived stress, droughts represent extreme hydrologic conditions that should be evaluated in any long-term management plan.

Ground Water Quality

Nearly all active ground water originates as rain or snow that infiltrates through the vadose zone to the water table or saturated zone. Because most water vapor that becomes precipitation occurs as a result of evaporation, it typically contains low concentrations of dissolved solids. Consequently, the chemical composition of natural ground water is primarily a result of physical, chemical, and biological processes that occur as water interacts with geologic materials as it moves downward through the vadose zone (in recharge areas) and flows (as ground water) to areas of discharge. Some of the more important processes include weathering of rock and soil, mineral dissolution and precipitation reactions (including for example, oxidation and reduction, ion exchange, and adsorption), and interactions between water and air. The types and concentrations of dissolved constituents in ground water are net effects of chemical reactions that have dissolved material from solid phases, altered previously dissolved constituents, or removed dissolved constituents by precipitation or other processes (Hem 1989). Biological activity and numerous physical processes influence these chemical processes.

Commonly, precipitation that infiltrates to the subsurface moves vertically through a thickness of unsaturated (vadose) zone before reaching the water table. In the vadose zone, carbonic acid (H_2CO_3) is generated as water interacts with soils and oxygen. Carbonic acid typically lowers the pH of the water slightly. As ground water flows from recharge areas to discharge areas, residence time increases and continuing rock-water interaction results in an increase in total dissolved solids in the downgradient direction. As a result, ground water contains a wide variety of dissolved inorganic constituents in various concentrations. The concentration of TDS in ground water varies from less than 500 to more than 100,000 mg/L. The TDS of seawater is 35,000 mg/L. The EPA secondary drinking-water standard is 500 mg//L.

Most of the chemical constituents that are dissolved in ground water occur in ionic form. Ions that have negative charge (excess electrons) are referred to as anions. Ions that have a positive charge (excess of protons) are referred to as cations. Common major, minor, and trace dissolved inorganic constituents are listed in table 7. In intermediate and regional ground water flow systems the dominant anion often changes from bicarbonate to sulfate to chloride. Types and concentrations of dominant cations vary depending on the mineralogy and chemical composition of the rock or sediment and the dominant chemical reactions.

Dissolved organic constituents and dissolved gases also occur in ground water but concentrations are usually low. Dissolved organic matter is ubiquitous in natural ground water and is thought to be primarily fulvic and humic acids. The most abundant dissolved gases in ground water are nitrogen (N_2), oxygen (O_2), carbon dioxide (CO_2), methane (CH_4), and hydrogen sulfide (H_2S). The first three are atmospheric gases and the last two are products of biogeochemical processes that occur in anaerobic subsurface zones. Other minor dissolved gases include radon (radioactive), argon, helium, and neon.

Hardness is a water-quality property that has had widespread interest for centuries. Hardness refers to the effects observed in the use of soap with some types of water or to the encrustations left by heating some types of water. Hardness is further defined as the content of metallic ions that react with sodium soaps to form a scummy residue. Because hardness results primarily from the presence of calcium and magnesium, it is typically reported as the total concentration of Ca^{2+} and Mg^{2+} expressed in terms of an equivalent concentration of $CaCO_3$. The designation of "soft" and "hard" water is somewhat arbitrary. Table 8 presents a commonly used classification developed by Durfor and Becker (1964).

Table 7. Major, minor, and trace dissolved inorganic constituents in ground water.

Major dissolved constituents (> 5 mg/L)	Minor dissolved constituents (> 0.01–10mg/L)	Trace dissolved constituents (< 0.1 mg/L)	
Carbonic acid (H_2CO_3)	Boron (B)	Arsenic (As)	Chromium (Cr)
Chloride[1] (Cl)	Carbonate[1] (CO_3)	Cadmium (Cd)	Phosphate (PO_4)
Sulfate[1] (SO_4)	Fluoride[1] (F)	Zinc (Zn)	Copper (Cu)
Bicarbonate[1] (HCO_3)	Iron[2] (Fe)	Lead (Pb)	Silver (Ag)
Calcium[2] (Ca)	Nitrate[1] (NO_3)	Manganese (Mn)	Selenium (Se)
Magnesium [2] (Mg)	Potassium[2] (K)	Aluminum (Al)	Radium[3] (Ra)
Sodium[2] (Na)	Strontium[2] (Sr)	Antimony (Sb)	Uranium[3] (U)
Silicon[2] (Si)		Barium (Ba)	Thorium[3] (Th)

[1] anions

[2] cations

[3] radioactive

Table 8. Hardness classification based on equivalent concentration of CaCO₃ (mg/L).

Description	Hardness
Soft	0–60
Moderately hard	61–120
Hard	121–180
Very hard	> 180

Alkalinity and acidity are other important properties of natural ground water. These properties refer to the capacity of ground water to neutralize an acid or base. Chemical reactions related to rock-water interaction result in low dissolved concentrations of hydrogen (H^+) and hydroxyl (OH^-) ions, which contribute significantly to acidity and alkalinity, respectively. Alkalinity of ground water is defined as the capacity for solutes it contains to react with and neutralize acid. Acidity is defined as the quantitative capacity of aqueous media to react with and neutralize a base. In most natural waters, the alkalinity is largely produced by dissolved carbon dioxide species (CO_2), bicarbonate (HCO_3^-), and carbonate (CO_3^{2-}). Hydroxide, silicate, and borate are important noncarbonate contributors to alkalinity. Alkalinity is most often reported as an equivalent amount of $CaCO_3$ (mg/L). The pH of water is a measure of the concentration (activity) of H^+ ions. Sources of acidity in natural ground water include low pH rain and snow, dissolved CO_2, solution of volcanic gases or gaseous discharges in geothermal areas, and the oxidation of sulfide minerals and ferrous iron. Acidity is reported as meq/L or mg/L of H^+.

As discussed previously, the chemistry of natural ground water is generally influenced greatly by the geological materials through which it flows. As a result, ground water in similar geologic materials tends to exhibit similar chemistry. In saturated sedimentary rock sequences, characterized by active ground water flushing through well-leached rocks, the ground water tends to be low in TDS with bicarbonate as the dominant anion. In rock sequences characterized by intermediate and regional-scale ground water flow systems, ground water circulation is relatively slow, and residence times are relatively long. In these hydrogeological settings, TDS concentrations in ground water tend to be higher with sulfate and chloride as the dominant anions. Ground water in carbonate rock aquifers tends to be higher in calcium, magnesium, and bicarbonate and lower in sodium, potassium, chloride, and sulfate. Ground water contained in crystalline igneous and metamorphic rocks is commonly soft and slightly acidic, with low concentrations of TDS and high concentrations of dissolved silica. It is important to note that ground water in karst aquifers can often be an exception to these general tendencies, since water flow may be quite rapid.

Ground water quality in glacial deposits is quite variable because of the large variability in mineralogy of glacial deposits. Glacial deposits that overlie the North American Pre-Cambrian Shield commonly contain soft, slightly acidic water with TDS concentrations less than 100 mg/L. Sodium, calcium, and magnesium are the common cations, and bicarbonate is the dominant anion. Ground water in glacial deposits overlying the interior plains of the United States commonly contains high concentrations of TDS. Sodium, magnesium, calcium, and sulfate occur in major concentrations. Ground water in shallow fluvial deposits is generally low in TDS and slightly acidic if derived from infiltration of precipitation.

Ground Water/ Surface Water Interactions

Ground water and surface water both originate as precipitation. From the moment that water from precipitation reaches the soil surface, its chemistry begins to change. Water that infiltrates surface soils and is underground for long periods tends to develop considerably different water-quality characteristics (chemical composition, temperature, and microbiological quality) than water that flows overland. Ground water and surface water interact (join and mix) at many locations in most watersheds; consequently, their flow rates, chemistries, temperatures, and microbiological qualities are often neither uniform nor estimable by simple extrapolation downstream or downslope. Ground water that originated as infiltrating precipitation today may be discharged months or years later to a stream or lake that also contains water recently contributed by precipitation. Conversely, surface water that originated as runoff from recent precipitation may be lost by seepage downward through the streambed to mix with ground water of much greater age.

Ground water and surface water interact on many physical scales and over a wide range of time periods. Some of these interactions may be observed and measured directly, while others may be detected and evaluated only by indirect methods or surrogate measures. The interactions of significant interest include (1) ground water supply to the baseflow of perennial streams and full flow of some ephemeral streams; (2) ground water supply of flow to springs, seeps, and cave systems; (3) streamflow supply of recharge to the ground water system; (4) ground water flow into, and ground water recharge from, reservoirs, lakes, ponds and lagoons; and (5) ground water controls on landforms and stream morphology. Observations and measurements of these interactions may be used to provide key inputs and constraints for watershed models and ecological assessments, thereby greatly improving their reliability and usefulness.

Ground water and streams may interact in a variety of ways. Ground water may flow directly into a stream through seeps or springs in the streambanks or streambed. Surface water may be lost by seepage from a stream channel to underlying ground water. Ground water and surface water may exchange repeatedly along the length of a stream or cyclically over time in a given stream reach in response to changing water-table and/or runoff conditions. Streams may disappear into the ground, and reappear elsewhere, especially in karst (limestone) terrain and in fractured rock settings.

The water regime of a stream is defined in terms of the presence of running water in the channel. Perennial streams flow year-round and are generally supported by abundant ground water discharge during dry periods. Some drainages are intermittent, containing perennial water only in certain segments fed by springs or ground water and dry for long distances during dry periods. Many drainages in semiarid areas and most desert drainages are ephemeral, containing running water only seasonally, usually in response to rainfall, and not necessarily every year.

Ground water contributes to streamflow under at least some conditions in most physiographic and climatic settings. The proportion of stream water that is derived from ground water varies across physiographic, climatic, and seasonal settings. Knowledge of the amount of ground water recharge and discharge to streams and other surface-water bodies is important in quantifying the total ground water available in an area. In areas where streams primarily lose water to ground water, such as the arid Southwest, discharge of ground water may supply the drainage at its head, while in downstream areas infiltration of streamflow may be the major source of recharge to the ground water system. Ground water discharge to streams can be estimated by measuring streamflow during "baseflow" periods, when streamflow is almost entirely supported by ground water inflow to the stream channel. The baseflow of a stream is that portion of streamflow in the channel that has been contributed by ground water inflow to the stream. Baseflow may constitute a small portion or a majority of the streamflow. The average proportion of baseflow to the total streamflow ranges from a few percent annually to almost all of the streamflow in the channel annually.

Streamflow can be measured by several methods. The most common method currently used by the USGS involves obtaining point velocity measurements at predetermined locations along a cross-section of stream channel and multiplying these velocities by the area represented by each velocity measurement to get volumetric discharge. The velocity measurements may be made by either wading in small streams or by taking measurements in boats or from bridges and cableways. This method provides a measurement of stream discharge at one point along a stream at a single point in time. Continuous measurements are obtained at gaging stations by continuously measuring stream stage along a section of the stream, and applying a uniquely determined stage-discharge relation for that site. Streamflow obtained at these sites are often presented graphically as a "hydrograph." Real-time stream-discharge data are available on the Internet for many streams at the USGS Web site at http://water.usgs.gov. Historic stream-discharge data can be obtained from individual water science centers of the USGS. Historic data can be used to develop flow-duration curves for gaged reaches, which can be used to help develop bounding estimates of baseflow in perennial streams.

Case Study: Contribution of Metal Loads to Daisy Creek from Ground Water, Custer National Forest, MT

Water quality and aquatic habitat in Daisy Creek on the Custer National Forest has been adversely affected by drainage from the McLaren Mine, as well as by natural weathering of pyrite-rich mineralized rock. Specific surface and subsurface sources of metals to the creek were identified by a synoptic sampling and tracer injection study. Knowledge of the main sources and pathways of metals and acid to Daisy Creek has aided resource managers in planning and conducting cost-efficient remediation activities.

Acid drainage from the McLaren Mine affects the water quality of Daisy Creek, an alpine headwater tributary of the Stillwater River. Water quality and aquatic habitat have been severely affected by drainage from mining as well as by natural weathering of pyrite-rich mineralized rock (fig. 43). Effective planning for remediation requires detailed knowledge of the sources of metals and how the metals from these sources enter the stream. Metal-loading studies have been useful in characterizing water quality in historical mining areas and identifying surface as well as subsurface metal sources and pathways. The USGS in cooperation with the Forest Service conducted a constant-rate tracer injection synoptic sampling study to quantify the principle sources of metal loads to Daisy Creek.

In August 1999, a sodium chloride tracer was added to the stream for 29.5 hours to provide a hydrologic context for synoptic sampling of metal chemistry in the stream and its inflows. Detailed profiles of metal loads along Daisy Creek were developed from streamflow data (obtained by tracer injection) and metal-concentration data (obtained by synoptic water-quality sampling) collected at many closely spaced sites. These profiles helped to identify reaches of Daisy Creek where most of the metal loading occurs.

Inflows to the stream can be divided between visible surface inflows, which were sampled, and subsurface inflows, which were not sampled, but the effects of both types of inflows on the stream were quantified. Substantial loads were attributed to both sources (fig. 44). About 54 percent of the total copper load was contributed by surface inflows. Copper loading from ground water inflows was also substantial, contributing 46 percent of the total dissolved copper load to Daisy Creek.

The upper 270 feet of Daisy Creek are relatively unaffected by historical mining activity and resulting water-quality impacts (fig. 44). Once the tributaries draining the McLaren Mine and related ground water inflows are encountered, however, significant impacts of acidity and elevated metals are encountered in Daisy Creek. The principal observable impacts from the McLaren Mine occur in the subreach from 270 feet to 611 feet downstream. The subreaches from 611 feet to 5,475 feet receive little surface water inflow, but ground water inflow into Daisy Creek continues to provide copper loading
.

Flow through the shallow subsurface appears to be a major copper transport pathway from the McLaren Mine and surrounding mineralized bedrock to Daisy Creek during baseflow conditions. These results indicate that remediation of large visible inflows could still leave ground water-derived metal concentrations in Daisy Creek at levels that may adversely affect aquatic life.

For additional information see Nimick and Cleasby (2001).

Figure 43. Iron oxyhydroxide and associated heavy metals from acidic inflows degrade the water quality of Daisy Creek, Park County, MT.

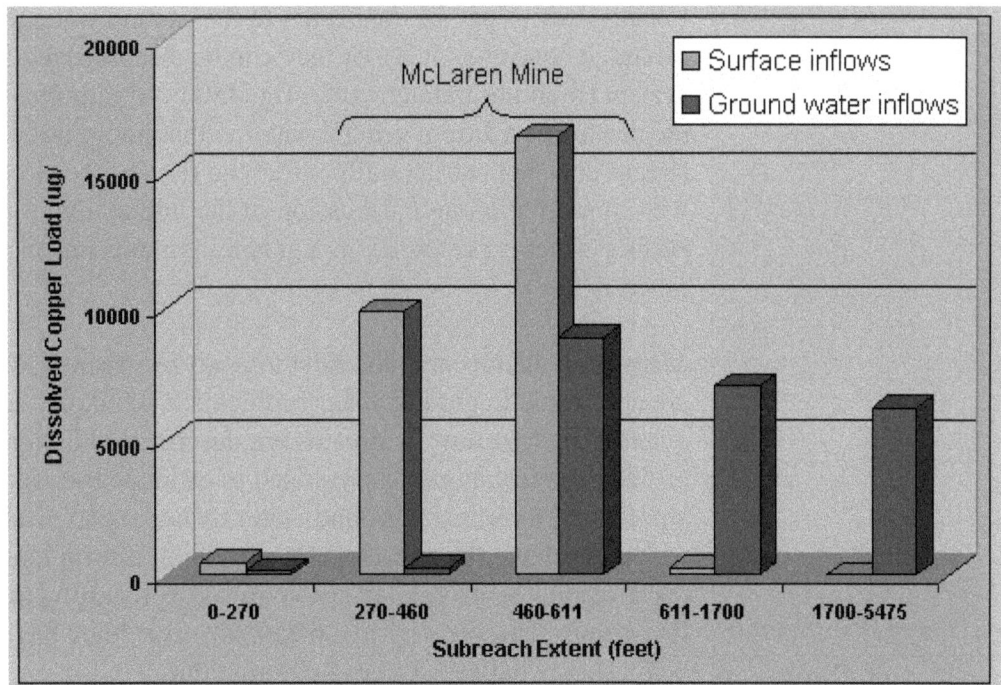

Figure 44. Sources of dissolved copper to subreaches of Daisy Creek, including relative contributions of copper from surface water and ground water sources. Copper loading occurs primarily from surface inflow in the upper reaches, while ground water contributes substantial loads in the lower reaches.

Gaining and Losing Stream Reaches

Some shallow ground water has a water level that lies above the elevation of the water surface in an adjacent stream channel. In such cases, ground water seeps through the stream bank and bed to discharge into the stream, which is referred to as a "gaining stream or reach" (fig. 45A). Where shallow ground water has a water level that lies below the elevation of the water surface in an adjacent stream, water may seep out of the channel through the stream bank and bed to locally recharge the ground water. In such a situation, the channel is referred to as a "losing stream or reach" (figs. 45B and C). Many streams have reaches of both types, gaining in some and losing in others. Most mountain streams have gaining reaches from their headwaters on downstream to a mid-valley location, where they may have losing reaches. Farther downstream, near the mouths of the streams, additional gaining reaches are frequently found; these are typically the discharge zones of shallow ground water flow systems.

The flow directions between ground water and surface water can change seasonally as the elevation of the ground water table changes with respect to the stream-surface elevation. They can change over shorter timeframes when rises in stream surfaces during storms cause recharge to the streambank. Under natural conditions, ground water makes some contribution to streamflow in most physiographic and climatic settings. Thus, even in settings where streams are primarily losing water to ground water, certain reaches may receive ground water inflow seasonally or under particular hydrologic conditions.

Losing streams can be connected to the ground water system by a continuous saturated zone (fig. 45B), or they can be disconnected from the ground water system by an unsaturated zone (fig. 45C). An important feature of streams that are disconnected from ground water is that pumping of ground water near the stream does not substantially affect the flow of the stream near the pumped well. A more thorough discussion of the interaction of ground water and surface water is presented, in a generally nontechnical format, by Winter and others (1998).

Many graphical techniques exist to analyze streamflow data, and several are applicable to ground water problems. Techniques and procedures for quantifying baseflow in streams are described in Fetter (2001) and McCuen (1998). The techniques can be used to estimate hydraulic properties of an aquifer and to estimate ground water recharge and discharge in a basin. These techniques have the advantage of providing information over a wide area (as compared to an aquifer-test analysis), integrating the effects of climate, topography, and geology in a basin; however, they have the disadvantages of being somewhat subjective and nonunique.

Flow-duration curves are cumulative frequency curves that show the percentage of time during which specified discharges of streams were equaled or exceeded in a given period of time (Searcy 1959). Comparison of flow-duration curves can provide valuable insights into the drainage characteristics of different streams or of different reaches of the same stream. Steep curves

A   GAINING STREAM

Flow direction

Unsaturated zone

Water table

Saturated zone

B   LOSING STREAM

Flow direction

Water table

Unsaturated zone

C   LOSING STREAM THAT IS DISCONNECTED FROM THE WATER TABLE

Flow direction

Unsaturated zone

Water table

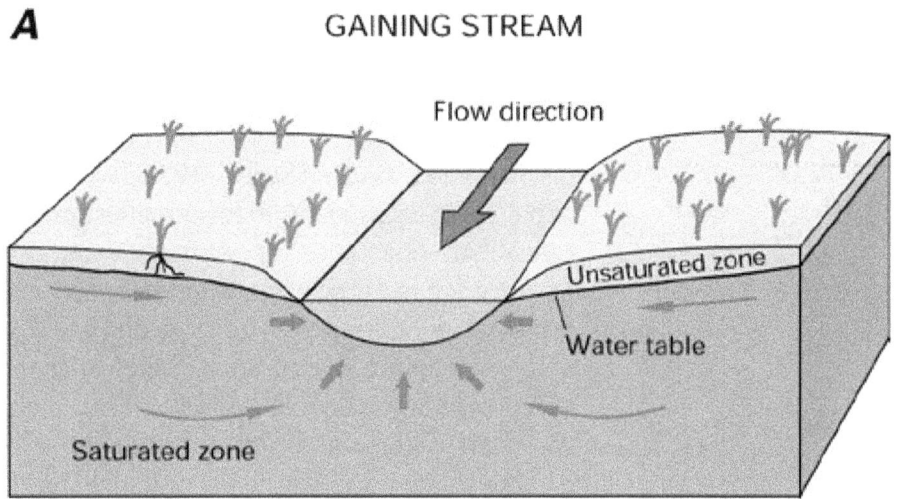

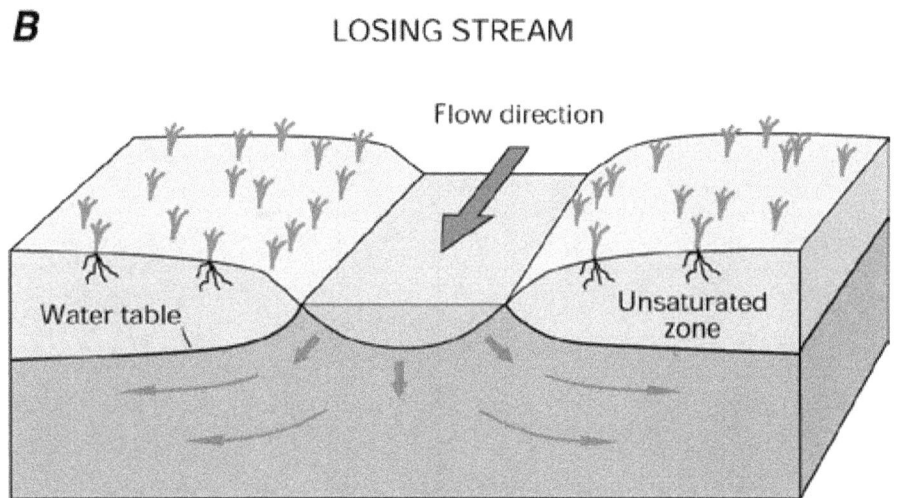

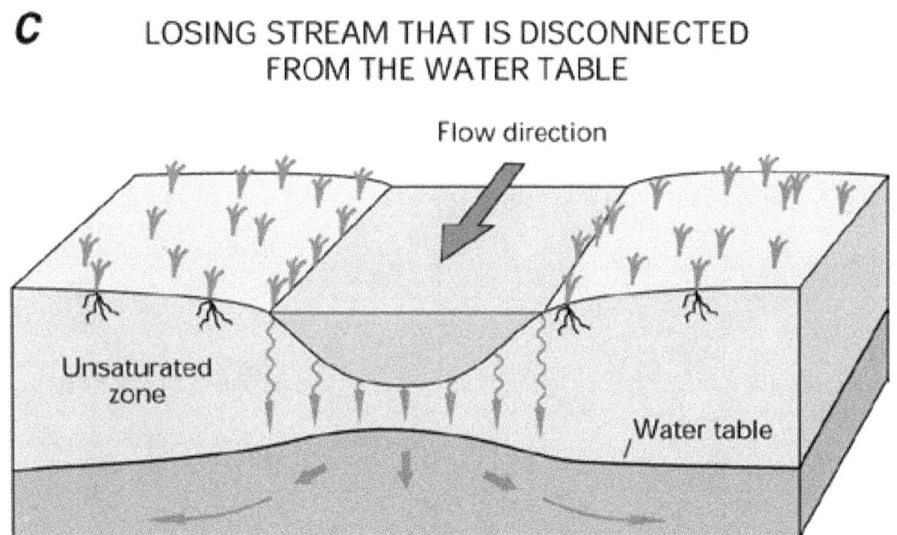

Figure 45. Interaction of streams and ground water (after Winter and others 1998).

125

indicate a high degree of runoff; flat curves indicate a high degree of surface or subsurface storage in the basin. Because the distribution of low flows is controlled chiefly by the geology of the basin, the lower end of the curve is a valuable means for studying the effects of geology on the ground water discharge to a stream (Searcy 1959). Many studies have used a flow-duration curve value as a substitute for direct estimates of mean baseflow. Some researchers select the 90 percent flow-duration value (the flow that is equaled or exceeded by 90 percent of the flow on record) as a conservative estimator of ground water discharge (Rutledge and Mesko 1996), but individual basins are highly variable. Flow duration values of as low as 40 percent have been found to represent a reasonable estimate of mean ground water discharge. Figure 46 shows an example of a flow-duration curve for Rio Camuy, near Hatillo, PR. An independent evaluation of ground water discharge for Rio Camuy at this site indicated a baseflow of about 72 cfs, which approximately corresponds to the 60 percent flow-duration value on this curve (Tucci and Martinez 1995).

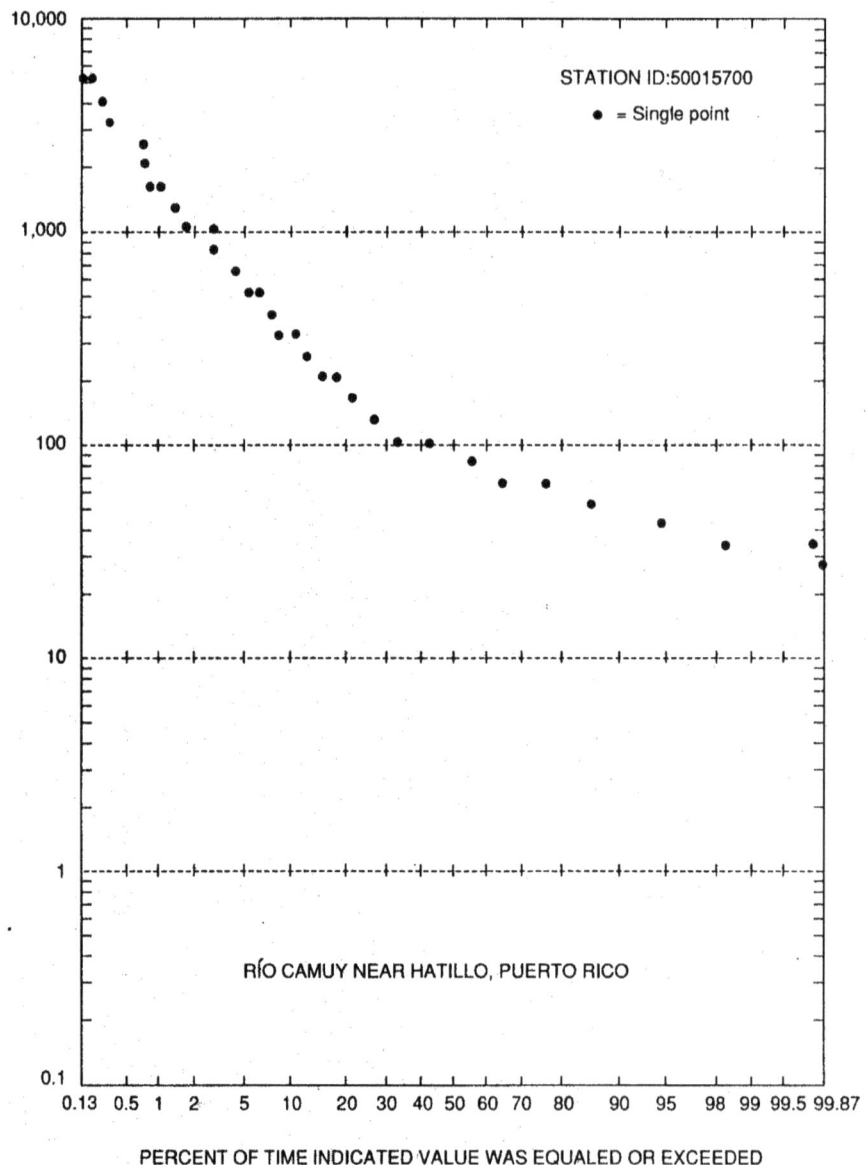

Figure 46. Flow-duration curve for the Rio Camuy near Hatillo, PR (Tucci and Martinez 1995).

Streamflow recession methods, also referred to as hydrograph-separation methods, can provide information not only on baseflow to streams but also on transmissivity and storage values for a basin. These methods characterize the portions of the hydrograph following a recharge event, which are represented by a sharp increase in streamflow followed by a decline (or "recession") (fig. 47). The method was described in detail by Rorabaugh (1964), and computer programs to apply the method are available (Rutledge 1998, 2000). Several techniques for assessing the quantity and quality of ground water discharging to streams are presented in table 9.

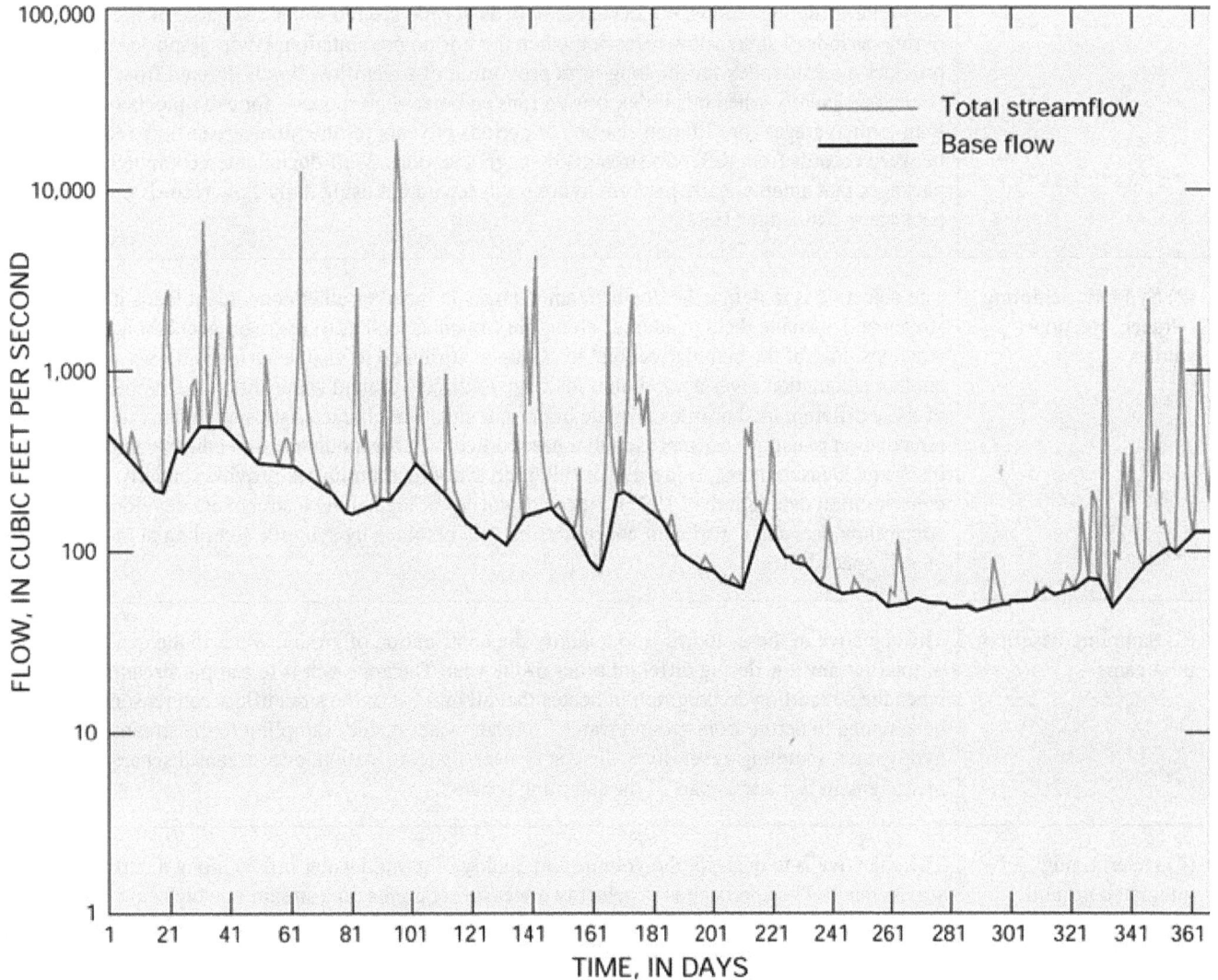

Figure 47. Streamflow and baseflow hydrographs for the Homochitto River in Mississippi (Winter and others 1998).

Table 9. Types of studies that evaluate water-flow and water-quality interactions between ground water and surface water.

Type of study	Explanation
Ground Water Contributions to Stream Flow and Quality	
(1) Hydrograph separation	The objective is to divide the total streamflow hydrograph into two parts: (1) Storm runoff or quick-response flow that is related to storms and (2) ground water, which may augment storm runoff during storms, but occurs mainly as normal ground water discharge to streams during periods of streamflow recession when there is no precipitation. Hydrograph separation provides a useful index for the long-term proportion of streamflow that is derived from ground water, particularly when this index is used for comparative purposes—for example, between long-term averages for different seasons or periods of years for the same streamflow record or between records from different streamflow-gaging stations. Well-documented computer software packages that automatically perform hydrograph separation using daily flow records include the package by Rutledge (1993).
(2) Synoptic sampling – tracer injection studies	The objective is to define the downstream changes in metal or other constituent loads in the stream and attribute them to sources along the stream as well as to instream geochemical reactions. Part of the cumulative total load can be attributed to visible surface inflows, and another calculation gives a maximum load due to diffuse ground water inflows. Comparisons of these different load profiles provide important chemical characteristics of streams useful for remediation planning. An approach that has worked well for mountain watersheds combines discharge measurements, using dye or salt, with synoptic sampling to provide spatially detailed concentration data (Kimball 1997). Detailed profiles of load along a stream are developed from streamflow data and constituent concentration data obtained by synoptic sampling at many closely spaced sites.
(3) Sampling baseflow of streams	The objective of these studies is to quantify the contribution of ground water to the quality of total streamflow during different times of the year. The approach is to sample streamflow when the streamflow hydrograph indicates that all or most of the streamflow can reasonably be assumed to derive from ground water. To relate water quality sampling to the streamflow hydrograph, sampling generally is done at or near a gaging station, or a stream-discharge measurement is made as part of the sampling process.
(4) Determining integrated ground water inflow along stream reaches	The objective is to quantify the volume and quality of ground water inflow along a particular stream reach. The approach is to select two measuring points on a stream at which flow and water quality are measured. The ground water contribution to flow and water quality along the stream reach are determined by difference. These studies are more locally focused than assessments of the baseflow of streams.
(5) Evaluating discharging ground water	The objective is to determine the quality of shallow ground water that soon will discharge into a stream. The approach is to sample using streambed and streambank piezometers and to compare the ground water quality with stream water quality. Another possibility is direct sampling of ground water discharge to streams by means of seepage meters. Sampling from shallow streambed piezometers can be used in reconnaissance surveys to characterize ground water quality. The design of these surveys is guided by knowledge of flow patterns in the shallow ground water flow system and land use near the streams.

Table 9. — Cont.

Type of study	Explanation
(6) Assessing spring or seep water	Springs are points of concentrated ground water discharge. They represent an opportunity to sample ground water discharge directly. Although often difficult to determine, the contributing area of a sampling site is a useful concept for springs. The sampled water quality from a spring may vary for some constituents, depending on where and how the spring is sampled—for example, as ground water from a piezometer immediately upgradient from the orifice or as surface water after discharge. A seep is an area where ground water oozes from the earth in small quantities. Therefore, seeps can be viewed as low-discharge end members of springs.
(7) Measuring surface water capture from shallow pumping wells located near a surface water body	Generally, the objective is to determine the proportion of the water pumped from a well field that is derived from surface water at different pumping rates. An additional objective may be to determine if pumping-induced movement of surface water through the shallow ground water system to the well results in changes in the original quality of the surface water; for example, one can determine whether the concentrations of key constituents from the surface water are decreased or eliminated before the stream water reaches the pumping well. Tools of analysis for this type of study and bank storage/overbank-flooding studies include water-mixing models, analysis of isotope data, and local-scale simulation of the ground water flow system.
Determining Interactions Related to Increases in Stream Stage	
(1) Bank storage	The objectives of bank-storage studies include determining (1) the movement of stream water into the ground water system during periods of rising stream stage and (2) the volume, time of release, and quality of former surface water, possibly mixed with original ground water, that returns to the stream during periods of falling stream stage. These studies rely on determining the quality of the surface water and shallow ground water near the stream.
(2) Overbank flooding	The objective of overbank-flooding studies (similar to the objectives for bank storage studies) is to determine the volume and quality of surface water that recharges shallow ground water from the flooded land surface, as well as surface water that enters the shallow ground water system through the stream banks and bed. Given sufficient overbank flooding, parts of the underlying surficial hydrogeological unit may become completely saturated to the land surface.

Hyporheic Zone and Floodplain Mixing

In certain circumstances, surface water and ground water may mix and remix rapidly over short distances. Hyporheic water is stream water that flows through shallow unconfined aquifers and is returned to the stream over relatively short time periods. It ranges from 100 percent for recent stream-source water in downwelling locations, to some mixture of longer residence time "ground water" and recent stream-source water. A popular convention is to limit the spatial extent of the hyporheic zone to areas where the water in the unconfined aquifer is composed of 10 percent or more recent stream-source water (Triska and others 1989). The following are among the hydrologic features that lead to hyporheic exchange flows:

129

- Any change in the longitudinal profile of the stream (pool-step or pool-riffle sequences).
- Changes in aquifer thickness or width.
- Changes in saturated hydraulic conductivity.
- Presence of multiple channels (channel splits around islands, secondary channels, floodplain spring brooks).
- Buried relic channels that create longitudinally continuous preferential flow paths.
- Channel meander bends (channel sinuosity).
- Interactions between streamflow and channel bed forms.
- Bank storage or overbank flooding and infiltration of flood water.
- Entrainment of stream water into mobilized bed sediments during floods.

Ground water and surface water may exchange over short distances along a stream because of streambed slope changes in high-gradient step-pool streams, and between meanders in lowland valley streams. In step-pool streams, ground water may flow into the stream channel at the upstream ends of the pools and return as stream water recharging the ground water at the downstream ends of the pools. This sequence of flows creates a unique ecological environment within the streambed sediments and their immediate surroundings. The water that washes back and forth through these sediments is much richer in oxygen and nutrients than that found deeper in the subsurface. This especially active envelope of sediments around and including the streambed can support unique biota that have evolved in and inhabit the hyporheic zone.

The meandering of streams offers opportunities for stream water and ground water to mix. Often, the direction of shallow ground water flow is roughly coincident with the predominant course of the stream. This assumption is reasonable for small undeveloped valleys, but less so for large valleys and where development has led to significant use of ground water. If flows are coincident with high-stage conditions in the stream channel, water moves from the stream to the ground water. For relatively short duration events, only bank storage of stream water occurs during the high stage; the water stored in the streambanks is released back into the stream as the high stage subsides. For relatively long duration events, the water that is moving continuously from the stream into the streambanks is pushed through the streambank and farther into the adjacent sediments, where it blends with the local ground water and is carried downgradient as part of the ground water flow system. Consequently, only some of the water lost by the stream may return to the channel as ground water inflow after the high stream stage subsides.

Overbank floods provide additional opportunities for the mixing of stream water and ground water. Stream water spreads out over the floodplain during overbank flooding and infiltrates the floodplain sediments to mix with ground water. Because overbank flooding may occur infrequently and be of very limited duration, the influx of oxygen and nutrients may be too short-lived to evolve and sustain unique biota. Certain riparian plants, however, require occasional inundation by overbank flooding for proper establishment and growth. Although the specific role of the influx of nutrients in floodwaters may not be known as yet for most riparian plant species, ecological principles suggest that the role is not incidental.

Influence of Wells on Streams

Ground water pumping can substantially affect the quantity of surface waters, including not only downstream water supply for human consumption but also the maintenance of instream-flow requirements for fish habitat and other environmental needs. Long-term reductions in streamflow can affect vegetation along streams (riparian areas) that serve critical roles in maintaining wildlife habitat and in enhancing the quality of surface water. Pumping-induced changes in the flow direction to and from streams may affect temperature, oxygen levels, and nutrient concentrations in the stream, which may in turn affect aquatic life in the stream.

Figure 48 illustrates the following discussion on the source of water to wells (from Alley and others 1999). Under natural conditions (A), recharge at the water table is equal to ground water discharge to the stream. Assume a well is installed and is pumped continuously at a rate, Q1, as in (B). After a new state of dynamic equilibrium is achieved, inflow to the ground water system from recharge will equal outflow to the stream plus the withdrawal from the well. In this new equilibrium, some of the ground water that would have discharged to the stream is intercepted by the well, and a ground water divide (a line separating directions of flow) is established locally between the well and the stream. If the well is pumped at a higher rate, Q2, a different equilibrium is reached, as shown in (C). Under this condition, the ground water divide between the well and the stream is no longer present, and withdrawals from the well induce movement of water from the stream into the aquifer. Thus, pumping reverses the hydrologic condition of the stream in this reach from ground water discharge to ground water recharge. Note that in the hydrologic system depicted in (A) and (B), the quality of the stream water generally will have little effect on the quality of ground water; however, the loss of ground water to the stream could have an effect on water quality in the stream. In the case of the well pumping at the higher rate in (C), however, the quality of the stream water can affect the quality of ground water between the well and the stream, as well as the quality of the water withdrawn from the well. Although a stream is used in this example, the general concepts apply to all surface water bodies including lakes, reservoirs, wetlands, and estuaries.

The factors that influence the location of areas contributing water to wells can be categorized as dependent either on the ground water system or the well (Franke and others 1998, Reilly and Pollock 1993). The ground water factors that affect the paths of water movement in ground water systems are (1) the hydrogeological framework, (2) system boundary conditions, (3) aquifer properties, and (4) other transient effects, such as rainfall. The well factors are the location of the well, the depth of the screened or open-hole section of the well, and pumping rates.

131

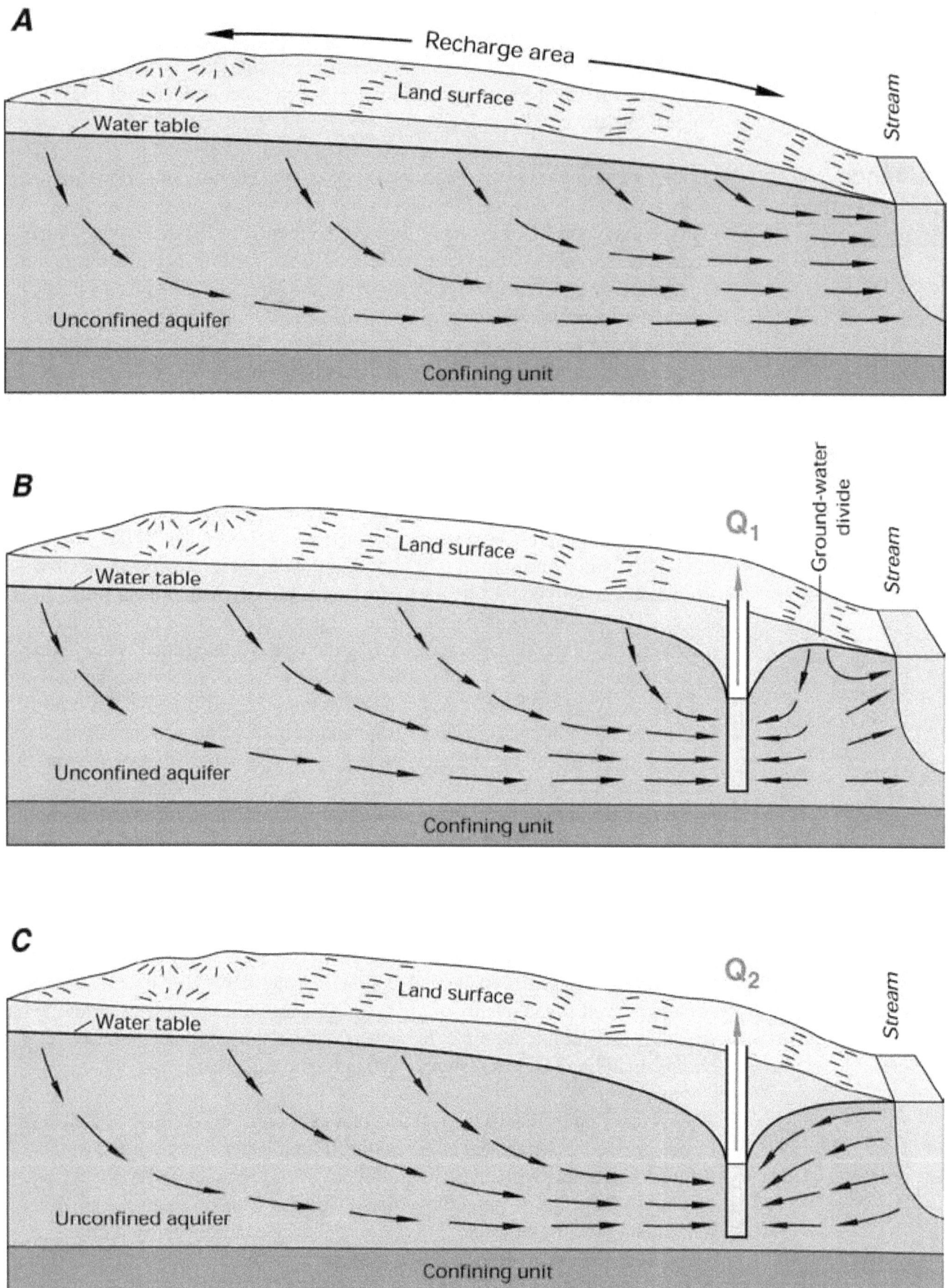

A

Recharge area

Land surface

Water table

Unconfined aquifer

Confining unit

Stream

B

Land surface

Water table

Unconfined aquifer

Confining unit

Q_1

Ground-water divide

Stream

C

Land surface

Water table

Unconfined aquifer

Confining unit

Q_2

Stream

Figure 48. Effects of pumping from a hypothetical ground water system that discharges to a stream (after Heath 1983).

132

The adjustments to pumping of an actual hydrological system may take place over many years, depending on the physical characteristics of the aquifer, the degree of hydraulic connection between the stream and aquifer, and the locations and pumping histories of wells. Reductions of streamflow as a result of ground water pumping are likely to be of greatest concern during periods of low flow, particularly when the reliability of surface water supplies is threatened during droughts.

Characterizing a ground water flow system involves definition of the aquifers and confining units that comprise the system, as well as quantification of the amount of ground water present in the system at any one time and the amount of ground water that is entering and leaving the system at that time. Definition of the system is scale dependent. For example the ground water system of interest in a problem involving a leaking underground storage tank can be very different from the system of interest when trying to quantify the amount of ground water that flows through a national forest.

Springs

Springs are important sources of hydrogeological information. They occur because the hydraulic head in the aquifer intersects the land surface. The distribution, flow characteristics, and water quality of springs can provide as much, or more, information about an aquifer system as a well. Springs are relatively small riparian ecosystems that are maintained by water flowing from the ground (Hynes 1970). The classic definition is from Meinzer (1923, 48): "A spring is a place where, without the agency of man, water flows from a rock or soil upon the land or into a body of surface water." Spring ecosystems include aquatic and riparian habitats that are similar to those associated with rivers, streams, lakes, and ponds. They are distinctive habitats because they provide relatively constant water temperature, depend on subterranean flow through aquifers, and on occasion provide refuge habitats that support species that occur only in springs (Hynes 1970, Erman and Erman 1995, O'Brien and Blinn 1999).

Springs are replenished by precipitation that percolates into aquifers by seeping into the soil and entering fractures, joints, bedding planes, or interstitial pore space. Springs occur where water flowing through aquifers discharges at the ground surface through fault zones or fractures, or by flow along an impermeable layer (fig. 49). They can also occur where water flows from large orifices that occur when water creates a passage by enlarging fractures or joints by dissolving carbonate rock. Characteristics of regional and local geology influence spring occurrence and flow rates.

TYPES OF SPRINGS

Fetter (2001) identifies five types of springs: (1) depression springs, (2) contact springs, (3) fault springs, (4) sinkhole springs, and (5) fracture springs. Depression springs form in low topographic spots where the water table reaches the surface. Where permeable rocks overlie rocks of much lower permeability, a contact spring may result. Such a lithologic contact between rock of contrasting permeability is often marked by a line of springs. Faulting

133

Figure 49. Comal Springs, near San Antonio, TX, discharges ground water from the highly productive Edwards aquifer. (Photo by Robert Morris, USGS.)

can form a boundary to ground water flow and force water in the aquifer to discharge as a fault spring. Sinkhole springs are formed where water dissolves the limestone beneath the surface and creates a sinkhole. If the artesian pressure in the subterranean solution cavities is high enough to reach the surface, a sinkhole spring is formed. Fracture springs form where ground water flowing along a fracture or joint intersects the land surface.

Springs can also be classified as gravity springs and artesian springs, with thermal springs classed as a type of artesian spring. Water that moves along an elevation gradient emerging at the surface creates gravity springs. Depression springs, contact springs, and fracture springs are different types of gravity springs. They are the result of ground water discharging from a permeable rock unit in contact with impermeable rocks or rocks having lower permeability. Fracture springs, for example, are often the result of fractured basalt or limestone overlying an impermeable rock stratum, and water flows along the outcrop of the two units. The temperature of the water will approximate the mean annual atmospheric temperature of the location. If movement of water occurs through passages that are open to the circulation of air, cooling to as much as several degrees below mean annual temperature will occur. If water is not in contact with circulating air and the depth to the water table is several hundred feet, the water will be a few degrees warmer than the mean annual temperature (generally about 1 degree for each 100 feet in depth).

Artesian springs occur where the potentiometric level of the ground water flow system is above land surface and water flows at the land surface under artesian pressure, or where water is forced to the surface from deep sources by thermal and pressure gradients. They usually occur at lower elevations in mountainous areas, especially along mountain fronts. Aquifer outcrop springs and fault springs are the two main types of artesian springs. Thermal springs are usually a variation of artesian spring that connect to deep-seated thermal sources, and they are classed as volcanic springs or fissure springs (Milligan and others 1966). Temperatures of thermal springs can be greater than 100° C. Fault-related springs can also be thermal if they are from a deep source of water. This type of spring is common in the Great Basin (UT, NV and adjacent States), where mountain blocks are faulted along the margins, allowing water from deep sources to rise along the fault. Devil's Hole, NV, is an example.

Karst springs can be classified by mode of ground water recharge. The three major recharge modes are (1) diffuse through permeable material producing network conduit/cave patterns; (2) authigenic through many discrete sources such as sink holes, producing dendritic conduit/cave patterns; and (3) allogenic through a few major inflow points such as sinking streams, producing braided conduit/cave patterns. Springs are natural ground water discharge points, while sinks can be ground water recharge or discharge points.

Springs can be regional (long flow paths that may connect more than one surface water basin) or local discharge points (short flow paths). Local springs are comparatively small, can be low flow and low temperature, and are typically from shallow aquifers. The discharge from these springs often fluctuates either seasonally or in greater cycles, sometimes in response to local precipitation. Local aquifers are quickly recharged and water movement through them is comparatively rapid, resulting in water that is low in mineralization. Springs supported by local aquifers are more likely to periodically stop flowing than springs supported by regional aquifers. Springs at higher elevations generally display greater fluctuations in flow rates, and dry more frequently than regional springs or springs at lower elevations; however, they are generally less susceptible to impacts from dewatering at mining operations or from pumping wells.

Springs fed from regional aquifers typically have large discharge, and are discharge points for aquifers covering hundreds of square miles (fig. 50). In the Great Basin, the majority of springs with high discharge rates occur in intermontane basins of the carbonate rock province and are often closely associated with limestone outcrops (Mifflin 1988). Regional springs are typically of nearly constant discharge, and can be more mineralized than local springs because of their long flow paths. Their temperatures can be cold or warm depending on the depth of circulation. Seasonal and annual variations in discharge from regional springs are usually limited, and they are comparatively stable aquatic environments. Regional springs rarely stop flowing, even during long droughts, but they can be affected by pumping from the regional aquifer (Dudley and Larson 1976).

Figure 50. Ground water from a large regional limestone aquifer discharges at Crystal Spring in Ash Meadows National Wildlife Refuge, NV. (Photo by Pat Tucci, USGS.)

PHYSICAL ENVIRONMENT AND WATER CHEMISTRY OF SPRINGS

Springs occur in many sizes, types of discharge points, and location with respect to topography. They occur at the highest elevations of mountainous areas and they occur in valley floors. Many springs on public land are small, provide limited aquatic habitats, and are intermittent in flow. They sometimes support limited amounts of riparian vegetation. Some small springs, however, provide aquatic habitat, are permanent, and support high species diversity over large riparian areas. Springs can be categorized by the morphology of their discharge area. Limnocrenes are springs where water flows from large deep pools, helocrenes are marshy bogs, and rheocrenes flow from a confined channel (Hynes 1970). It is often difficult, however, to categorize springs because morphology can involve a combination of features from more than one of these categories.

Springs may occur singly or in groups that can include dozens of habitats in various sizes and morphologies. Many springs are tributaries to rivers, lakes, or streams. A few are even the major source for a river or lake. Many springs are isolated from other surface waters and frequently flow a short distance on the surface before drying. Springs in dry regions may stop flowing seasonally or during droughts. Some groups of springs can support wetland areas with unique habitat and species; examples are Ruby Marsh in northeastern Nevada, Ash Meadows in southern Nevada (fig. 50), Fish Springs in northwestern Utah, and San Bernardino Ranch in southern Arizona (Hendrickson and Minckley 1984, Dudley and Larson 1976). Some springs support fens that are at middle to low elevations in the watershed, usually in large open areas or parks, such as in South Park in south-central Colorado. Some springs are the source for low-order streams high in watersheds.

Riparian vegetation may be narrowly restricted to immediate boundaries of the aquatic habitat, or it may extend outward for substantial distances. Narrow riparian areas are typically dominated by sedges, grasses, and woody phreatophytes such as willows and mesquite. Wider riparian ecosystems are generally associated with spring provinces where water seeps outward from aquatic habitats, which saturate and create hydric soils. In these provinces, riparian ecosystems are characterized by marsh vegetation or expansive mesic alkali meadows.

Physical and chemical conditions of springs vary (Hynes 1970). They can be cold (near or below mean-annual air temperature), thermal (5° to 10° C above mean annual air temperature) (van Everdingen 1991), or hot (more than 10° C above mean annual air temperature) (Peterken 1957). The temperature of spring water is also an indicator of the flow path of water discharging to the spring and its recharge area. Shallow circulating ground water has temperatures generally within a few degrees of the mean annual ambient air temperature (Mifflin 1988). Higher temperatures are usually indicative of deeper, regional circulation, although some cool regional springs exist. Thermal springs may gain their heat when water comes in contact with or in close proximity to recently emplaced igneous masses, such as at Steamboat Springs, NV; Yellowstone National Park; and Geyser, CA (Wood and Fernandez 1988), or through the higher temperatures encountered at large depths caused by the natural geothermal gradient.

Springs may be highly mineralized, especially thermal springs and regional springs that have a very long flow path. Dissolved oxygen (DO) concentrations are primarily a function of temperature and pressure; as temperature increases, the DO concentration decreases (Hem 1989). As a result, DO concentrations are frequently very low (less than 2 parts per million) in hot springs and high (greater than 5 ppm) in cold springs. However, DO can also be substantially affected by the nature of the geologic materials along the flow path. For example, a flow path that involves materials with high organic content will generally have low DO concentrations. Electrical conductance may range from very low (near 0 micromhos per centimeter) to very high (greater than 10,000 micromhos per centimeter). Local low-flowing springs may freeze during winter while the larger and warmer regional springs generally do not.

FIELD OBSERVATIONS OF SEEPS AND SPRINGS

Springs and seeps provide a means of assessing ground water quality and of helping to determine ground water flow patterns. It is essential that flow from a spring is identified as to the geologic formation from which the water discharges. Springs may be caused by bodies of perched ground water, water under artesian pressure, or outcrops of the main water table. Gains or losses in baseflow of streams mark reaches affected by ground-water discharge or recharge. The following are among the basic data collection requirements for springs:

1. Elevation of the spring.
2. Uses of the spring water (stock watering, domestic, unused).

3. Permanence of flow (perennial or seasonal).
4. Discharge of the spring, including date and time of measurement.
5. Chemical characteristics of the water.
6. Type of spring (perched, contact, fracture, and so on).
7. Source aquifer.

Examination of water quality in the field is an important part of hydrogeological studies; especially when investigating spring sources. Certain properties of natural water, especially pH and DO, are so closely related to the environment of the water that they are likely to be altered by sampling and storage, and a meaningful value can be obtained only in the field. Other properties that should be sampled while in the field are specific conductance, redox potential, and temperature.

When conducting spring investigations, any geological outcrops at the spring need to be evaluated to determine the hydrogeological setting for the spring. Assess how water is being recharged into the ground water system, how it moves, and what mechanism forces the water to the surface at that particular point. Features such as fractures, faults, sand-and-gravel layers overlying impermeable bedrock, and silt or shale layers that impede downward flow of water should be noted. Before development of a spring is attempted, the type of spring must be determined to properly design a collection system that will result in a reliable water source without damaging the natural condition or ecological values of the spring. A classification of the spring can often be made from geological and topographical observations.

Ground Water Exchange in Reservoirs, Lakes, and Ponds

The evaporation of water from reservoirs, lakes, and ponds often maintains a water level that is somewhat lower than the local ground water table. In such circumstances, the primary productivity of the surface water body may be greatly enhanced by the nutrients carried by the ground water flowing into it (Kenoyer and Anderson 1989). Another potential influence on primary productivity is that the temperature of ground water is usually fairly constant, so that it is colder in the summer and warmer in winter than the water in the receiving surface-water body. The amount of ground water flowing to a water body may be estimated by mapping the slope of the ground water table, performing water-budget analysis (Winter 1995), conducting seepage studies (Carr and Winter 1980, Paulsen and others 2001), applying chemical mass-balance methods (Krabbenhoft and others 1990a, Lerman and others 1995, Sacks 2002, Stauffer 1985,), or performing numerical modeling (Krabbenhoft and others 1990b). In some instances, the inflows from surface streams and runoff to a surface-water body, usually one with no surface outlet, may be sufficient to maintain its level at an elevation that is higher than the local ground water table. In those cases, seepage from the surface-water body is a recharge source for local ground water (Winter 1981).

Sensitive Hydrogeological Settings

Karst Terrains

Many professionals consider the most sensitive aquifers to be those that are composed of karst limestone or dolomite. Dissolution of portions of a soluble rock body by water flowing through the pores and fractures generates preferential pathways of flow, which can vary in size from lengthy but small diameter solution cavities to cave systems and large caverns. Because the rock itself is often highly porous, karst aquifers offer the possibility of enormous withdrawals of ground water; however, the consequences of excessive withdrawals from karst aquifers can be highly destructive. A sinkhole is one possible result. The ready movement of large amounts of ground water within karst aquifers, coupled with the presence of preferential pathways of flow, can make ground water contamination spread quickly and often in unpredictable patterns.

Karst is a general term for a wide range of landscape settings in which the underlying rocks have been modified by solutional processes. Ground water interactions with stream water in karst settings include springs, sinkholes, swallows, and resurgences. The flow of ground water through karstic limestone formations occurs through the pores of the bulk rock, through fractures in the bulk rock, and through solution cavities and channels, including cave streams, pools, and waterfalls. Ground water emerges as springs, seeps, and wetlands of various kinds, including fens and marshes. Streams that flow over karst terrain may swell or shrink in size sporadically, in response to passing over springs and sinks. Streams may disappear completely into sinkholes or swallows (the land surface entry points of solution channels or cavities), only to reappear by resurgence farther downslope. Tracer studies have shown that the flow paths within karst limestone can be circuitous and multibranched, so that it is not uncommon for swallowed streamflow to reappear miles away, on the opposite side of a ridge, or at several widely separated locations. Similarly, tracer studies have indicated that the water in some streams in karst terrain contained ground water from various springs that originated in widely separated areas.

Focused recharge and discharge can be readily identified in many karst systems from remote sensing and mapping of geomorphic features such as sinkholes, stream networks, and vegetation patterns defining fracture traces. The National Research Council (2004) suggests the following field components for assessing the hydrogeology of a karst area:

1. Develop a long-term water balance for the watershed.
2. Measure discharge and geochemical parameters at all major springs.
3. Measure travel times and residence times using environmental isotopes, geochemistry, temperature, and tracer tests.
4. Install monitoring wells and meteorological stations and monitor continuous water levels and geochemistry.

5. Map in detail the topography, soils, karst features, and vegetation that correlates with discharge zones and seeps.
6. Estimate stream-hydrograph separation using both physical and chemical parameters to discern baseflow, stormflow, and old and new water components (Kendall and McDonnell 1998).

Additional information on characterization of karst and fractured rock hydrogeologic systems can be found in American Society for Testing and Materials (ASTM) standard D5717-95e1 (ASTM 1996).

Unconsolidated Deposits

Unconsolidated deposits comprise the most common and most accessible aquifers in the United States. The unconsolidated-rock aquifers occur as alluvium, colluvium, and glacial drift deposits. These aquifers are typically composed of sand or sand and gravel, often intermixed with finer-grained sediments. They are usually unconfined aquifers, but may also occur as partially confined or confined aquifers. Because unconsolidated aquifers are generally shallow and well connected to surface water, knowledge of ground water/surface water interactions is critical to understanding these aquifers. Those same characteristics also make unconsolidated aquifers often highly susceptible to contamination.

Alluvial aquifers generally occur along rivers and streams and were deposited as coarse-grained sediments by streams. Their extent in area may be restricted to a zone on either side approximately parallel to the stream or it may be quite extensive, especially along major rivers. An example of an extensive aquifer is the Mississippi Aquifer in western Tennessee and eastern Arkansas. Alluvial aquifers generally are tens to hundreds of feet thick. The distribution of coarse-grained sediments, which are the most productive parts of the aquifers, is controlled somewhat by the type of stream that deposited the sediments. In streams that are somewhat confined in area and have a relatively steep gradient, coarse-grained sediments may be distributed throughout the aquifer. In larger, meandering streams, the coarse-grained sediments tend to be associated with sand and gravel bars distributed between finer grained sediments. Alluvial aquifers are usually well connected to the nearby streams, which can provide a source of recharge to the aquifer. Because of their shallow, unconfined nature, these aquifers are susceptible to contamination from human activities at the surface.

The High Plains Aquifer is major alluvial aquifer that underlies some National Grasslands. It is one of the largest aquifers in the United States. It was formed from sediments eroded from the Rocky Mountains to the west. It is a highly productive, thick, generally unconfined aquifer, which has undergone significant water-level declines because of irrigation pumpage.

Some alluvial aquifers are buried beneath more recent stream or glacial sediments. These aquifers most often occur within bedrock channels carved by ancient rivers in Northeastern and Midwestern States. An example is the Teays Aquifer of Ohio, Indiana, and Illinois (Sharp 1988).

Valley-fill aquifers, sometimes termed basin-fill aquifers, occur in the Western United States and are the most important aquifers in the Basin and Range physiographic province. These aquifers, which often are adjacent to NFS lands, were deposited as a combination of alluvium and colluvium as the basins subsided relative to the surrounding mountain ranges. They tend to be coarser grained and most productive along the basin margins and near modern stream channels. They tend to be finer grained and less productive near the basin centers, but this general pattern can be altered somewhat by the structural history of the basin (Anderson and others 1992). In some basins, evaporite deposits occur within the valley-fill aquifer, and these deposits can degrade the quality of the ground water. Ground water generally occurs under unconfined conditions, but confined conditions can occur where extensive fine-grained sediments are present. Valley-fill aquifers can be very productive, and some wells can yield more than 1,000 gallons/minute. Because of the typically arid environments in which these aquifers occur, recharge to them is small. Much of the water pumped from them is removed from storage, causing large water-level declines. In some basins, large ground water withdrawals have resulted in land subsidence.

Glacial aquifers are generally derived from coarse-grained sediments associated with glacial outwash and ice-contact stratified deposits associated with fast-moving glacial meltwaters. Some sandy tills and loess (aeolian silt) deposits can provide adequate water to domestic wells. In mountainous areas, the productive glacial aquifers are confined to valley bottoms and sides. In areas subjected to continental glaciation, the productive aquifers may occur along the surface or may be buried by sediments deposited by subsequent glaciation or other processes. Hydraulic properties and thickness of glacial aquifers are highly variable, depending on the type of glacial deposits and subsequent modification to those deposits (Stephenson and others 1988). Glacial aquifers are generally shallow and unconfined, but ground water can occur under confined conditions where the aquifers occur beneath glacial lake deposits.

Characterizing unconsolidated aquifers is generally straightforward because much of the theoretical basis for quantification of ground water flow was developed from studies of these types of deposits. Aquifer tests and computer modeling are well-suited to analysis of unconsolidated aquifers because porous-media flow is commonly a reasonable assumption. Heterogeneity of unconsolidated deposits, however, greatly complicates the characterization of these aquifers, particularly for small-scale problems such as ground water contamination; for example, braided-stream or glacial-outwash deposits can vary greatly within short distances both horizontally and vertically.

These variations can make correlation of units within these deposits almost impossible. In addition, many of the fine-grained tills and lacustrine deposits associated with glacial aquifers developed fractures as a result of unloading following ice retreat or periglacial freeze-thaw action. Due to the high variability in these systems, large numbers of boreholes and wells, completed at different depths, may be required for unconsolidated aquifer characterization.

Geophysical methods may help in characterization for large-scale problems; for example, gravity and seismic surveys have been used successfully to map the extent of buried alluvial aquifers. Electrical and electromagnetic geophysical methods can be used to map the extent of fine-grained materials within the unconsolidated aquifers. Basic knowledge of the geomorphic and sedimentologic characteristics of unconsolidated deposits can be used to map the location of the various depositional facies and estimate the productive portions of these aquifers.

Volcanic Terrains

Volcanic rocks retain porosity associated with lava-flow features and pyroclastic deposition. Hydraulic conductivity can be quite high, but ash beds, intrusive dikes, and sills may be barriers to ground water flow. Flow features such as vertical contraction joints and stream gravels buried between successive flows contribute to overall permeability and produce some of the most productive aquifers. Flow distribution, timing, volumes, and rates can be affected by stratigraphic differences in texture, jointing and fracture patterns and spacing, contact relationships between lithologic units, and the presence or absence of lava tubes. Ground water geochemistry can be directly affected by venting volcanic gases and ground water circulation driven by geothermal systems. The Columbia Lava Plateau is the largest sequence of basalt flows and interbedded sediments in the United States. Ground water is replenished by precipitation, runoff from adjacent mountains, and excess irrigation water applied to the surface.

Fractured-rock Settings

Much of the NFS lands are located in mountain-dominated terrain. The landscape is rugged and composed of exposed igneous, metamorphic, and sedimentary bedrock. A weathered zone of soil a few meters to tens of meters thick may exist. Sources of recharge largely involve diffuse infiltration of precipitation, including melting snowpack. Discharge occurs locally as focused spring flows and seeps, as diffuse inflow to streams, and as transpiration in riparian areas. Ground water flows through pores and fractures. In rock formations with large numbers of fractures that are highly interconnected, ground water flow can be very similar to that through porous sediments. A predominant flow direction exists, and responses to pumping or intersection with a lake or stream are predictable. In rock formations with few or poorly connected fractures, ground water flow occurs in a far less predictable manner. Discrete fracture flows may be independent of ground water flow through the bulk rock. Consequently, the seepage of ground water from, and the seepage of streamflow into, fractured rock can result in the mixing of chemically dissimilar waters at flow rates that are very difficult to predict. Some suggested methods for studying the recharge and discharge characteristics in mountainous hydrogeological settings are described in National Research Council (2004).

The majority of NFS land is underlain by fractured-rock aquifers, and demands are increasing on ground water on and around the NFS. These types of aquifers occur in the igneous, metamorphic, and sedimentary rocks that form the mountain uplifts and in sedimentary rocks that flank uplifts. The occurrence and flow of ground water in these types of aquifers is controlled by the spacing, aperture size, orientation, and connectivity of permeable "preferential" pathways that occur within discontinuities created by structural processes related to uplift and mountain building (Caine and others 1996). Types of discontinuities that facilitate ground water flow include joints and fractures, foliation, faults, shear zones, geological contacts, and bedding planes. It should be noted that some structures, such as dikes and faults, may also function as barriers to ground water flow. The quantitative aspects of ground water flow in fractured rocks are not well understood, particularly at the fracture scale; however, in many fractured-rock settings, the watershed or surface drainage basin can be an appropriate, natural unit within which to characterize and manage surface-water and ground water resources.

Heath (1988) divided North America into 28 ground water regions based on, among other things, the nature of the water-bearing openings of the dominant aquifer or aquifers. Eleven of these regions are underlain by mountainous areas dominated by fractured-rock hydrogeologic settings. Six of the regions (Western Mountain Ranges, Columbia Lava Plateau, Sierra Madre Occidental, Sierra Madre Oriental, Sierra Madre Del Sur, and Faja Volcanica Tansmexicano) occur along the western edge of North America. Three regions (Northeastern Appalachians, Appalachian Plateaus, and Piedmont and Blue Ridge) occur along the eastern edge of North America. The other two are the West Indies and the Hawaiian Islands.

Hydrogeological settings in the 11 ground water regions in mountainous areas are typically characterized by steep slopes on the sides of ridges and mountains, and thin soil or regolith overlying moderately to highly fractured and/or folded bedrock. Exceptions are the Northeastern Appalachian and Appalachian Plateau ground water regions, where the regolith is thick. In some bedrock types, the upper 10 to 100 feet are commonly highly weathered and, when saturated, comprise a significant water-bearing unit separate from the bedrock. In areas where snowmelt is a significant seasonal event, interflow is often a dominant process in this zone during the snowmelt flux. Below this zone, ground water flow occurs predominantly in individual fractures, fracture zones, faults, fault zones, and other structural discontinuities (Gerhart 1984). The rock matrix in igneous and metamorphic bedrock typically plays a minor role in ground water flow and usually has low porosity and permeability. As a result, ground water flow in these settings is highly preferential and controlled primarily by the spacing, orientation, hydraulic properties, and connectivity of the permeable discontinuities (Forster and Smith 1988b).

Because of the steep topography in mountainous and upland areas, hydraulic gradients along preferential pathways can be very high, causing relatively high ground water flow velocities. Ground water flow along preferential pathways is generally toward valley bottoms, where ground water discharges at seeps and springs, which are very common in fractured-rock settings, or to an intermittent or perennial stream in the valley bottom. Strong upward gradients are common in bedrock underlying mountain valley bottoms. In mountainous watersheds, the topographic drainage basins are not always coincident with ground water flow divides. Because of low porosity and storage, seasonal ground water levels in fractured-rock aquifers commonly vary from 10s to 100s of feet. As a result, the location of ground water divides can shift seasonally because more and different flow paths are available to ground water when water levels are high than when water levels are low. Hydrogeological conditions in mountain watersheds often result in distinct ground water flow systems that are temporally and spatially dynamic.

Inadequate collection, interpretation, and use of fracture-scale hydraulic data continue to be general deficiencies in fractured-rock hydrogeological investigations. In such settings, the relationship between seasonal interflow in a surficial water-bearing zone and recharge to an underlying fracture flow system is poorly understood. A number of factors control ground water flow through fractures or other discontinuities, including fracture aperture and length and the degree of roughness and nonparallelism of the fracture walls. The hydraulic conductivity through a fracture is directly proportional to the aperture width and inversely related to normal stress and depth. Fracture permeability is affected by rock temperature, cementation, in-filling, and chemical and physical weathering. In fractured-rock settings it is important to distinguish between the hydraulic conductivity of a fracture, the hydraulic conductivity of the rock matrix and the hydraulic conductivity of a rock mass. As the development of ground water increases, a better understanding of the quantitative aspects of ground water flow at the fracture scale will be essential to adequately manage the spatial and temporal withdrawal of ground water for human use. This will also be essential for characterizing contaminant transport in fractured-rock settings.

As more attention is being focused on the hydrogeology of fractured-rock settings, a wide variety of tools are being used to characterize (1) ground water storage and flow and development; (2) ground water/surface water interaction; (3) chemical, isotopic, and biological quality of ground water and surface water; and (4) contaminant transport in these settings. Given the complexity, it is advisable to use a number of different tools and data sets (multiple lines of evidence) for evaluation. Some of the more appropriate characterizations tools include aquifer tests, evaluation of drill core and drillholes, isotopes, and geophysics.

As the population density has increased in these fractured-rock settings, the development of ground water resources for domestic, municipal, and commercial uses has increased significantly. Concurrent with this growth has been an increase in anthropogenic contamination of ground water. The issues of population pressures and contamination are becoming increasingly important in and around the NFS. As a result of the increasing stress on water resources, ground water scientists and water-resource managers have recognized the need to develop a more appropriate approach to characterizing the occurrence, movement, and chemistry of ground water in fractured rock. A concurrent need exists to develop more appropriate ways to characterize contaminant transport in fractured rocks and to select, design, and operate remedial technologies at ground water contamination sites. As the overall understanding of ground water occurrence and flow in fractured rock has improved, a new conceptual model has evolved that is more appropriate for characterizing water resources in these types of hydrogeological settings. Ground water scientists have been forced to move beyond conventional approaches that have traditionally been used for unconsolidated deposits and consolidated rock aquifers where porous media flow is dominant. Other, more thorough, discussions of these topics are provided by the National Research Council (1996), the National Ground Water Association (2002), and ASTM standard D5717-95e1 (ASTM 1996).

Ground Water Investigation Methods

Ground Water Resource Inventories and Evaluations

Ground water inventories can help provide the basis for selecting suitable areas for major land uses, identifying areas that need more intensive investigation, evaluating various land-management alternatives, and predicting the effects of a given activity on resource health or condition. The resultant maps, data, descriptions, and management interpretations provide basic ground water resource information necessary for ecological assessments, project planning, watershed analysis, forest plan revisions, and implementation and monitoring of forest plans. The information provided can be used for activities such as assessing resource conditions, conducting environmental analyses, defining and establishing desired conditions, and managing and monitoring natural resources. Ground water inventories and monitoring programs will necessarily involve various levels of detail, focus, and spatial extent depending on the geographic location of a national forest and the specific resource issues that that national forest is dealing with. The basic elements in a ground water inventory are shown in figure 51.

The discussion that follows explains how various strategies, field methods, and data analyses are useful for accomplishing ground water inventories and evaluations. Although each ground water inventory will be more or less unique, general guidelines for a successful inventory can be followed.

ELEMENTS OF A GROUND WATER INVENTORY

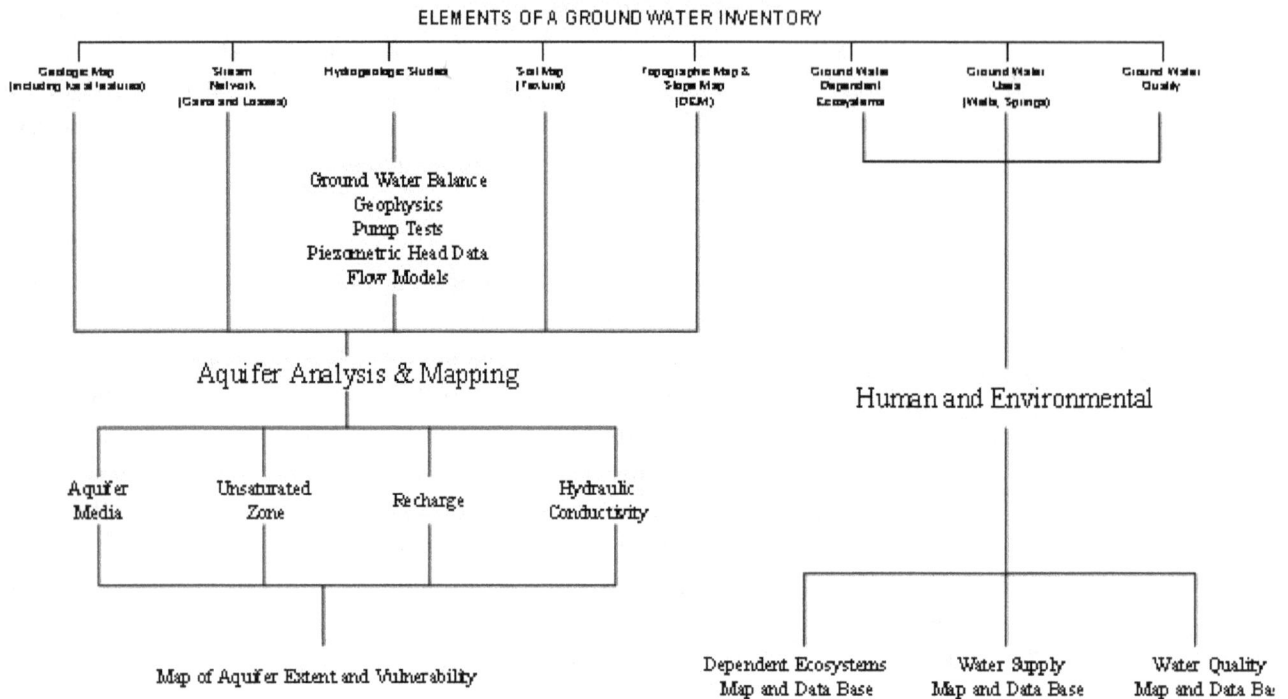

Figure 51. Elements of a ground water inventory and assessment.

Aquifer Delineation and Assessment

Aquifer assessments can be used by land-management agencies to define the overall usefulness of an aquifer and/or its susceptibility to contamination or hydraulic disruption. Ground water assessments of various kinds are needed in many Federal, State, and local water-management programs. An assessment should include the identification and location of sustainable sources of drinking water, State pesticide management plans, underground injection of waste, and confined animal feeding operations. A National Research Council (1993) publication summarizes the broad array of definitions and approaches that are used by government as well as private and academic organizations in assessing the vulnerability of ground water to contamination. The National Research Council (1993) defines vulnerability as "the tendency or likelihood for contaminants to reach a specified position in the ground water system after introduction at some location above the uppermost aquifer."

Depending on specific objectives and available resources, assessments can be designed to include individual wells or entire aquifer systems. They can target one contaminant, or contamination in general. They can focus on hydraulic disruption. The effectiveness of individual assessments will be linked to the degree to which the important physical/chemical processes have been identified and accounted for, the manner in which uncertainty is addressed, and the extent to which the original science and management objectives are met.

The vulnerability of ground water to contamination depends on intrinsic susceptibility as well as the locations and types of potential sources of contamination, the relative locations of wells, and the fate and transport

146

of potential contaminant(s). The intrinsic susceptibility of a ground water system depends on its geologic setting, the aquifer properties including hydraulic conductivity, porosity, and hydraulic gradients, and on the associated sources of water and stresses for the system. Key elements are recharge, interactions with surface water, travel through the unsaturated zone, and well discharge. Intrinsic susceptibility assessments do not target specific natural or anthropogenic sources of contamination but instead consider only the physical factors affecting the flow of water to, and through, the ground water resource. Karst aquifers typically have a high intrinsic vulnerability because of the ease and speed with which contaminants can enter and move within the system (Zwahlen 2003). Some volcanic aquifers are similarly vulnerable.

Assessments of the vulnerability of ground water to contamination range in scope and complexity from simple, qualitative, and relatively inexpensive approaches to rigorous, quantitative, and costly ones. Tradeoffs must be carefully considered among the competing influences of the cost of an assessment, the scientific defensibility, and the amount of acceptable uncertainty in meeting the objectives of the water-resource decision maker. Subjective rating methods focus on policy or management objectives. Relative degrees of ground water vulnerability are usually delineated as low, medium, and high. These classes are common endpoints for all subjective rating methods. These broad classes are appropriate for the "index" methods described below, but not for the more costly and involved statistical and process-based methods.

Index methods and closely associated "overlay methods" assign numerical scores or ratings directly to various physical attributes to develop a range of vulnerability categories. The index method is one of the earliest and most commonly used categorical rating methods (National Research Council 1993). The most widely used index method is DRASTIC, which is an acronym for the seven factors considered in the method: Depth to water, net Recharge, Aquifer media, Soil media, Topography, Impact of vadose zone media, and hydraulic Conductivity of the aquifer (Aller and others 1985, 1987). The point rating system for DRASTIC was determined by the best professional judgment of the original method developers. The DRASTIC method has been used to produce maps in many parts of the United States (Durnford and others 1990). The maps have a variety of scales, including national (Kellogg and others 1997, Lynch and others 1994), statewide (Hamerlinck and Ameson 1998, Seelig 1994), and individual counties and townships (Regional Groundwater Center 1995, Shukla and others 2000). The index method is popular for ground water vulnerability assessments because it is relatively inexpensive and straightforward, uses data that are commonly available or estimated, and produces an end product that is easily interpreted and incorporated into decision-making processes.

Figure 52 shows how the DRASTIC method can be applied to NFS land. This coverage was constructed using GIS layers for topography, slope aspect, and geology. Depth to water was estimated from water well information stored in a Statewide well database. Hydraulic conductivity, recharge, and soil thickness were estimated by consulting a hydrogeologist and soil scientest familiar with the area. Index value compuations for the various hydrogeological settings are presented in table 10. Geologic units were combined into hydrogeologic settings based on similar hydrogeological properties, as suggested by Aller and others (1985, 1987).

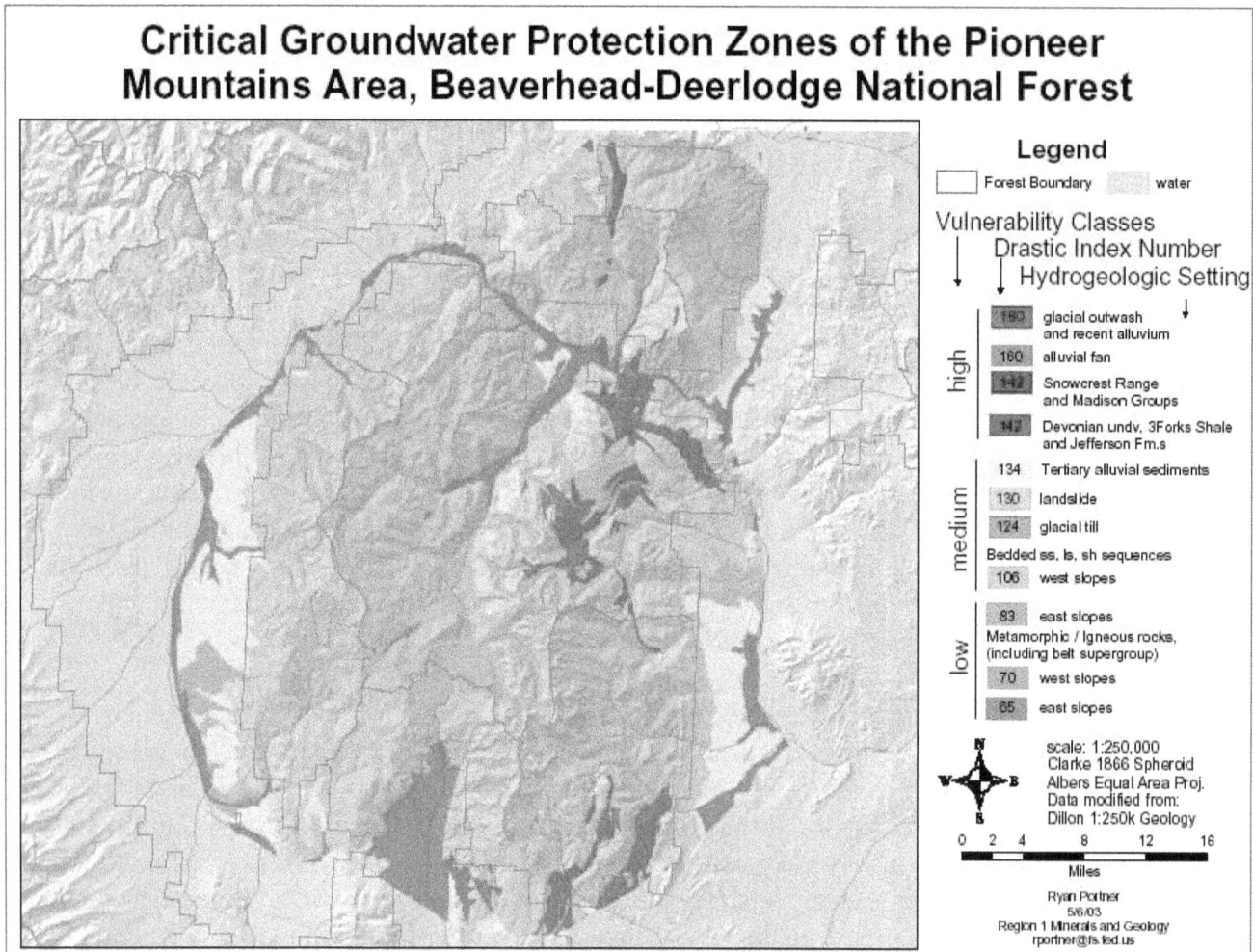

Critical Groundwater Protection Zones of the Pioneer Mountains Area, Beaverhead-Deerlodge National Forest

Figure 52. Vulnerability map of a portion of the Beaverhead-Deerlodge National Forest constructed using the DRASTIC method. Aquifers are rated from high to low vulnerability based on hydrogeological factors.

Hydrogeological Mapping

Geological mapping for hydrogeological purposes involves standard geological mapping procedures. Aerial photographs (1:24,000 scale or less) are used to delineate geological contacts. Photo data are combined with field checks to correlate map units, characterize rock units, measure stratigraphic sections, and measure strike and dip of formations. Of primary interest in hydrogeology is the ability of the various rock units to store and transmit water and act as aquifers.

Understanding geological conditions is the cornerstone of any ground water evaluation. Geology forms the physical framework for the flow of ground water. Primary and secondary porosity, storage properties, and transmitting properties are largely a function of the geological materials present. Stratigraphy affects local and regional ground water flow. Structural features,

Table 10. DRASTIC computation matrix showing methods for computing index values for various hydrogeological settings in the Pioneer Mountains, Beaverhead-Deerlodge National Forest, MT.

Map units	DRASTIC Index	Aquifer media	Rating	Depth to water (ft)	Rating	Recharge (in)	Rating	Soil	Rating	Topography (% Slope)	Rating	Vadose zone	Rating	K (gpd/ft²)	Rating
Qo	180	glacial outwash	8	5-15	9	4-7	6	gravel	10	2-6	9	sand & gravel	8	700-1000	6
Qm	124	glacial till	5	15-30	7	4-7	6	sandy loam	6	6-12	5	(s & g) with silt and clay	6	1-100	1
Ql	130	landslide	8	50-75	3	7-10	8	silty loam	4	2-6	9	s & g with silt and clay	6	300-700	4
Qf	160	alluvial fan	8	30-50	5	7-10	8	sandy loam	6	2-6	9	s & g	8	700-1000	6
Qtg, Tbz	134	alluvial gravels, Bozeman group	8	30-50	5	7-10	8	silty loam	4	2-6	9	s & g with silt and clay	6	100-300	2
Ym, Ymm, Kbgg, Kgtd, Tkg	65	metamorphic igneous (east slopes)	3	100+	1	0-2	1	thin or absent	10	18+	1	metamorphic igneous	4	100-300	2
	70	metamorphic igneous (west slopes)	3	75-100	2	0-2	1	thin or absent	10	18+	1	metamorphic, igneous	4	100-300	2
Cu,Kk,Ks, Pmu,IPmu,Pp,	83	bedded ss, ls, sh sequences (east slopes)	6	75-100	2	0-2	1	sandy loam	6	12-18	3	sandstone, limestone, shale	6	100-300	2
Tc,Tvu,Tru	106	bedded ss, ls, sh sequences (west slopes)	6	30-50	5	2-4	3	sandy loam	6	12-18	3	sandstone, limestone, shale	6	100-300	2
Mmm	142	limestone	9	100+	1	4-7	6	silty loam	4	12-18	3	limestone	9	1000-2000	10

*Map Units: Qo - Quaternary glacial outwash; Qm - Quaternary glacial moraine; Ql - Quaternary landslide deposits; Qf - Quaternary alluvial fan; Qtg - Quaternary alluvial gravels; Tbz - Tertiary Bozeman group; Ym - Precambrian Missoula group; Ymm - Precambrian Missoula group; Kbgg - Cretacious biotite grandiorite and granite; Kgtd - Cretacious granite and diabase; Tkg - Tertiary-Cretacious granite; Cu - Precambrian Cherry Creek series; Kk - Cretacious Kootenai Fm; Ks - Cretacious sediments, undifferentiated; Pmu - Permian, undifferentiated; IPmu - Triassic Permian, undifferentiated; Pp - Permian Phosphoria Fm; Tc - Tertiary colluvium; Tvu - Tertiary volcanics, undivided; Tru - volcanics, undifferentiated; Mmm - Mississippian Madison Mission Canyon Fm.**

such as the folding and fracturing of rock by tectonic processes, may alter directions of ground water flow compared to horizontal sediments by changing the inclination of permeable sediments and confining units. Displacement of sediments by faulting may either provide zones of increased permeability through fracturing or create aquifer boundaries when impermeable strata block the flow of water through permeable strata. Secondary fracture porosity results primarily from tectonic stresses.

Geological maps are the basis for interpreting the movement of ground water. Distinctions between unconsolidated and consolidated, and permeable and impermeable rock units are made on a qualitative basis, using rock type, structure, and knowledge about depositional environments. Interpretations of hydraulic conductivity can be made using such information, and estimates of potential ground water movement through the rock unit. Geological maps showing rock type and genesis are more useful for hydrogeological purposes than maps that classify rock units only as to their stratigraphic age. In addition to bedrock maps, those showing surficial geology are also very useful. In fractured-rock settings, maps that show geological structures such as faults, folds, joint orientation, strike and dip of beds, and cross-sections are useful for hydrogeological purposes because geologic discontinuities frequently are preferential ground water flow paths.

Hydrogeological mapping requires the systematic and integrated appraisal of soils, geomorphology, geology, hydrology (including meteorology), geochemistry, and water chemistry as they affect the occurrence, flow, and quality of ground water. It is also important to understand the hydrogeological setting as whole, including (1) surface water hydrology, (2) other nonfractured-rock aquifers that occur within the setting, and (3) data on meteorological and other water-budget elements in the watershed.

The character and distribution of soils and landforms are major considerations in hydrogeological mapping in humid areas where unconfined aquifers develop in unconsolidated materials and lie relatively near the land surface. In such settings, the water table generally follows the land surface, but with more subdued relief. Recharge areas are generally located in upland areas, and ground water divides tend to coincide with surface watershed boundaries. Valley bottoms and floodplains with perennial streams represent discharge areas. For all areas, soils and topography are the primary features that determine how much precipitation infiltrates into the ground to recharge ground water, and how much runs off to surface streams. In general, highly permeable soils and flat topography favor infiltration; less permeable soils and steep slopes promote surface runoff. However, steep, forested slopes with near-surface exposure of bedrock can serve as focused recharge areas for associated aquifers (Potter and others 1995).

Although the focus of hydrogeological mapping is the saturated ground water system, the occurrence and flow of ground water must be understood in the context of the larger hydrological cycle, which includes atmospheric water,

water in the vadose zone (unsaturated ground water), and surface water. Such understanding is especially important for unconfined aquifers, which are intimately connected to the hydrological cycle. Complete characterization of unconfined aquifers requires consideration of infiltration of precipitation, the effects of evapotranspiration, and the relationship between the ground water and surface water systems. Potentiometric surface mapping is one of the most important aspects of hydrogeological characterization. Confined aquifers that are distant from areas of surface recharge can be considered effectively isolated from the hydrological cycle, provided that they are highly confined. Such an assumption greatly simplifies analysis of a ground water flow system. A potentiometric-surface map is one of the most basic and useful tools available for characterization of ground water flow systems. A water-table map depicts the elevation of saturated ground water in an unconfined aquifer; a piezometric (pressure) surface map depicts the pressure potentials of confined aquifers. Either type of map is called a potentiometric-surface map. In practice, the terms water table, potentiometric, and piezometric are all often used interchangeably.

REMOTE SENSING

Interpretation of aerial photographs for hydrogeological purposes generally has two purposes: (1) location of potential sites for drilling water supply wells, and (2) analysis of regional or local ground water flow systems. Methods employed in such investigations include (1) analysis of soil patterns that may reflect on infiltration potential, drainage characteristics that suggest rock type, and soil/rock permeability (permeable soils will have good drainage, reflected in drainage patterns that are course textured or even absent); (2) lineament analysis; (3) mapping and interpretation of joints and fractures; (4) land form analysis, which gives suggestive evidence of the kind of geologic material making up the landform; and (5) observation of vegetation patterns or types that provide inferences about the presence or preferential movement of water or its chemical quality. Other related uses of aerial photographs in the assessment of hydrogeology include the interpretation of the geological history of an area, using landforms, channel geometry, and the identification of fluvial or lacustrine sediments and bedrock contacts.

Aerial photography is an essential element of many geological or hydrogeological studies. Considerable information on ground water conditions can be obtained from stereo pairs of low-level black and white (panchromatic) or color air photos. The pairs provide a three-dimensional image of the topography when they are viewed through a stereoscope. Patterns of vegetation, variations of gray or color tones in soil and rock, drainage patterns, joint patterns, and linear features (landforms, fault traces) allow preliminary interpretations of geology, soils, and hydrogeology. Historic aerial photography can also be useful in documenting preexisting physical conditions and monitoring the progress of cleanup operations at hazardous waste sites.

Aerial photos of areas with near-surface bedrock often reveal linear features called fracture traces, which indicate zones of relatively high permeability in the subsurface. Fracture-trace analysis on aerial photographs can provide preliminary information on possible preferential movement of ground water or contaminants. Fetter (2001) provides a useful introduction to fracture-trace analysis. Sonderegger (1970) describes use of panchromatic, color, and infrared photography to locate fracture traces as an aid to the interpretation of the occurrence and movement of ground water in limestone terrain. Parizek (1976) thoroughly reviews the North American literature on fracture-trace and lineament analysis. If possible, aerial-photo analysis of fracture traces should be supplemented with surface analysis of bedrock fracture orientations. Tracing and analysis of drainage patterns on aerial photos using overlays can suggest various rock types and geological structures, based on characteristic drainage patterns and densities (fig. 53).

Color infrared photography is particularly useful for identifying ground water discharge areas or areas where contamination changes vegetation; for example, it can help identify a failed septic tank absorption system (Farrell 1985), areas where fertilizer has been applied, or areas of oil pollution. Thermal infrared scanning can detect ground water discharge into surface waters by sensing temperature differences in the ground water and surface water. Ellyett and Pratt (1975) considered this type of photography to be potentially the most useful remote sensing tool in the study of hydrogeological indicators. The use of thermal infrared imagery to measure soil moisture (Jackson 1986, Jackson

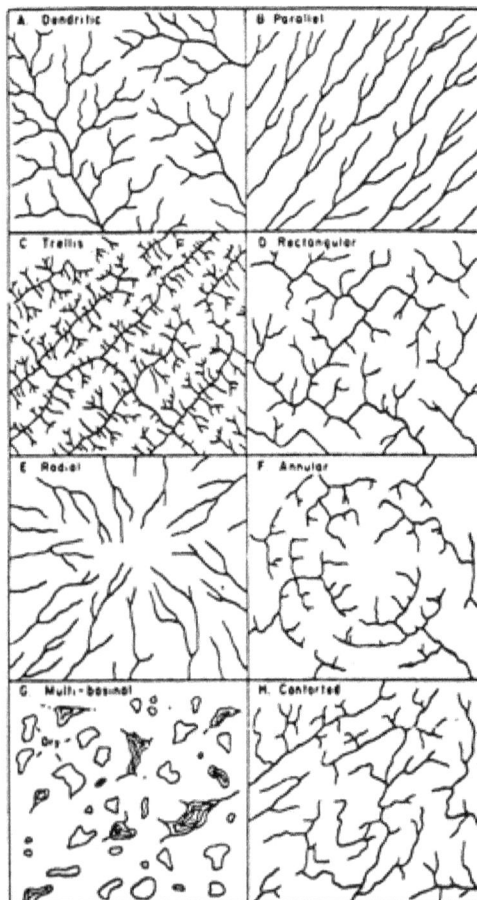

A. Dendritic occurs on rocks of uniform resistance to erosion and on gentle regional slopes.

B. Parallel occurs on steep regional slopes.

C. Trellis occurs in areas of folded rocks with major divides formed along outcrops of resistant rocks and valleys on easily eroded rocks.

D. Rectangular occurs in areas where joints and faults intersect at right angle.

E. Radial occurs on flanks of domes and volcanoes where there is no effect of differing rock resistance.

F. Annular occurs on eroded structural domes and basins, where resistant outcrops form major divides and weak rocks form valleys (a concentric type of trellis pattern).

G. Multi-basinal occurs in areas where the original drainage pattern has been disrupted by glaciation, recent volcanism, limestone solution, or permafrost.

H. Contorted occurs in areas of complex geology where dikes veins, faults or metamorphic rocks control the pattern.
The fishhook patern of the main stream might also result from capture of a northeast flowing stream by the southward flowing main stream.

Figure 53. Interpretation of drainage patterns from aerial photos and corresponding rock types and structure.

and others 1982, Price 1980, U.S. Geological Survey 1982) and evaporation (Price 1980, U.S. Geological Survey 1982) is reasonably well established. The National Research Council (2004) evaluated remote sensing technology as a means of detecting shallow aquifers and concluded that it is not practical for measuring ground water depth directly.

Airborne geophysical methods such as side-looking radar (SLAR), airborne electromagnetic imaging, and aeromagnetics have not been widely used in ground water studies, but the potential exists for their use in regional water-quality studies. A special feature of SLAR is its ability to distinguish grain size in alluvium. This technique requires unvegetated surfaces, a condition that is more likely to occur in arid areas (Ellyett and Pratt 1975).

Black and white photographs are available from various State and Federal agencies for almost any location in the United States and are the least expensive type of photo to obtain. Other types of photography are available at greater expense and should be used for special applications or to expand the scope of the photographic interpretation of the study area. These include the following photo types:

- *True color photos that record all colors in the visible spectrum.*
- *Color infrared film that records yellows and reds as green and the near infrared (not visible to the eye) as red. Since vegetation reflects near-infrared radiation, this image is especially useful for observing vegetation patterns. Other types of images that record or display colors differentially (false color) can be created.*
- *Ultraviolet (UV) photography uses special film and filters to record UV energy. Oil and carbonate minerals are fluorescent in UV bands when they are stimulated by sunlight. A disadvantage of UV wavelengths is that they are scattered in the atmosphere and result in a low-contrast image, especially when dust or haze is present.*
- *Multiband or multispectral images use multiple lenses and filters to record simultaneous exposures of different portions of the visible and near-infrared spectra of the same area on the ground. Images can also be recorded electronically using a multispectral scanning system.*

Because aerial photography is basic to preparation of geological maps, its use is crucial for hydrogeological studies if published geological maps are not available. Contracts for aerial photography should be awarded only for areas that are not mapped and only when the project has high priority. Existing aerial photography can often be purchased, realizing significant cost savings. A hydrogeologist can make basic interpretations about hydrogeological conditions of an area using aerial photography along with field verification. A general geological map can be prepared for many study areas using the aerial photography. The lengthy time requirements and great expense required for preparation of geological maps restrict the use of map preparation to small project areas. Detailed geological mapping for large areas (for example, er

1:24,000) can often take years, and makes such efforts impractical for most hydrogeological studies. Satellite images can also provide useful geological and hydrological information (Salama and others 1994), but at a scale that is useful primarily for large regional assessments. Imagery can be purchased from several commercial vendors.

WELL AND BOREHOLE LOGS

Inventory data about hydrogeological conditions are available from well logs filed with State regulatory agencies. Well logs are drillers' descriptions of lithological units penetrated by drilling. Unfortunately, these logs are often only minimally descriptive and sometimes describe rock types inaccurately; however, well logs are very descriptive if a geologist was present during drilling and supervises the logging, or the driller knew various rock types and could identify subtle changes in drilling that translate into changes in rock type. These logs also include specific capacity test data that can be used to infer relative transmissivity of the formation (Theis and others 1963, Bradbury and Rothschild 1985). Regardless of the quality of the logs, well logs can provide important information about the subsurface and, where available, should be incorporated into any ground water investigation.

Hydrogeological studies may require installation of monitoring wells. Often, such installations involve the detailed logging of the associated borehole by a geologist. In addition, there may be other boring logs or detailed soil pit information generated by other ground water investigations conducted in the area. These detailed logs can provide much needed information during the early stages of an investigation.

Design of a Ground Water Monitoring Network

Ground water monitoring networks are should be designed for the specific purpose for which they are established. For example, a network may be used to (1) monitor long-term effects of climatic changes on ground water systems, (2) monitor the effects of a new well field adjacent to a national forest on ground water levels within the national forest boundaries, or (3) monitor contaminant movement within national forest boundaries from a landfill or other pollution source. Each of these networks has different design considerations.

Often overlooked in ground water investigations is the need for an observation network to collect other types of hydrological data, in addition to ground water levels and water-quality data (Taylor and Alley 2001). Because meteorological data aid in the interpretation of water-level changes in observation wells, rain gages should be included as part of a network. Where observation wells are located in aquifers that have a strong hydraulic connections to streams or lakes, hydrological data, such as stream discharge and stage or lake stage, are important to examine the interaction between ground water and surface water (ASTM 1996). In addition, water-use data, such as rates and volumes of extracted ground water, can greatly enhance the interpretation of trends observed in water levels.

Contaminant detection is generally the most important aspect of a water-quality program, and must be considered in network design. False negative contaminant readings because of the loss of chemical constituents or the introduction of interfering substances that mask the presence of the contaminants in water samples can be very serious. Such errors may delay needed remedial action and expose either the public or the environment to an unreasonably high risk. False positive observations of contaminants may call for costly remedial actions or more intensive study, which are not warranted by the actual situation. Thus, reliable sample collection and data interpretation procedures are central to an optimized network design.

The ideal observation network consists of monitoring wells constructed specifically for that network, as well as instrumentation to collect ancillary hydrological data such as rainfall and streamflow. Budgetary constraints, however, may require the use of existing observation, domestic, or other wells for all or part of the network.

Extreme care must be taken in the selection of existing wells for use in the networks; for example, water-supply wells are drilled for maximum capacity, and may be completed so that they tap more than one aquifer or water-bearing unit. Such wells may provide data of minimal value in a study of contaminant movement from a pollution source. If existing wells must be used, all available well-construction information should be obtained so that the usefulness of the well for the network can be evaluated. If a pumping well is to be used for the network, both the pumping level and the static water level in the well should be obtained.

Use of Existing Wells

Identification of existing wells that are suitable for sampling may be divided into four steps: (1) identifying all the wells that exist in the area of interest; (2) identify those wells that are screened only in the hydrogeological unit targeted for sampling; (3) applying a screening process to the wells identified in step 2 to determine which wells meet the explicitly defined suitability criteria for sampling; and (4) evaluating the spatial distribution of wells that are suitable for sampling, not only in map view but also relative to the depths of the screened intervals of these wells within the hydrogeological unit. This evaluation is accomplished most efficiently by plotting available wells on a map, showing depth of screened interval below the water table and/or depth of screened interval relative to total thickness of the hydrogeological unit.

Criteria for wells that are suitable for sampling may vary for different projects; therefore, the first step in defining suitable wells is to list explicitly (and document subsequently in the monitoring project database) a set of criteria that must be present, and information about the well that must be available to meet the minimum-acceptable criteria for sampling. These same criteria are a starting point in developing specifications for newly constructed project wells. The criterion for a monitoring well that must be met is that it must yield water from, and only from, the particular zone (hydrostratigraphic unit) that is targeted for sampling.

A second criterion, the well type (primary purpose for which the well was constructed), relates to existing wells and is a key consideration in judging their suitability for sampling to meet project objectives. Wells may be divided into two major size categories: (1) high-capacity wells and (2) low-capacity wells. Sometimes the type of well to be sampled is an explicit part of the project objectives; for example, interest may be only in low-capacity domestic wells or wells constructed to a certain depth or screened in a particular zone. It is essential that well construction information be available for any well that is to be sampled. Key information includes screened interval, total depth, casing material, filter pack, and surface seal design.

A third criterion involves construction of the well. Key considerations include (1) length of intake interval (well screen or open hole)— project objectives may not be served by very long well screens or long open-hole intervals in bedrock wells because these create uncertainties in the actual water source; (2) the type of casing and screen material—results of sampling for metals may be compromised by metal casing, and sampling for some volatile organic compounds (VOCs) may be compromised by polyvinyl chloride (PVC) casing, particularly if casing joints are glued; and (3) methods and materials used to drill, complete, and develop the well—contaminants could be introduced into the strata or change the chemical environment in the vicinity of the well bore (Brobst 1984).

In addition to these criteria for selecting existing wells suitable for sampling is the availability of detailed information about these wells. The process of evaluating the suitability of existing wells for sampling begins as part of the assembling and evaluating of existing geologic and water-quality information in the different hydrogeological units of interest. An important prerequisite for screening existing water-quality data is the existence (preferably in an electronic database) of basic information on well location, well-construction details, and at least one water level when the well was not being pumped. Large numbers of otherwise suitable wells may have to be eliminated as candidates for sampling because essential information about the wells is not available.

Installation of New Monitoring Wells

Following the determination of the number and spatial distribution of existing wells that are suitable for sampling, as well as the number of wells to be sampled, the next question is whether a subset of the existing suitable wells can be selected that will meet project objectives. If not, new project wells are needed either for all samples in the study, or to merely "fill in the gaps" where wells do not exist, or suitable wells are not available. The obvious advantages of drilling new project wells include (1) selection of the well location and access to the well for sampling, (2) designation of the screened interval of the well within the hydrogeological unit, (3) control over specific construction features of the well, (4) possible assurance of long-term availability of the well for sampling; and (5) control over the collection of detailed geologic and hydrologic information during well construction. The principal disadvantages of drilling new project wells are the potentially large additional drilling costs for the project, and the increased time and cost associated with difficulties in obtaining permission and legal easements to drill wells in desired locations. Additional time is also required for completing all required environmental assessments.

A type of monitoring well that often justifies special design and construction is one that will be used for long-term monitoring and trend analysis. Because of the large costs to collect and analyze the samples over a period of many years, reliability of the data is imperative, and a well specifically constructed for this purpose would likely be cost-effective.

Effective design and construction of a monitoring well require considerable care and at least some understanding of the hydrogeology and subsurface geochemistry of the site. Preliminary borings, well drilling experience, and the details of the operational history of a site can be very helpful. Common monitoring well design criteria include depth, screen size, gravel-pack specifications, and yield potential. These considerations differ substantially from those applied to production wells. The simplest, small-diameter well completions that will permit development, accommodate the sampling gear, and minimize the need to purge large volumes of potentially contaminated water are preferred for effective routine monitoring activities. Helpful references include Barcelona and others (1983), Scalf and others (1981), Wehrmann (1983), Aller and others (1991), Lapham and others (1997), and Driscoll (1986).

Well Placement

The placement and number of wells in a network, as well as the placement and number of rain gages or stream and lake gaging stations, depends on the purpose of the network and the hydrogeologic complexity of the study area. Ideally, the wells chosen for an observation network provide data representative of the topographic, geologic, climatic, and land-use environments present in the area of interest. The more varied these environments are within that area, the more wells will be needed. Subsurface geophysical techniques can be very helpful in determining the optimum placement of monitoring wells under appropriate conditions and when sufficient hydrogeological information is available (Evans and Schweitzer 1984). The placement and number of wells will also depend on the degree of spatial and temporal detail needed to meet the objectives of the program. Both the directions and approximate rates of ground water movement must be known to satisfactorily interpret the chemical data. An understanding of the variability or distribution of hydraulic conductivity, in both the vertical and horizontal dimension, allows one to isolate the major zones of water transmission and, therefore, to select the proper depths of wells and the position and length of well screens. The same is true for offsite, upgradient, and downgradient monitoring wells. With this knowledge, it also may be possible to estimate the nature and location of pollutant sources (Gorelick and others 1983). Well placement should be viewed as an evolutionary activity that may expand or contract as the needs of the program dictate.

For contaminant monitoring, wells should be placed near the area of suspected contamination pathway, as well as upgradient of the site. Initial investigations need to be conducted to determine the flow system before monitoring wells can be effectively installed. If several aquifers are present and of interest, observation wells will be required that monitor each individual aquifer. Individual sampling can be accomplished by using multiple wells, each completed in an individual aquifer and isolated from the others at a site, or by using multiple screened intervals isolated from other aquifers with packers or some other isolating medium, such as bentonite or cement. Wells completed at multiple depths may also be needed where there are vertical head gradients within a single aquifer (such as near a lake or stream that receives ground water discharge), or where contaminant migration may be along preferential flow paths (such as fractures or sand lenses). Where multiple-completion wells are required, care should be taken to physically isolate each zone of interest. Lapham and others (1997) provide detailed discussions of well design, well completion, and well development that provide optimal information for water-quality studies. Generally, the placement of nested piezometers in closely spaced, separate boreholes of different depths generally is the preferred method to determine vertical head differences and the potential for vertical movement of contaminants, while monitoring wells with appropriately located screens are used to determining the lateral movement of contaminants in the saturated zone. One must also consider whether vadose-zone monitoring is required. Nested lysimeters can be used to detect contaminants in the vadose zone, but great care must be taken to ensure that the collected samples are representative and not affected by preferential flow paths and sorption and volatilization of the contaminant.

The length and position of well screens also must be predicated on the nature of the contaminant; for example, if the contaminants are miscible with the liquid phase, it may be possible to use only one well per sampling point. It also may be possible to use only one well if the transmissive zone is very thin. If the contaminants are immiscible with the liquid phase (sinkers or floaters), the well screen must be located accordingly. Selection of a length of well screen depends on the vertical scale of investigation, and on the thickness and properties of the hydrogeological unit of interest. The longer the screened (or open) interval relative to aquifer thickness, the less likely differences in water quality at specific depth intervals will be able to be distinguished. Mixing of waters within the screened interval can lead to constituent concentrations that do not necessarily represent the maximum or minimum concentrations of those constituents at any point. For this reason, relatively short screens are generally used if the objective is to investigate water quality at discrete intervals and to define chemical stratification within the aquifer. If determining the vertical distribution of water quality in an aquifer is a data-collection objective, installing wells at different depths, each with a relatively short screen length, is often the most effective design.

Screen lengths for monitoring wells typically range from 2 to 20 feet. As a general rule, screen lengths of 20 feet or less generally are appropriate for most resource assessment studies, while screen lengths of 5 feet or less generally are better suited for studies to determine fate, transport, and geochemistry of ground water constituents. A screen length of 5 feet might be too long if information suggests that marked vertical differences in the distribution of hydraulic head or water quality occur on the order of a few feet or less. The length of the open interval also depends on the scale of the investigation. For example, a 20-foot-long screen is too long for an investigation of a 5-foot-thick contaminant plume; however, such a screen might be considered too short in an investigation of the general water quality of an aquifer that is several hundred feet thick. The following are additional factors to consider when deciding on screen length:

- A short screen generally provides measurements of hydraulic head and ground water quality that more closely represent point measurements in the aquifer than measurements provided by a long screen.
- Samples taken from wells with long screened intervals could exhibit smaller concentrations or a higher frequency of samples with undetectable concentrations (leading to a "false negative" assessment) in comparison with samples taken from wells with short screened intervals.

A long well screen also can induce mixing of waters of different chemistry in comparison to a short well screen because of vertical flow along the screened interval resulting from differences in head (well-bore flow). Well-bore flow can occur even in homogeneous aquifers with very small vertical head differences. Well-bore flow might contribute to aquifer contamination by providing a pathway for contaminant movement from contaminated to uncontaminated zones along the screened interval(s).

The selection of a particular drilling technique for observation-well construction depends on the geology of the site, the expected depths of the well, the requirements for subsurface lithologic samples, and the suitability of drilling equipment for the contaminants of interest. Available drilling methods include auger; rotary, using water-based fluids or air; cable-tool, jet-wash and jet-percussion; coring; direct push; and vibration. The advantages and disadvantages of each method are described in detail by Lapham and others (1997). Regardless of the technique used, every effort should be made to minimize subsurface disturbance. For environmental applications, the drilling rig and tools should be steam cleaned to minimize the potential for cross-contamination between formations or successive borings.

CASE STUDY: LANDFILL EVALUATION, WEST YELLOWSTONE, GALLATIN NATIONAL FOREST, MT

Leaching of contaminants from unlined landfills is a common problem. This case study documents a hydrogeological investigation of a landfill on Forest Service land used by the city of West Yellowstone, MT. The investigation detected a plume of heavy metals and VOCs migrating from the landfill toward the nearby Madison River, resulting in a decision by the city to close the landfill and truck all of its trash to a distant certified landfill.

In 1971, the city of West Yellowstone obtained a special-use permit from the Gallatin National Forest to locate and operate a Class II (household wastes) landfill on national forest land 4 miles north of the city and east of U.S. Highway 191. It replaced an older dump 1 mile away. The site is on a flat alluvial terrace formed by the Madison River. The geological setting is medium- to coarse-grained, highly permeable, obsidian sand that is about 600 feet thick and derived from volcanic activity originating in Yellowstone National Park. Annual precipitation in the area averages 40 inches, mostly as snowfall from October through May. Elevation is about 6,700 feet.

By most accounts, the landfill was well-operated and maintained by the private contractor to whom the city leased the site. In 1976, the Forest Service zone sanitary engineer asked the Soil Conservation Service (SCS) to investigate and evaluate the landfill. The SCS drilled two monitoring wells and described the landfill's trenches as being about 25 feet deep with about 25–35 feet of obsidian sand separating the garbage from the water table. The ground water flow was estimated to be southwesterly toward the Madison River about 3,700 feet away. The potential for leachate from the landfill to ultimately reach the river and Hebgen Lake was considered to be high.

Monitoring results from the two wells in 1976 revealed high levels of carbon dioxide, iron, manganese, lead, mercury, cadmium, biological oxygen demand, chemical oxygen demand, and a trace of 2,4-dichlorophenoxyacetic acid (2,4-D), indicating leachate contamination. A decision was made to drill seven more 4-inch diameter, shallow wells, 44–62 inches deep, between the landfill and the river. A background well was located inside Yellowstone National Park upgradient of the landfill. The state office of the SCS performed the drilling in 1976–77. The water table depth ranged from 42 to 60 feet below the land surface, and was shown to fluctuate 6 to 8 feet annually.

Water-quality samples were collected at each well in 1977–78, 1980–81, and 1985. Results revealed a leachate plume in ground water, moving steadily toward the Madison River. The Solid Waste District sampled for 10 parameters in 1989, 1990, 1992, and added VOCs in 1995 and 1996. Low-level concentrations of several VOCs were detected, but none exceeded human health standards. Dissolved metals (iron, manganese, zinc) and specific conductance values indicated that the leachate plume was continuing to move toward the river.

In 1982, dumping of household wastes at the landfill ceased, but Class III waste (trees, cars, demolition debris) dumping continued until 1988 when a soil cap was constructed. A solid waste transfer facility was built in 1983–84 and continues in operation. A 1994 special-use permit required testing of VOCs twice at the closed landfill site. These tests were done, and no additional testing is planned.

Old, unlined landfills can leach heavy metals, VOCs, and other constituents into ground water and eventually into surface water bodies hydrologically connected to the ground water. Hydrogeological investigations are necessary to quantify the magnitude and extent of ground water contamination and to determine the direction and rate of ground water movement. Cooperation of local, State, and Federal agencies produces better results than going it alone.

Water-level Monitoring

Water-level measurements from observation wells are the principal source of information about the hydrologic stresses acting on aquifers and how these stresses affect ground water recharge, storage, and discharge (Taylor and Alley 2001). The frequency of water-level measurements from the network is an important design consideration. Water levels may be measured continuously using floats and paper strip charts, at frequent intervals ranging from seconds to hours using transducers and data loggers, or periodically (daily, weekly, monthly) by obtaining manual measurements on site. Again, the purpose of the network determines the frequency of measurements that is required. Although influenced by budgetary considerations, the frequency of measurements should be determined to the extent possible with regard to the anticipated variability of water-level fluctuations in the observation wells and the amount of detail needed to fully characterize the hydrological behavior of the aquifer or the rate of contaminant movement (Fig. 54).

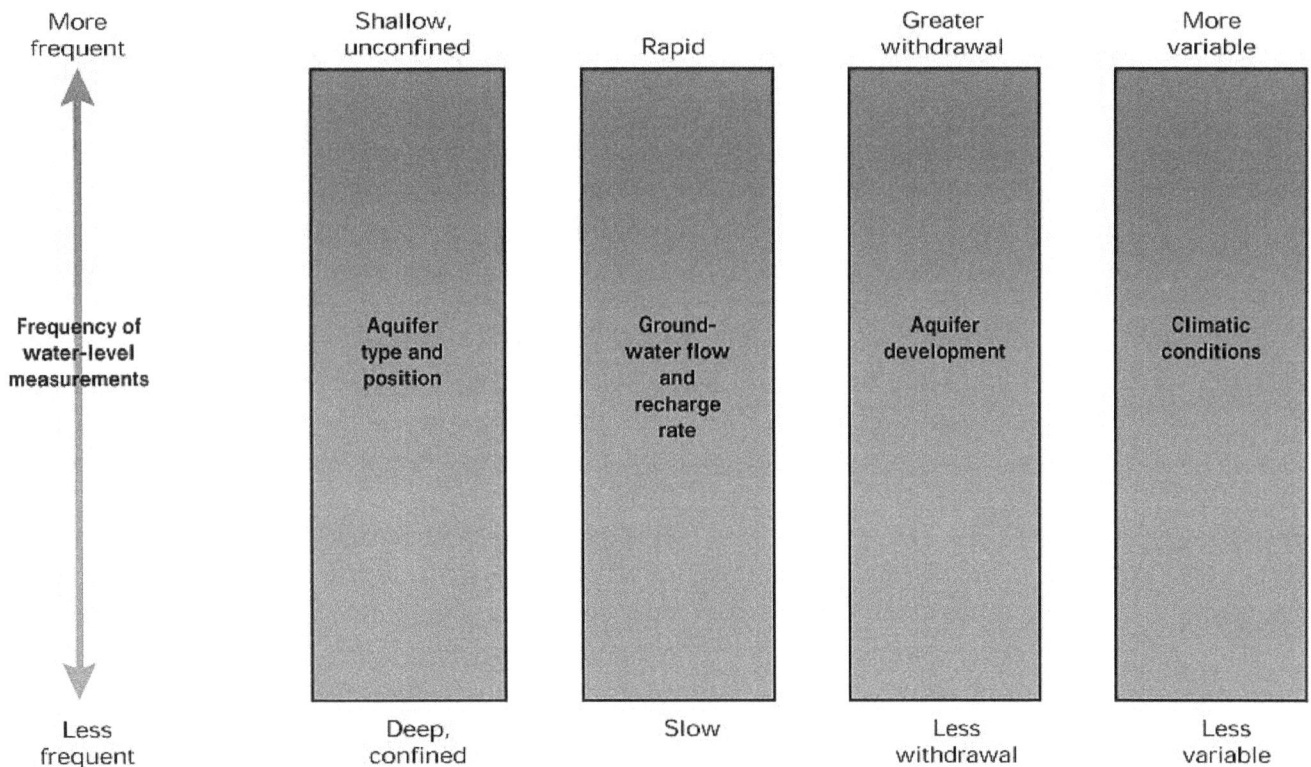

Figure 54. Common environmental factors that influence the choice of frequency of water-level measurements in observation wells (Taylor and Alley 2001).

Synoptic water-level measurements are a special type of periodic measurement in which water levels in wells and nearby surface waters are measured within a relatively short period and under specified hydrologic conditions. Synoptic measurements provide a "snapshot" of aquifer water levels. These measurements commonly are taken when data are needed for mapping the altitude of the water table or potentiometric surface, for determining hydraulic gradients, or for defining the physical boundaries of an aquifer. Regional synoptic measurements made on an annual or multiyear basis can be used as part of long-term monitoring to complement more frequent measurements made from a smaller number of wells.

QUALITY ASSURANCE

Good quality assurance helps to maintain the accuracy and precision of water-level measurements, ensure that observation wells reflect conditions in the aquifer being monitored, and provide data that can be relied on for many uses (Taylor and Alley 2001). Therefore, field and office practices that will provide the needed levels of quality assurance for water-level data should be carefully considered and consistently employed. Some important field practices that will ensure the quality of ground water-level data include the establishment of a permanent datum (a reference point for water-level measurements) for observation wells, periodic inspection of well structures, and periodic hydraulic testing of the well to ensure its communication with the aquifer. The locations and elevations of the wells should be accurately surveyed initially and checked periodically (annually or every other year). Existing wells selected for use in the network should be carefully inspected, with a downhole video camera if necessary, to ensure that no construction defects are present that might affect the accuracy of the water-level measurements. Water levels are typically measured to within 0.01 foot or 1 millimeter.

To help ensure quality, both paper and electronic files should be established containing information for each observation well. The files should include a physical description of well construction, location (both horizontal and vertical) in an appropriate coordinate system, coordinate system datum, and results of hydraulic tests. Recent water-level measurements should be compared with previous measurements made under similar hydrological conditions on a regular basis to identify potential anomalies in water-level fluctuations that may indicate measurement equipment malfunction or a defect in well construction. Ongoing data evaluations can help identify issues early to allow correction or equipment repair prior to the end of the study.

WATER-LEVEL MONITORING EQUIPMENT

The type of water-level monitoring equipment needed depends on several factors, such as study objectives, depth to the water table, required accuracy, type of well to be monitored (pumping or observation well) and frequency of measurements. Manual-measurement equipment tends to be used only when periodic (monthly, quarterly, or annual) measurements are needed. Continuous-measurement equipment is most useful when short-term (minutes, hours, days) measurements are needed.

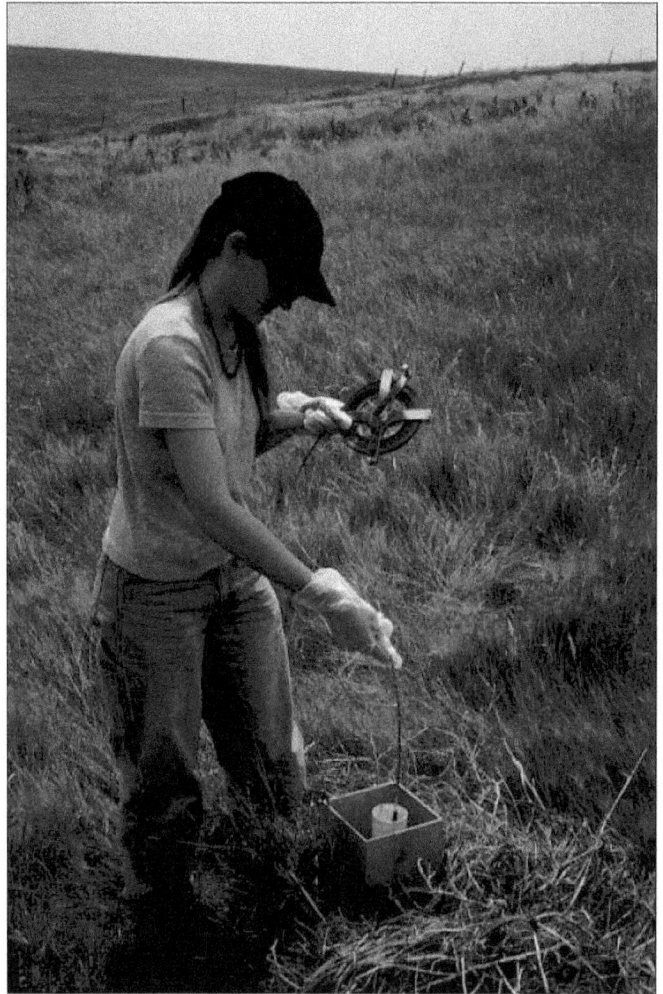

Figure 55. Measuring water levels in an observation well with a steel tape. (Photo by Heather S. Eppler, USGS.)

Manual-measurement equipment includes steel or electric tapes. Steel tapes with attached "poppers" (Fig. 55) are most useful at shallow depths, generally less than 200 feet, and can be accurate within 0.01 foot. Electric tapes are used at shallow to intermediate depths, generally less than 1,000 feet, but can be less accurate at greater depths or in wells with cascading water. Correction factors for both types of tapes (particularly electric tapes) may be required for measurements greater than 1,000 feet, because of stretching of the tape at those depths (Garber and Koopman 1968).

Common continuous-measurement equipment includes floats with strip charts and pressure transducers with data-logger systems. Strip charts and floats monitor water levels truly continuously (barring equipment failures, such as hung-up floats), but transducer and data-logger systems can be programmed to monitor water levels at any desired interval. In order to save battery power or data storage space, data loggers can even be programmed to measure only when water levels change by a specified value, over a specified time interval, or when also connected to a rain gage, during or immediately following storm events. The data are recorded and stored electronically, and can be transmitted by radio, phone, or satellite to an office miles away. Such capability is especially useful for monitoring in remote locations, and for early identification of potential equipment problems prior to significant loss of data.

163

Figure 56. Downloading data from an automatic water-level recorder. (Photo by Michael D. Unthank, USGS.)

Several manufacturers produce downhole pressure transducer-data logger packages to measure water level changes in wells (fig. 56). These units generally have small diameters (5/8 inch) and short lengths (1–2 feet), which permit them to be installed inside the well casing and left in place for long-term monitoring. They have no exposed wires or other indications that anything is inside the well, which minimizes vandalism. The units can be set to record water levels every few seconds if necessary (during pump tests for example), or once or twice per day, or even monthly for long-term monitoring. Data is downloaded into either a laptop computer or a handheld device. Data are then transferred to a database through data management software or analyzed with graphic analysis software.

Ground Water Quality Sampling and Analysis

Water-quality sampling and field analysis for Forest Service ground water studies, inventories, and investigations should conform to techniques and protocols established in the *National Field Manual for the Collection of Water-Quality Data (National Field Manual)* by the USGS (1997 to present). The *National Field Manual* describes protocols and provides guidelines for personnel who collect data used to assess the quality of the Nation's surface water and ground water resources. A chapter of the *National Field Manual* addresses field-trip preparations, including selection of sample-collection sites for studies of surface water quality. It also covers site reconnaissance and well selection for studies of ground water quality, and the establishment of field files for a sampling site. Each chapter of the *National Field Manual* is published separately and revised periodically. Newly published and revised chapters are announced at http://water.usgs.gov/ under "New Publications of the U.S. Geological Survey."

The *National Field Manual* is targeted specifically for field personnel to (1) establish and communicate scientifically sound methods and procedures; (2) provide methods that minimize data bias and, when properly applied, result in data that are reproducible within acceptable limits of variability; (3) encourage consistent use of field methods for the purpose of producing nationally comparable data; and (4) provide citable documentation for USGS water-quality data-collection protocols. **Formal training and field experience are needed to correctly implement the protocols described in this manual.**

Sampling protocols addressed in the USGS *National Field Manual for the Collection of Water-quality Data* include the following:

- Preparations for water sampling.
- Selection of equipment for water sampling.
- Cleaning of equipment for water sampling.
- Collection of water samples.
- Processing of water samples.
- Field measurements.
    - Temperature.
    - Dissolved oxygen.
    - Specific electrical conductance.
    - pH.
    - Oxidation-reduction (redox) potential by the electrode method (also known as ORP or Eh).
    - Alkalinity and acid neutralizing capacity.
    - Turbidity.
- Biological indicators.
    - Five-day biochemical oxygen demand (also known as BOD_5).
    - Fecal indicator bacteria.
    - Fecal indicator viruses.
    - Protozoan pathogens.
- Bottom-material samples.
- Safety in field activities.

Whether the goal of the monitoring effort is inventory or detection of specific contamination, the information gathered in sampling must be of known quality and must be well documented. High-quality chemical data collection is essential in ground water monitoring programs. Each monitoring program, however, has unique needs and goals that are fundamentally different from surface-water investigative activities. The reliable detection and assessment of subsurface contamination require minimal disturbance of geochemical and hydrogeological conditions during sampling. The technical difficulties involved in "representative" samplings are well documented (Wood 1976).

Gillham and others (1983) published a very useful reference on the principal sources of bias and imprecision in ground water monitoring. Their treatment is extensive and stresses the minimization of random error, which can enter into well construction, sample collection, and sample handling. They further stress the importance of collecting precise data over time to maximize the effectiveness of trend analysis, particularly for regulatory purposes. Accuracy also is very important, because the ultimate reliability of statistical comparisons of results from different wells may depend on differences between mean values for selected constituents from relatively small numbers of replicates; therefore, systematic error must be controlled by selecting proven methods for establishing sampling points and collecting samples to ensure known levels of accuracy. Collecting representative samples that are free of artifacts and errors is a function of the degree of detail needed to characterize subsurface hydrological and geochemical conditions and the care taken to minimize disturbance of these conditions (EPA 1993a). Each well or boring represents a potential conduit for short-circuited contaminant migration or ground water flow, which must be considered a potential liability to investigative activities.

Filtration of samples being analyzed for contaminants has received considerable attention in the literature in the past several years because contaminants can be sorbed onto colloidal particles moving through an aquifer. Analysis of filtered ground water samples might underestimate the amount of a contaminant that is actively moving through an aquifer if colloidal transport is occurring. Analysis of unfiltered samples, however, may overestimate the amount of contaminant in the aquifer as a result of changed physical and chemical conditions in and immediately surrounding the well. Given the changing status of regulatory thinking on this issue, consultation with the appropriate regulatory agency or water-quality expert is recommended to determine whether samples should be filtered. Filtering samples for total organic carbon, total organic halogens, or other organics is inadvisable because the increased handling required may result in the loss of the chemical constituents of interest (EPA 1991). Filtering of ground water samples that are to be analyzed for metals, however, may be appropriate. To minimize the problems associated with sample filtration, any filtration should be conducted in the field as soon after sample collection as possible.

Preparations for sampling involve extensive planning and logistical support to ensure that the sampling effort is conducted properly and will provide legally and scientifically defensible data. Planning may involve months of personnel time. Activities include (1) selection of wells to be sampled, if existing wells are to be used, or selection of drilling locations for placing monitoring wells for new monitoring locations; (2) ordering of suitable equipment to obtain the desired sample parameters and associated supplies, including health and safety gear; (3) training of field personnel in sampling protocols; (4) implementation of the quality assurance/quality control (QA/QC) program in connection with the training of field personnel, including analysis of equipment blanks and possibly other types of QA/QC samples; (5) selection of the lab and establishing procedures for receipt of samples and turnaround times for lab analysis; (6) establishing methods of shipping and ensuring that holding times are not exceeded, that proper sample preservation methods are employed, and that proper chain of custody control is maintained; (7) visits to all wells to confirm permission to sample, obtain access to wells (physical access and access into the well bore with sampling equipment), establish exact location of wells, and anticipate any sampling problems that might arise with new personnel on the site. In some cases, some of the time and expense associate with proper sampling may be addressed through the use of well-trained, experienced contractors.

To collect sensitive, high-quality concentration data, investigators must identify the types and magnitudes of errors that may arise in ground water sampling. Table 11 presents a generalized diagram of the steps involved in sampling and the principal sources of error.

Table 11. Generalized diagram of the steps involved in sampling and the principal sources of error.

Sampling activity	Sources of error
Establishment of sampling points	Improper well construction or placement.
Field measurements	Instrument malfunction; operator error; poor field conditions.
Sample collection	Improper protocols – cross contamination, sample exposure, degassing, oxygenation.
Preservation/storage	Improper protocols – handling, labeling errors, wrong preservative; matrix interference.
Transportation	Delay beyond holding times; sample loss.
Field and trip blanks, standards	Contamination; operator error; matrix interferences.

Other factors controlling ground water sampling errors are the contamination of the subsurface by drilling fluids, grouts, or sealing materials; the sorptive or leaching effects on waters samples from well casing; pump or sampling tubing materials; and the effects on the solution chemistry from oxygenation, depressurization, or gas exchange caused by the sampling mechanism. Two of the most critical elements of a monitoring program are establishing both reliable sampling points and simple, efficient sampling protocols that will yield data of known quality.

SELECTION OF
GROUND WATER-
QUALITY INDICATORS

Whether a program is focused on background/existing conditions, land-use impacts, or compliance monitoring, a key element is the selection of the properties, elements, and compounds (indicators) to be measured. Selection

167

of indicators for monitoring programs should be based on their relevance to important water-quality issues, such as human or aquatic health protection, the need for gaining an understanding of important geochemical processes, as well as the existence of appropriate analytical methodologies. Because of differences in the importance of water-quality issues in various regions of the country and because of the potential for significant differences in the objectives of monitoring programs, no one set of indicators is suitable or appropriate for all Forest Service monitoring programs.

Indicators appropriate for ground water-quality monitoring should meet two general criteria. First, a parameter should be a candidate for monitoring because it fulfills any of or all the following criteria:

- The parameter is potentially toxic to human health and the environment, livestock, and plants; for example, pesticides, VOCs, trace elements, and nitrate.
- The parameter impairs the suitability of the water for general use; for example, hardness, iron, manganese, taste, odor, and color.
- The parameter is of interest in surface water and may be transported from ground water to surface water; for example, ammonia, nitrite, and nitrate.
- The parameter is an important "support variable" for interpreting the results of physical and chemical measurements; for example, temperature, specific conductance, pH, major ion balance, depth to the water table, and selected isotopes.

Second, analysis of the candidate indicator should be made using well-established analytical methods at appropriate minimum-detection and reporting levels necessary to achieve the objectives of study. In general, only published analysis protocols established or recognized by EPA, USGS, ASTM, or States should be used on Forest Service projects. An additional important source of portentially applicable analysis prototcols is Standard Methods http://www.standardmethods.org. Based on these criteria, the following general groups of indicators should be considered for ground water-monitoring programs:

- Field measurements (temperature, specific conductance, pH, Eh (redox potential), dissolved oxygen, alkalinity, depth to water).
- Major inorganic ions and dissolved nutrients (total dissolved solids (TDS), Cl, NO_3, SO_4, PO_4, SiO_2, Na, K, Ca, Mg, NH_4).
- Total organic carbon (also known as TOC) [Barcelona 1984].
- Pesticides.
- VOCs.
- Metals and trace elements (Fe, Mn, Zn, Cd, Cu, Pb, Cr, Ni, Ag, Hg, As, Sb, Se, Be, B).
- Bacteria.
- Radionuclides.
- Environmental isotopes (H, O, N, C, S).

The steps for selecting specific indicators for ground water monitoring are discussed below.

1. *Analyze existing information.* The first step in the process is to use existing information to determine whether a recently documented occurrence of the indicator(s) exists. In many areas, large amounts of water-quality data have been collected by many organizations to address a wide range of objectives. Much of these data can be obtained from the EPA STOrage and RETrieval (STORET) database (http://www.epa.gov/storet/), the USGS National Water Information System (NWIS) database (http://waterdata.usgs.gov/nwis), and specific State ground water and/or water-quality databases.

2. *Determine whether the constituent or contaminant is likely to occur in the ground water system.* This step assesses the likelihood that specific indicators, which have no documented occurrence and have not been found in samples collected from the aquifer system, will be present. This assessment should take into account what is known about the potential sources of the contaminant(s) of interest, the physical and chemical properties of the contaminants, and knowledge of the local hydrogeology and the susceptibility of the aquifer to contamination. Franke (1997) provides detailed lists of indicators that could be considered for monitoring in areas with different types of land use and sources of contaminants.

3. *Test and validate constituent occurrence.* A screening sampling could be conducted to determine if the constituent of interest is present in selected wells in the aquifer system to be sampled. The number of wells to be assessed in such a screening survey should be determined on the basis of the size of the study region and the complexity of the hydrogeological setting. Using the results of this survey, the investigator should refine the list of constituents to include in subsequent sampling of the system. As knowledge of the occurrence of different constituents in different environmental settings improves, the uncertainty associated with the understanding of indicator occurrence, as well as the need for extensive verification, should decrease.

WELL PURGING AND SAMPLING

It is generally accepted that water present within the well casing is not likely to be representative of the formation water and needs to be purged prior to collection of ground water samples. The water in the screened interval, however, may indeed be representative of the formation, depending on well construction and site hydrogeology. Wells are purged for the following reasons: (1) the presence of the air interface at the top of the water column results in an increased oxygen concentration within the well and surrounding materials, (2) the loss of volatiles up the water column, (3) the leaching from or sorption to the casing, filter pack, seal or fill, of constituents of interest, (4) the changes to the aquifer flow field from the physical presence of the well, and (5) the presence of stagnant water within the well casing. Low-flow purging has been found to minimize the amount of purging needed while obtaining a representative aquifer sample.

Low flow refers to the velocity with which water enters the pump intake and is transmitted to the formation pore water in the immediate vicinity of the well screen. It does not necessarily refer to the flow rate of water discharged at the surface, which can be affected by flow regulators or restrictions. Water-level drawdown provides the best indicator of the stress imparted by a given flow rate for a given hydrological situation. The objective is to pump in a manner that minimizes stress (drawdown) to the system to the extent practicable, taking into account established site sampling objectives. Typically, flow rates on the order of 0.1—0.5 liters/minute are used; however, rates depend on site-specific hydrogeology. Some extremely coarse-textured formations have been successfully sampled in this manner at flow rates up to 1 liter/minute. The effectiveness of using low-flow purging is closely linked with proper screen location, screen length, and well construction and development techniques. The reestablishment of natural flow paths in both the vertical and horizontal directions is important for correct interpretation of the data. For high-resolution sampling needs, screens 1 meter in length or less should be used.

Most of the need for purging has been found to be caused by passing the sampling device through the overlying casing water, which causes mixing of this stagnant water with the dynamic water in the screened interval. In addition, this action causes disturbance to sediment collected in the bottom of the casing and the displacement of water out into the formation immediately adjacent to the well screen. These disturbances and impacts can be avoided by using dedicated sampling equipment, which precludes the need to insert the sampling device prior to purging and sampling.

Low-flow purging using portable or dedicated systems should be implemented with a pump intake located in the middle or slightly above the middle of the screened interval. Placement of the pump too close to the bottom of the well will cause increased entrainment of solids that have collected in the well over time. These particles are present as a result of well development, prior purging and sampling, natural colloidal transport and deposition, and changes to aquifer redox conditions within the well. Placement of the pump at the top of the water column for sampling is only recommended in unconfined aquifers that are screened across the water table, where the top of the water column is the desired sampling point. Low-flow purging has the advantage of minimizing mixing between the overlying stagnant casing water and water within the screened interval.

The water in the interval can be isolated from the overlying stagnant casing water by using low-flow minimal drawdown techniques. If the pump intake is located within the screened interval, most of the water pumped will be drawn directly from the formation into the well with little mixing of casing water or disturbance to the sampling zone; however, if the well is not constructed and developed properly, zones other than those intended may be sampled. At some sites where geologic heterogeneities are sufficiently different within the screened interval, the higher conductivity zones may be preferentially sampled. This is another reason to use short screened intervals, especially where high spatial resolution of the aquifer is a sampling objective.

Water-quality indicator parameters should be used to determine purging needs before sample collection in each well. Stabilization of parameters such as pH, specific conductance, dissolved oxygen, redox potential (ORP or Eh), temperature, and turbidity should be used to determine when stagnant water in the well is purged and sampling can begin. In general, the order of stabilization is pH, temperature, and specific conductance, followed by dissolved oxygen and turbidity. Performance criteria for determination of stabilization should be based on water-level drawdown, pumping rate, and equipment specifications for measuring indicator parameters. Special devices such as inline flow cells are available that continuously measure the above parameters.

Sampling aquifers that have low hydraulic conductivity presents unique difficulties. The traditional approach of purging the well of several well casing volumes may be ineffective because the drawdown is so rapid that perhaps only one casing volume can be obtained in a reasonable time. The approach used for low-yielding wells has been to pump the well until the water column is evacuated, wait for the well to recover, then repeat, if possible, or directly sample the water. This approach poses several concerns:

- The time required for sufficient recovery of the well may be excessive (perhaps days), affecting sample chemistry through prolonged exposure to the atmosphere.
- Purging below the top of the screen (or open section) may cause "jetting" or cascading in the well screen (or open section) as the well recovers, resulting in a change in dissolved gases and redox state and ultimately affecting the concentration of the analytes of interest through the oxidation of dissolved metals and possible loss of VOCs if they are present.
- Draining water from the filter pack around the well screen can trap air in the pore spaces, causing lingering effects on dissolved gas levels and redox state.
- Increased sample turbidity can result from the stress on the formation and stirring up of any settled solids in the bottom of the well.

Low-flow sampling as described above is an alternative for sampling aquifers having low hydraulic conductivity. Use of these methods will minimize the pitfalls in traditional well purging and permit sampling of low-yield formations where traditional methods would not be effective. In addition, use of small-diameter tubing and the smallest pump chamber (or bladder) volume minimizes the sampling system volume and the water displaced by the equipment.

A plan for QA/QC requires the establishment of a sampling protocol that is designed to minimize sources of error in each stage of the sampling process, from sample collection to analysis to reporting of analytical data. Key elements include (1) development of a statistically sound sampling plan for spatial and temporal characterization of ground water (EPA 1989); (2) installation of a vertical and horizontal sampling network that allows for the collection of samples that are representative of the subsurface; (3) use of sampling devices that minimize disturbance of the chemistry of the formation water; (4) use of decontamination procedures for all sampling equipment to minimize cross-contamination between sampling points (ASTM 1990); (5) collection of QA/QC samples (trip blanks, field blanks, and duplicates); and (6) processing, preserving, and transporting samples to maximize the integrity of the samples. Additional QA/QC procedures must be followed in the laboratory. As requirements for precision and accuracy increase, the type and number of appropriate QA/QC samples will increase.

QA/QC measures are activities undertaken to demonstrate the accuracy (how close to the real result) and precision (how reproducible the results are) of monitoring. QA generally refers to a broad plan for maintaining quality in all aspects of a program. This plan should describe how the monitoring effort would be undertaken. It should include proper documentation of all procedures, training of participants, study design, data management and analysis, and specific QC measures. QC consists of the steps to be taken to determine the validity of specific sampling and analytical procedures. The final quality assessment is the estimation of the overall precision and accuracy of the data after the analyses have been run.

The following are among the internal checks that should be performed by project field and laboratory staffs:

- *Trip blanks.* A trip blank is a sample bottle filled with deionized water under laboratory conditions that is exposed to all conditions experienced by the sample bottles and samples throughout the sample event. It is processed like any of the other samples. It is used to identify sample contamination that may have occurred through ambient exposure associated with the sampling event, shipping, and storage.
- *Field blanks.* A field blank is a sample bottle filled with deionized water under field conditions using the same or similar equipment used to collect the rest of the samples. It is processed like any of the other samples. It is used to identify errors or contamination in sample collection and analysis.
- *Negative and positive plates (for bacteria).* A negative plate results when the buffered rinse water (the water used to rinse down the sides of the filter funnel during filtration) has been filtered the same way as a sample. This material is different from a field blank in that it contains reagents used in the rinse water. There should be no bacteria growth on the filter after incubation. The purpose is to detect laboratory bacteria contamination of the sample. Positive plates result when water known

to contain bacteria (such as waste-water treatment plant influent) is filtered the same way as a sample. There should be plenty of bacteria growth on the filter after incubation. It is used to detect procedural errors or the presence of contaminants in the laboratory analysis that might inhibit bacteria growth.

- *Field duplicates*. A field duplicate is a duplicate sample collected by the same team or by another sampler or team at the same place, at the same time as a sample. It is used to estimate the precision of sampling and laboratory analysis.
- *Lab replicates.* A lab replicate is a sample that is split into subsamples at the lab. Each subsample is then analyzed and the results compared. They are used to test the precision of the laboratory measurements. For bacteria, they are used to obtain an optimal number of bacteria colonies on filters for counting purposes.
- *Spike samples.* A known concentration of the indicator being measured is added to the sample. This step should increase the concentration in the sample by a predictable amount. It is used to test the accuracy of the method.
- *Calibration blank.* A calibration blank is deionized water processed like any of the samples and used to "zero" the instrument. It is the first "sample" analyzed and used to set the meter to zero. It is different from the field blank in that it is "sampled" in the lab. It is used to check the measuring instrument periodically for "drift" (the instrument should always read "0" when this blank is measured). It can also be compared to the field blank to pinpoint where contamination might have occurred.
- *Calibration standards*. Calibration standards are used to calibrate a meter. They consist of one or more "standard concentrations" (made up in the lab to specified concentrations) of the indicator being measured, one of which is the calibration blank. Calibration standards can be used to calibrate the meter before running the test, or they can be used to convert the units read on the meter to the reporting units (for example, absorbance to milligrams per liter).

The following external checks may be performed by field staff and a second laboratory. The results are compared with those obtained by the project lab.

- *Split samples.* A split sample is a sample that is divided into two subsamples in the field or at the lab. One subsample is analyzed at the project lab and the other is analyzed at an independent lab. The results are compared.
- *Outside lab analysis of duplicate samples.* Either internal or external field duplicates can be analyzed at an independent lab. The results should be comparable with those obtained by the project lab.
- *Knowns.* The quality-control lab sends samples for selected indicators, labeled with the concentrations, to the project lab for analysis prior to the first sample run. These samples are analyzed and the results compared with the known concentrations. Problems are reported to the quality-control lab.

- *Unknowns.* The quality-control lab sends samples to the project lab for analysis for selected indicators, prior to the first sample run. The concentrations of these samples are unknown to the project lab. These samples are analyzed and the results reported to the quality-control lab. Discrepancies are reported to the project lab, and a problem-identification and problem-solving process follows. In general, many of the lab-specific portions of the QA/QC program can be satisfied by using a laboratory certified by formal EPA or State environmental lab certification programs. It is recommended that only certified labs be used for most water quality analyses.

Monitoring Network Cost Considerations

The total cost of an observation network includes the costs to plan, design, and construct the physical network, and the costs to operate and properly maintain the network for a specified period of time. Observation networks that include sites in remote areas can greatly add to the costs because of access restrictions. Managers must balance the needs and scope of the study with the budgetary constraints of the agency.

Construction costs are usually the largest costs encountered in observation networks. Drilling of observation wells can cost tens to hundreds of thousands of dollars for deep wells with multiple completions. Obtaining a continuous core sample provides a great deal of detailed geological information, but also can greatly increase the cost of a well. Simple, shallow wells or piezometers, however, can be relatively inexpensive, particularly if they are installed by hand in unconsolidated sediments. Well completion and development, borehole-geophysical logging, monitoring equipment, and construction of shelters for that equipment, however, add to the cost of monitoring wells. Use of existing wells may save on construction costs, but may not provide the information needed to correctly interpret the information obtained from those wells. The reader should consult a local contractor to ascertain the costs for the particular geographic area.

The cost of operation and maintenance of an observation network depends on a number of factors, such as measurement frequency, number of wells to be monitored, physical distance to and between wells and well access, and the frequency of well maintenance and testing. Manual measurements that are obtained infrequently (quarterly or annually) mainly involve relatively inexpensive equipment and personnel time, but may result in information that is difficult to interpret. If the network consists of many wells over a large distance that must be measured frequently, then self-operating automated systems (such as a transducer and datalogger) may actually be more cost-effective, despite the higher equipment costs. A regionally distributed network in which wells are in remote locations can be expensive to operate and maintain because of travel time and costs, as well as vehicle costs. Maintenance costs for self-operating systems can also be considerable. Periodic site visits to automated wells are required to check on equipment operation, download data, and upgrade or repair equipment. Monitoring equipment shelters often are favorite targets for vandalism. A local contractor should be contacted to ascertain the costs for a particular geographic area.

Aquifer Testing Techniques

Knowledge of the hydraulic properties of the subsurface systems being studied often is necessary for valid interpretation of ground water-quality data. A large-scale aquifer test can provide an overall estimate of the hydraulic conductivity and storage of water-bearing units within several hundred feet or more of a pumping well. A large-scale test usually involves measuring the response of an aquifer system to pumping by measuring changes in water levels in observation wells in the vicinity of the pumping well. Analysis of the response to other hydraulic stresses, such as injection of water into the system, also is possible. Typically, wells for an aquifer test consist of one large-diameter (4 inches or greater) pumping well that is associated with observation wells that can be of smaller diameter in which drawdown is measured as pumping proceeds. The larger diameter normally is required for the pumping well to ensure that pumping can be fast enough to cause measurable drawdown in the outlying observation wells. A diameter as small as 2 inches or less can be suitable for measuring water levels and can be used in an array of wells from which drawdown is to be determined.

The hydraulic properties that can be determined from an aquifer test depend on the onsite test conditions and installations (see table 12). The most commonly determined hydraulic parameters are the transmissivity (T), hydraulic conductivity (K), and storage coefficient or storativity (S).

Transmissivity is a measure of the ease with which the full thickness of the aquifer transmits water. *Hydraulic conductivity* is a measure of the ease with which a unit thickness of the aquifer transmits water (see appendix II for a chart of representative hydraulic conductivity values for different geological materials). Hydraulic conductivity measurements provide a basis for judging the hydraulic connection of the monitoring well and adjacent screened formation to the hydrogeological setting. These measurements also allow an experienced hydrogeologist to estimate an optimal sampling frequency for the monitoring program (Barcelona and others 1985). Hydraulic conductivity is most effectively determined under field conditions by aquifer testing methods, such as pump testing or slug testing. The water-level drawdown can be measured during water withdrawal. Alternatively, water levels can be measured after the static water level is depressed by application of gas pressure or elevated by the introduction of a slug of water. These procedures are rather straightforward for wells that have been properly developed (EPA 1991).

Traditionally, hydraulic conductivity has been estimated by collecting drill samples, which were then taken to the laboratory for testing. Several techniques involving laboratory permeameters are routinely used. Falling-head or constant-head permeameter tests on recompacted samples in fixed wall or triaxial test cells are among the most common. The relative applicability of these techniques depends on both operator skill and methodology because calibration standards are not available. The major problem with laboratory test procedures is that the determined values are based on remolded samples rather than on undisturbed materials. Work done to date with laboratory tests on "undisturbed" samples suggests that laboratory-determined values of hydraulic conductivity are three to six orders of magnitude smaller than values determined by in situ aquifer testing for unconsolidated, fine-grained material (Melby 1989). Therefore, considerable care must be exercised when evaluating laboratory-derived hydraulic conductivity information.

Storage coefficient (storativity), specific yield (S_y), effective porosity (n_e), and drainable porosity are all terms that express information about the storage capacity of an aquifer. Storage capacity is a measure of the interconnected void space of an aquifer medium. The storage coefficient of an unconfined aquifer is approximately equal to the effective porosity, and typically has values of 0.05 to 0.30, or 5–30%. The storage coefficient of a confined aquifer is typically much smaller than that of an unconfined aquifer, typically ranging from 10^{-5} to 10^{-3}. Storage coefficients are low in confined aquifers because they are not drained during pumping, and any water released from storage is the result of a combination of compression of the aquifer and expansion of the water that is being pumped. As a result, a small amount of water is released per unit change in head. Pressure is reduced in the aquifer, but the aquifer is not dewatered. Therefore, for equal changes in head in an unconfined aquifer vs. a confined aquifer, the unconfined aquifer will produce a greater volume of water.

Aquifer tests do not provide a direct analysis of hydraulic conductivity or effective porosity; however, hydraulic conductivity can be determined from an aquifer test where the saturated thickness of the aquifer is known. The effective porosity can be estimated as the storage coefficient from tests of an unconfined aquifer. The determination of storage coefficient with some confidence from an aquifer test requires analysis of the drawdown response in observation wells rather than in the pumping well. Drawdown response in the pumping well alone can be used to estimate transmissivity, but it is unreliable for determining the storage coefficient because the effective radius of the pumping well is not known.

Each aquifer-test method is commonly assumed to be limited to a relatively simple set of aquifer characteristics and boundary conditions as opposed to the complexity of actual sites. A method should be selected on the basis of (1) the hydrogeology of the test site, and (2) the field-test conditions. An additional set of criteria that affects the method(s) selected often involves the available budget, the project timeline, and the consequence of the results on future work. The hydrogeology of the test site—such as a nonleaky confined aquifer, a leaky confined aquifer, or an unconfined aquifer, and other natural conditions of the site—determines the applicable set of aquifer-test methods. The number and location of observation wells, if any, the instrumentation for measuring water levels, and the screened interval of the well and the capacity of the pump determine which aquifer-test methods can be applied to the data. These and other factors determine the physical constraints on stressing the aquifer and on determining the aquifer response, and may further limit the applicable aquifer-test methods.

One relatively simple method of determining aquifer characteristics is the slug test (table 12). A slug test at an observation well can provide an estimate of hydraulic conductivity of the aquifer in the immediate vicinity of the well screen. It is not a good method to characterize an aquifer at great distance from the well used for the slug test. A slug test is conducted in a single observation or monitoring well that usually is small in diameter (less than about 4 inches). Slug tests involve the instantaneous addition or removal of water from the well, commonly done by lowering a solid cylinder into the well, and withdrawing it, causing the water level to first rise and then drop in the well casing. Measurements of the recovery of the water level in the well from both falling head and rising head are used to determine hydraulic conductivity of the screened interval. Techniques for conducting slug tests can be found in many publications by the USGS.

Aquifer-test methods are numerous. The body of literature on these methods is extensive, and covers the selection, planning, design, and implementation of a test, and the analysis of results. Only a few of the many publications on aquifer-test methods are referenced below. A review of field procedures for conducting an aquifer test and a summary of the principal aquifer-test methods are provided in Bedinger and Reed (1988). Bedinger and Reed (1988) provide a glossary of terms and a syllabus of aquifer-test methods, classified by aquifer condition, control-well characteristics, recharge and discharge function, and boundary conditions. Description of the basic principles of well hydraulics and principal aquifer-test methods with examples of their application are described in Lohman (1972). Practical information related to aquifer-test planning and interpretation of aquifer-test data is given in Kruseman and deRidder (1990), and Driscoll (1986).

Geophysical Techniques

Geophysical methods applicable to ground water investigations are generally described in two broad categories: surface methods and borehole methods (EPA 1993b) [See appendix VII for more detailed information on geophysics in hydrogeological studies]. Borehole geophysical methods have the greatest utility in ground water studies, but their use is limited after wells are completed. Surface geophysical methods are used to interpret geological conditions and their possible controls on ground water. In addition, surface methods can be used to map contamination under some conditions. Recently, considerable technology and methodology have been developed for use in fractured-rock settings. Weight and Sondregger (2001) summarize geophysical techniques commonly used in hydrogeology.

Surface geophysical methods provide for the areal reconnaissance of geology and shallow ground water conditions (Zohdy and others 1974). Four techniques are widely applicable to a variety of geological settings, and can be useful in hydrogeological studies: (1) electrical resistivity, (2) electromagnetic conductivity, and (3) seismic refraction. These methods are generally employed in hydrogeological applications for four broad objectives: (1) evaluating ground water quality, (2) determining the depth to the water table, (3) determining the depth to the bedrock surface, and (4) evaluating subsurface lithology and physical properties. Surface geophysical methods can be useful in determining the surface location and orientation of potential water-bearing fractures. Electrical methods, including square-array and azimuthal resistivity surveys and electromagnetic surveys, are particularly useful for locating fractures (Lane and others 1995, Taylor and Fleming 1988, Slater and others 1998).

Borehole geophysics is the science of recording and analyzing continuous or point measurements of physical properties made in wells or test holes (Keys 1990). The terms borehole and downhole are used interchangeably to refer to such measurements. Most specific borehole geophysical techniques have long been in use by the petroleum industry, in which holes being logged are usually deep and filled with drilling muds or saline water. Many of these techniques are not suitable, or must be adapted, for use in freshwater aquifers, which are the focus of most near-surface hydrogeological investigations. Nevertheless, suitable borehole geophysical methods can greatly enhance the geological and hydrogeological information obtained from water supply or monitoring wells. The development of logging tools specifically designed for use in freshwater wells, such as the EM39 borehole conductivity meter (McNeill 1986), and high-precision thermal and electromagnetic borehole flowmeters (Paillet 1994, 2000) should contribute to greater use of downhole methods in the future.

Borehole and core logging can provide data on the geology of the borehole, individual fractures, and the fluid in the hole. Commonly used borehole logging methods include caliper, fluid, resistivity, and gamma logs. Optical and acoustic imaging methods and heat pulse flow meters are particularly useful for detecting and evaluating individual fractures. Newer technologies that are not yet in common use include digital borehole imaging, borehole radar, and seismic and resistivity tomography. It is important to keep in mind that many geophysical methods yield non-unique results that are best interpreted in combination with other lines of evidence, especially physical and geological sampling.

Table 12. Summary of aquifer test methods (National Academy of Sciences 1981).

Test	Reference	Major Items Required	Parameters Obtained	Comments
Pumping	USDI 1977; Lohman 1972, Stallman 1971; Walton 1962; Ferris & Knowles 1963; Ferris et al. 1962	Minimum of one observation well and preferably 4 or more; pump; power source; winch; tripod, mast or boom; discharge measuring device; stop watch; water level sounder	T, K, S	Yields parameter values averaged over a relatively large aquifer volume; most commonly used when accuracy and reliability is of high priority; best results in aquifers with good continuity and permeability provided by intergranular flow channels; can provide evidence of leakage through aquitards, directional permeability, and the presence of hydrogeologic boundaries. Relatively expensive, doesn't work well in very tight aquifers, requires a power source.
Drawdown/ specific capacity	USDI 1977; Lohman 1977; Walton 1970.	Same as above, but no observation wells are required.	T, K	Yields only rough estimates of T and/or K; storage coefficient or apparent specific yield must be estimated independently; conditions immediately adjacent to the well bore, well losses, etc., substantially affect results; in tight aquifers the effects of well-bore storage may be highly important. Relatively inexpensive; most useful in reconnaissance investigations.
Gravity injection	Same as above.	Supply of water (water truck or tank), injection hose or tubing, in-line flow meter, water-level measuring device, stop watch	T, K	Can be conducted on cased or open holes using the same equations as those for tests described above; conducted with constant head or with constant injection rate; best applications are with clean wells in poorly transmissive materials
Pressure pump-in	USDI 1977.	Inflatable or compression packers; pump; power source; pressure gages, stop watch; in-line discharge measuring device; storage capacity and source for water.	T, K	Usually conducted during exploration or reconnaissance investigations; permits determination of T and K in different intervals along the well bore; can be used above or below the water table or water level in the well; works best in consolidated aquifers or perforated well casing. Relatively expensive because it is usually conducted during the drilling operations using the contractor's rig and equipment.
Auger hole	Boast & Kirtham 1971.	Small pump or bail; stop watch; float	K	Applicable in cases of unconfined aquifers when the water table is within a few feet of ground surface; inexpensive, rapid, reliable
Recovery after any of the above tests.	Same as for test.	Same as for test.	T, K, S	Recovery should always be monitored following a drawdown/specific capacity test; usually yields more reliable values for T and K than the drawdown/ specific capacity test; has the additional advantage of providing an estimate of storage coefficient or apparent specific yield; because the rate of recovery is dependent upon the preceding pumping rate the results are effected by well-bore storage. Minimum expense in addition to that incurred during the pumping period and provides additional and more reliable information than the drawdown/specific capacity test.
Slug/falling head recovery	USDI 1977; Lohman 1972; Ferris & Knowles 1963; Kvorslev 1951; Covack & Papadopulos 1967; Bouver 1978	Equipment required depends upon the manner in which the slug is added or removed. Pump may be used but it is not required.	T, K	A specific type of recovery test; one of the simplest and least expensive of all tests; does not require a power source; yields values acceptably accurate for most purposes; analysis procedures available that account for aquifer storage only, well-bore storage only, or both. Applicable in both confined and unconfined aquifers.

*T = transmissivity, K = hydraulic conductivity, S = storage coefficient or specific yield

Source: National Acadamy of Sciences, 1981, p. 159

The characteristics of the borehole may constrain the type of borehole logging method that can be used, and therefore may be primary considerations when identifying borehole logging methods of potential value in a specific situation. These characteristics include the following:

- Whether a casing is present (electrical methods, for example, require uncased holes).
- If cased, the type of casing (borehole radar, for example, can be used with a PVC casing, but not with a steel casing).
- Borehole diameter must be large enough for the instrument of interest (some logs, such as dielectric and nuclear magnetic resonance logs, require borehole diameters that are considerably larger than are typically drilled for monitoring wells).
- Whether borehole fluid is present (electric logs, sonic logs, and any fluid characterization log require ground water or drilling fluid in the borehole).
- The radius of measurement of the specific method (radii can range from near the borehole surface for spontaneous potential and single-point resistance logs to more than 100 meters for borehole radar in highly resistive rock).
- Calibration (many logging methods require calibration for corrections of such factors as temperature, borehole diameter, and fluid resistivity).

The most commonly used borehole logging methods in hydrogeological and contaminated-site investigations include spontaneous potential, single-point resistance, fluid conductivity, natural gamma, gamma-gamma, neutron, sonic, caliper, temperature, and flow meter.

Ground Water Tracing Techniques

In recent years ground water tracing techniques have been used in a variety of hydrogeological settings to help characterize ground water flow systems. Tracing techniques have proven to be especially useful in fractured rock and karst settings and have been helpful for identifying and characterizing contaminant transport pathways and transport velocities. Ground water tracing techniques often require fewer assumptions about hydrogeological conditions than do hypothetical or numerical simulations; therefore, they can be more reliable. Tracer tests can be used to obtain empirical data related to ground water recharge, flow direction, flow rates, flow destinations, and flow-system boundaries. Tracer recovery data, when combined with ground water discharge data, can also provide quantitative data that can be useful for assessing the fate of contaminants in the subsurface. Several tracers can be used together, allowing several potential pathways to be evaluated simultaneously.

In general, tracing can be divided into two categories: label tracing and pulse tracing. Using tracers as labels allows for identification of specific waters or plumes. Pulse tracing involves sending an identifiable signal through part of a ground water flow system at concentrations significantly above background. Ground water tracers can be divided into two types: natural and artificial. In general, natural tracers are more applicable for label tracing, while artificial tracers are more suitable for pulse tracing. Important natural tracers include stable and radioactive isotopes, selected ions, selected field parameters (specific conductance and temperature) and selected microorganisms. Commonly used artificial tracers include organic fluorescent dyes, chlorofluorocarbons, gases, and salts (like chloride and bromide).

Tracing methods that include the deliberate or incidental introduction of a tracer into a stream or ground water flow system have been increasingly used in hydrogeological investigations in fractured-rock settings. Tracing techniques do not require the assumptions of a porous-media approach and can be used to delineate ground water flow paths, determine ground water flow velocities along the delineated flow paths, and help estimate mass loading to a stream along a preferential ground water flow path. Tracer recovery data, when combined with ground water flow data, can provide quantitative data for evaluating contaminant behavior and fate in the subsurface. An ideal artificial tracer should be (1) quantitatively detectable in very small concentrations; (2) found in low concentrations in the water to be traced; (3) not readily attenuated by the aquifer material, geochemical reactions, or biological degradation; and (4) nontoxic to humans and the ecosystem (Todd 1980). The application of surface-water tracing in combination with ground water tracing provides detailed information on ground water inflow zones to streams. Stream-tracing techniques, which include the continuous injection of a constant concentration of tracer, also provide very accurate stream discharge measurements based on dilution of the tracer (Kimball 1997). Discharge calculated in this manner includes that flow which is in the hyporheic zone, which is typically a significant hydrologic zone in many streams, especially those with high gradients. It should be noted that no way exists to quantify how the discharge estimate is influenced by hyporheic exchange flows. The only way to ensure that all hyporheic exchange flows are included in the discharge estimate is to locate the downstream sampling site on exposed, impermeable bedrock where all flow is forced into the surface stream channel.

Natural Tracers

Naturally occurring isotopes are the most common natural tracers used in ground water investigations. These include isotopes of common elements, such as carbon, oxygen, and hydrogen, in addition to isotopes of radioactive elements, such as tritium. Analyzing water samples for stable and radioactive isotopes can provide data for characterizing sources of ground water recharge. This step can be very helpful in delineating and characterizing preferential flow paths. The information can also be used to establish relative, or sometimes absolute, ages of ground water withdrawn from various depths and locations in an aquifer, which can greatly assist in the identification and differentiation of local, intermediate, and regional flow systems. It can also greatly assist in the estimation of exchange rates and flow directions at aquifer-stream interfaces and aquifer-lake interfaces. For detailed information on isotope geochemistry, processes affecting isotopic compositions, and isotopes in ground water hydrology, see Kendall and McDonnell (1998) and Clark and Fritz (1997).

Stable Isotopes. Stable isotopes have proven to be the most versatile natural tracers. For a given element, isotopic composition can vary because of partitioning or fractionation related to differences in reaction rates among the isotopes. Fractionation is typically proportional to the differences in isotopic mass for a given element of low atomic number. This property allows the ratios of isotopes of an element to become fingerprints of climatic and hydrologic conditions or serve as markers for different sources of that element. Variations in annual rainfall and snowmelt strongly affect the isotopic composition of waters. Data for ratios of stable isotopes of oxygen, hydrogen, sulfur, nitrogen and carbon can be especially useful for (1) characterizing ground water flow paths from areas of recharge to areas of discharge, (2) identifying mechanisms responsible for streamflow generation, and (3) testing flow path and water-budget models developed using hydrologic data.

Stable isotopes of oxygen and hydrogen are ideal tracers of water sources and movement because they are the two constituents of the water molecule and the ratios of each element tend to stay constant as long as the water has not experienced freezing or evaporation. Oxygen isotopes include ^{16}O, ^{17}O, and ^{18}O and hydrogen isotopes include protium (^{1}H), deuterium (^{2}H), and tritium (^{3}H).

Stable isotopes of sulfur, nitrogen, and carbon are also important in environmental studies. These isotopes are constituents that are dissolved in water or carried in the gas phase. Stable isotopes of oxygen and hydrogen behave conservatively because interactions with organic and geologic material along the flow path will have a negligible effect on the ratios of isotopes in the water molecule. The ratios of stable isotopes of dissolved sulfur, nitrogen, and carbon can be significantly altered by reactions with organic and geologic material. Thus, these solute isotopes may have limited use for tracing water sources and flow paths. Solute isotope data, however, can provide information on the reactions that are responsible for their presence in the water and the flow paths implied by their presence (Kendall and McDonnell 1998). In addition, they may be able to be used to identify the source of a contaminant plume should each potential source have a different isotopic composition.

Radioactive Isotopes. Tritium, a radioactive isotope of hydrogen (half-life = 12.43 years), is naturally produced in the upper atmosphere by bombardment of nitrogen by neutrons; however, large amounts of tritium were released to the atmosphere during the period of above-ground thermonuclear testing, which was at a maximum during the 1950s and was discontinued in 1963. Pre-bomb-testing concentrations of tritium in water have been determined to be close to the detection limit of 1 tritium unit. The presence of tritium above background levels in ground water is an indication that recharge occurred during or after the bomb testing period. In addition to radioactive decay, tritium in ground water is subject to significant attenuation through mixing with waters with less tritium; consequently, tritium concentrations usually cannot be used to obtain an "absolute" age of ground water.

Artificial Tracers

Artificial tracers are those introduced into the ground water flow system either purposely as part of a designed tracer test or inadvertently as a spill or other anthropogenic activity. To serve as a suitable tracer, a substance must be (1) nontoxic to humans and the ecosystem; (2) either absent from the ground water system or present at very low, near-constant levels; (3) soluble in water with the resultant solution having nearly the same density as water; (4) nonreactive; (5) easy to introduce into the flow system; and (6) unambiguously detectable in very low concentrations. Many organic dye and salt tracers have been approved for use in aquifers and streams that are used to obtain drinking water; however, some States, such as Wisconsin, have restricted the use of all artificial tracers in ground water.

Salts. Chloride, bromide, and lithium solutions are commonly used for both ground water and surface water tracing. These salts are very soluble, relatively inexpensive, conservative, and nontoxic at concentrations typically used for tracing. They are also easily detectable at low concentrations. Some ecological considerations for the use of salts in tracing studies are covered by Wood and Dykes (2002). Chloride occurs naturally in some ground water, often in the tens to hundreds of parts per million. Natural concentrations of bromide are usually much lower. Chloride has commonly been used to trace contaminant plumes that originate at landfills or other industrial facilities. Stream-tracing techniques, which include the continuous injection of a constant concentration of a salt, provide very accurate stream discharge measurements based on dilution of the tracer.

Organic Dyes. Fluorescent dyes are some of the most analytically sensitive, versatile, and inexpensive artificial water tracers available. Many references document the use of these dyes in stream and ground water tracing studies and their human and environmental toxicity (Field and others 1995, Smart and Laidlaw 1977). Fluorescent dyes commonly used in ground water investigations include uranine, fluorescein, rhodamine, eosin and phloxine, and sulpho-rhodamine B. Most fluorescent dyes work well in water with a nearly neutral pH. In acidic conditions, the fluorescence of some dyes is minimized; however, these dyes will fluoresce again if the pH of the sample is adjusted to more alkaline conditions. In addition, in ground water systems with substantial quantities of organic material, adsorption of the fluorescent dyes to the organic materials may limit their usefulness. Commercial grade, organic, fluorescent dyes can be purchased as liquid compounds or as powders. Uranine, eosin, and phloxine are FDA approved.

Chlorofluorocarbons. The chlorofluorocarbon (CFC) gases CFC-11, CFC-12 and CFC-13 were developed during the 1930s. These gases were chemically stable and safe and therefore found wide application, commonly as refrigerants. Unfortunately, waste CFCs accumulate in the atmosphere, where they are now thought to pose a serious hazard to stratospheric ozone. This problem has led to a very successful international action (Montreal Protocol) to reduce global CFC production. The known growth rates of atmospheric CFCs, their rapid mixing worldwide, their solubility in water, and their good chemical stability have enabled CFCs to become a useful tool for hydrologists to trace water movement in the oceans, in surface water, and in ground water. In the case of ground water, the method rests on the assumption that ground water at the water table will be in equilibrium with atmospheric air concentrations, including its CFC component. Once water moves into the saturated zone below the water table, it will not be able to acquire or lose any additional CFC gas to the atmosphere. The CFC concentration in the water will be characteristic of the atmospheric CFC level prevailing during its last contact with the atmosphere. This characteristic forms the basis of CFC dating of ground water on a time scale of 0 to 50 years. The steep increase in atmospheric CFC levels over time ensures that fairly precise dates can be obtained. In contrast, the input curves for tritium and radiocarbon are rather flat. The development of a reliable sampling and analytical procedure has ensured wide application of this technique. For information on dating of ground water using CFCs, see Plummer and others (1993) and Plummer and Friedman (1999).

Field Methods

The usefulness and appropriateness of a ground water tracer test depend on the questions to be answered by a particular hydrogeological investigation. Tracer tests are appropriate when ground water flow velocities are such that results will be obtained within a reasonable period of time, usually less than a year. The usefulness of tracer test results are highly dependent on proper test design (particularly determination of sampling locations) and execution, the nature of the tracer, the ability to detect the tracer at low concentrations, and correct interpretation of recovery data. Before conducting a tracer test, it is very important to use other geological and hydrological information to develop a fundamental understanding of the hydrogeological setting and the ground water flow system to be traced. This understanding can then be used to (1) determine the appropriate type of tracer, (2) determine the tracer injection location and method, (3) determine appropriate sample collection locations, and (4) determine which tracers should be included in the test design. It is always advisable to sample more locations rather than fewer locations. For artificial tracers, it is important to know precisely how much tracer mass is injected. This knowledge will allow for a determination of the percent of tracer mass recovered at a given sampling location. This quantitative aspect of tracing can be important in helping to evaluate the significance of any given ground water flow path.

Isotopes. Chapter 10 of Clark and Fritz (1997) includes an excellent discussion and comparison of sampling and analytical protocols and procedures for collecting water samples for isotopic analysis. Sample size, filtering, preservation, container type, holding times, and method of analysis vary quite a bit between different stable and radioactive isotopes. In general, isotopes of water (oxygen 18, deuterium, tritium) have simpler sampling protocols than isotopes of carbon (carbon 13 and carbon 14), sulfur (sulfur 34), dissolved gases (helium, argon 39, krypton 85), and uranium (uranium 234 and uranium

238). Water samples for isotopic analyses must be collected and stored in well-sealed bottles. Proper sealing and handling is necessary to prevent any additional fractionation.

Organic Dyes. The use of organic dyes as hydrogeological tracers requires specific field sampling and analysis procedures. Careful thought should be given to the selection of the proper dye and the method for introducing it into the ground water. Dyes should be selected based on their chemical and toxicity characteristics. Also, fluorescence is reduced in some dyes when dissolved in low pH waters or exposed to sunlight, and some dyes fluoresce better in cooler waters. It is also very important to carefully consider the best way to introduce the dye tracer into the ground water system. The following are four common methods:

- Inject into a well, making sure the well will take the desire quantities of water before introducing a dye tracer. The ability to take water can be ascertained by conducting a simple aquifer test.
- Inject into a stream, making sure the stream gradient is low and the reach of stream below the injection point is a losing reach.
- Inject into a constructed excavation, making sure the excavation will take the desire quantities of water in an appropriate time period.
- Inject into a sinkhole in karst terrain.

Protocols for collecting water samples that may contain the dye tracer are relatively simple. Sample containers and storage should minimize all exposure to light to prevent photo degradation of organic dyes. Samples do not require filtering or preservation, but water samples that may contain dye should be kept cool until analysis is complete. They should be analyzed within 2 weeks to minimize bacterial degradation of organic dyes.

Water samples can be collected by grab sampling or with an auto sampler, which is useful if many samples must be taken in a short time and from locations with difficult access. Once the water samples are collected, samples containing organic dyes should be analyzed on a spectrofluorometer to confirm the nature of the fluorescence and then samples containing all tracers should be analyzed with wet chemistry methods for dye concentration. It is important to conduct both types of analyses to confirm the presence of the dye that was injected. Sophisticated sampling of organic tracers can be achieved by using flow-through flourometers, which measure the fluorescence of the dye in water on a real-time basis. This type of sampling requires a power source and data loggers, but is indispensable in surface-water tracer studies.

Small bags of activated charcoal can also be used to detect organic dyes. Organic dyes will sorb onto charcoal if water that contains dye comes into con-tact with the charcoal. Charcoal bags are placed in water at sampling locations and then retrieved for analysis at selected time intervals. It is important to note that determining the travel time from an injection location to a given charcoal bag is constrained by the time interval between retrieval of the bags.

The most important rule of thumb for sampling is to collect samples often at many places. Collecting samples for organic dye analysis is relatively easy and inexpensive. Because it is not always possible to predict all locations where dye may be recovered, it is best to have more rather than fewer sampling locations. By collecting samples frequently, at important locations, the data can be used to construct breakthrough curves of recovery versus time. Detailed breakthrough curves can be used in rigorous analyses of the recovery data.

Analysis of Hydrogeological Data

This section presents techniques for assessing ground water flow conditions and hydraulic properties of aquifers. An equally important component for any hydrogeological investigation is analysis of water-quality data. Some techniques for analysis and plotting of water-quality data are detailed in appendix VI.

Analytical Methods

Analytical methods use exact closed-form solutions of the appropriate differential equations for particular sets of conditions and involve manually solving equations, such as Darcy's Law or the Theis equation, or generating solutions using curve-matching techniques. These approaches may be used either independently or in concert to develop solutions to complex problems. In contrast, numerical models apply approximate solutions to the same equations. Semianalytical models use numerical techniques to approximate complex analytical solutions, allowing a discrete solution in either time or space. Analytical methods are most useful in the analysis of aquifer test data, simplified aquifer system evaluation, and to assist in the design of numerical models.

Analytical models provide exact solutions, but employ many simplifying assumptions about the ground water system, its geometry, and external stresses to produce tractable solutions (Walton 1984). This approach places a burden on the user to test and justify the underlying assumptions and simplifications against the actual physical system (EPA 1991). For example, analytical models generally assume isotropic conditions and an infinite aquifer. These conditions may not exist in the problem at hand, and results may be inaccurate because of these constraints. The following are examples of the use of analytical models:

- Determining drawdown effects of pumping alluvial aquifers with relatively impermeable boundaries, as with mountain blocks bounding an alluvial valley floor. The use of image well theory provides for analysis of such a situation, and results in greater drawdown impacts than an infinite aquifer (Walton 1970).
- Determining drawdown effects at a well field with several wells pumping simultaneously.

- Determining the ground water flow rate to a finite line sink, as in the study of ditches, canals, strip mines or ground water flow to finite sections of rivers or streams. In this case, the head or drawdown at the line sink is known and the flow rate is unknown.
- Mounding of ground water beneath a water body such as a tailings pond.

Semi-analytical models can provide streamline and travel time information through the use of numerical or analytical expressions in space or time. This information is especially useful for delineation of wellhead protection areas (EPA 1994). Analytical element models are a relatively recent development in semi-analytical modeling of regional ground water flow. They use approximate analytical solutions by superposing various exact or approximate analytical functions, each representing a particular feature of the aquifer (Haijtema 1985, Strack 1989). A major advantage of these models compared to analytic models is greater flexibility in incorporating varying hydrogeology and stresses without a significantly increased need for data (van der Heijde and others 1988).

POTENTIOMETRIC MAPS   Developing a potentiometric map is not as straightforward as preparing a topographic map. An accurate potentiometric map requires enough well observations to develop contours of equal head that do not miss important features of the flow system. Considerable interpretation and judgment may be required in developing contours when well data points do not seem to fit into a coherent pattern; for example, if water-level data from wells are drawn from multiple sources, measurements in nearby wells may have been taken at different times of the year and may not be directly comparable. On the other hand, if all the data have been collected so as to minimize effects of short-term or seasonal fluctuations, examination of individual well characteristics may yield explanations for anomalous data points; for example, a single well data point that is far out of line with nearby wells may be tapping a different aquifer. If an anomalous well data point cannot be readily explained as being unrepresentative for any reason, then further field investigation may be required to determine whether any localized hydrogeological conditions are causing the anomaly.

The starting point for a potentiometric map is a base map. The base map identifies well locations, water-level elevations in the wells, and other surface hydrologic features, such as streams, rivers, and water bodies. Drawing equipotential contours requires some skill and judgment. Errors in contouring fall into two general categories: (1) failure to exclude data points that are not representative and (2) failure to take into account subsurface features that change the distribution of potentiometric head as a result of aquifer heterogeneity or boundary conditions. Following are six situations in which contouring errors might occur:

1. For water-table maps, failure to exclude measurements from wells cased below the water-table surface in recharge and discharge areas; for example, only well c in figure 57 gives an accurate reading of the water table surface.
2. For water-table maps, failure to adjust contour lines in areas of topographic depressions occupied by lakes. Figure 58a illustrates the incorrect and correct interpretations in this situation.
3. Failure to recognize locally steep gradients caused by fault zones. Figure 58b illustrates how conventional contouring methods erroneously portray the ground water flow systems on the two sides of a fault.
4. Failure to consider localized mounding or depression of the potentiometric surface from anthropogenic recharge or withdrawal. Pumping wells create a cone of depression around the well, with steepened hydraulic gradients. Agricultural irrigation, artificial recharge using municipally treated wastewater, and artificial ponds and lagoons usually cause a mounding of the water table. When the source of recharge is confined to a relatively small area, a localized mound develops with elevations increasing toward the center, rather than decreasing as in a pumped well. Area wide recharge will reduce hydraulic gradients compared to natural aquifer conditions. These features are especially significant when they are located near a ground water divide because small shifts in the location of a divide may have a major impact on the direction in which contaminants flow.
5. Failure to consider seasonal and other short-term fluctuations in well levels. If an aquifer experiences seasonal high and low water tables, well measurements are not comparable unless they are taken at the same time of year. Other factors, such as dramatic changes in atmospheric pressure and precipitation events, might reduce the comparability of well measurements even if the measurements are taken at about the same time of year.
6. Use of measurements from wells tapping multiple aquifers. Wells in which the screened interval includes multiple aquifers generally yield inaccurate water level or piezometric measurements because the measured head reflects the interaction between heads of the intersected aquifers. Figure 59 illustrates how the failure to differentiate measurements from wells completed in two aquifers, combined with a well that connects the two, results in an apparent depression in the potentiometric surface.

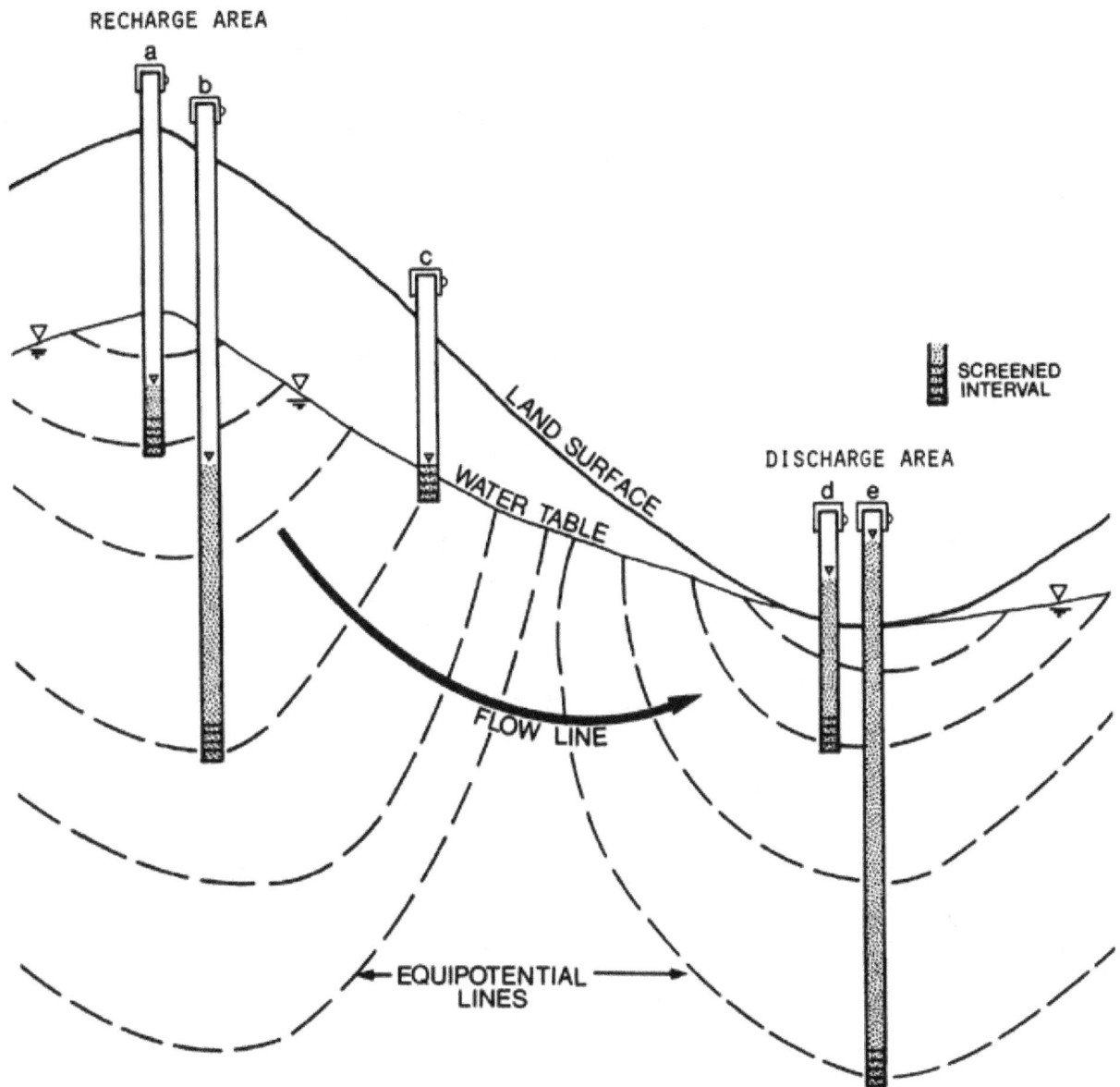

Figure 57. Cross-sectional diagram showing the water level as measured by piezometers located at various depths. The water level in piezometer c is the same as well b since it lies along the same equipotential line (after Mills and others 1985).

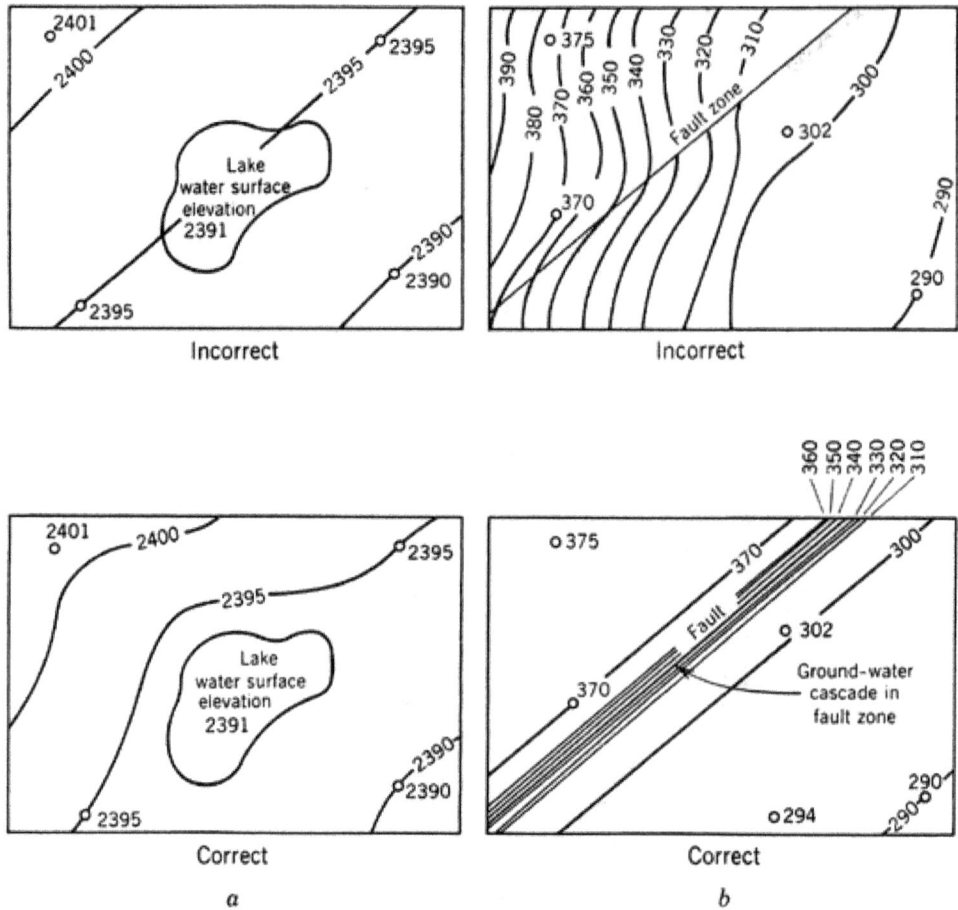

Figure 58. Common errors in contouring water table maps: (a) topographic depression occupied by lakes and (b) fault zones (Davis and DeWiest 1966).

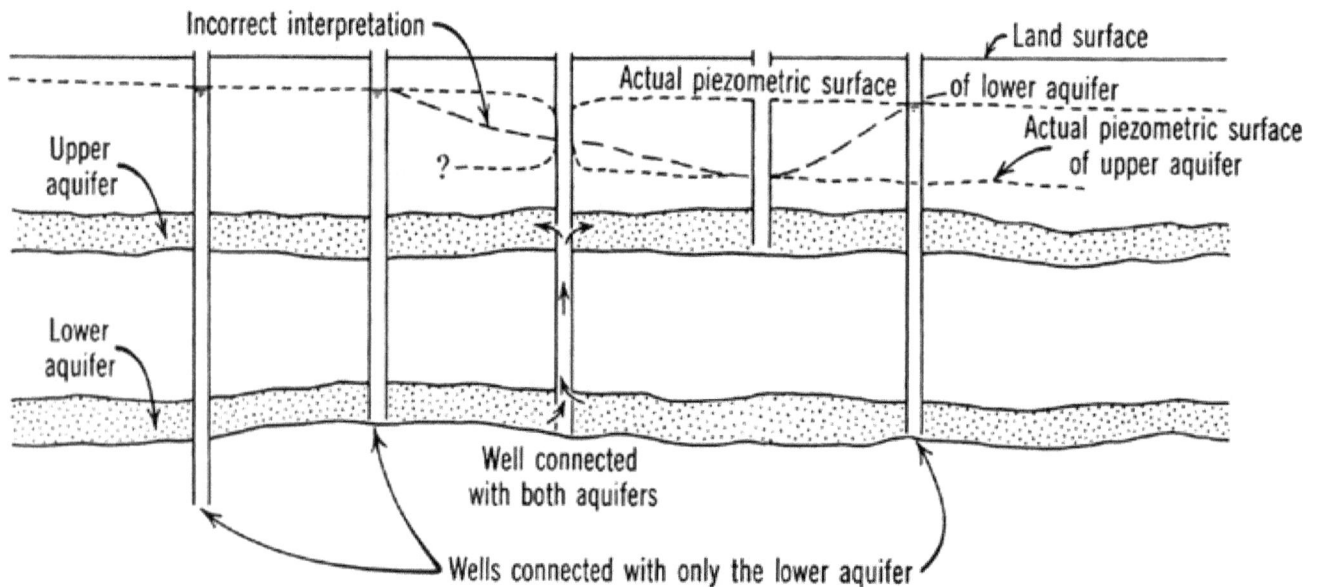

Figure 59. Error in mapping potentiometric surface because of mixing of two confined aquifers with different pressures (Davis and DeWiest 1966).

CALCULATING GROUND WATER FLOW

The quantity of ground water moving through a volume of rock can be estimated using Darcy's Law (Darcy 1856),

$$Q = KIA,$$

where Q is the quantity of ground water flow, K is the hydraulic conductivity, I is the hydraulic gradient, and A is the cross-sectional area of the aquifer of interest (saturated interval).

Note that the quantity of flow is directly proportional to the hydraulic gradient. This equation provides a rapid way to estimate the flow through an alluvial channel, for example.

FLOW NETS

A set of intersecting equipotential lines and flow lines, constructed according to a strict set of rules, is called a flow net. It can be a powerful analytical tool for the analysis of ground water flow (Freeze and Cherry 1979). A discussion of the rules governing the construction of flow nets is beyond the scope of this section, and the reader is referred to chapter 5 of Freeze and Cherry (1979) for a detailed description of flow net construction. Once a flow net is properly constructed, the amount of ground water flow through the area represented by the flow net, under steady-state conditions, can be calculated if the hydraulic conductivity of the aquifer is known. Figure 60 shows an example flow net for a simple system (modified from Freeze and Cherry 1979), in which ground water is flowing from the left side of the figure to the right side.

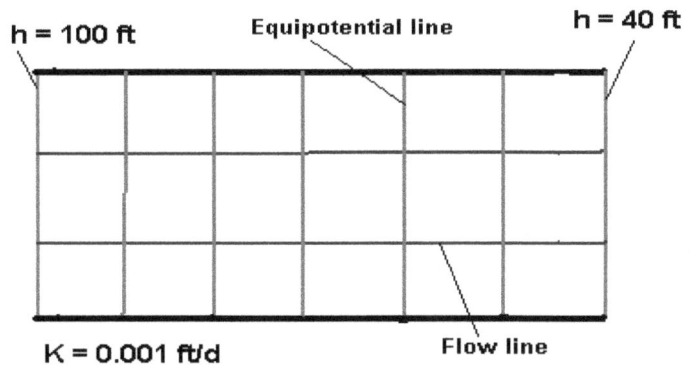

Figure 60. Example of a flow net for a simple flow system. m = 3, n = 6, H = 60 feet, K = 10^{-3} feet/day, so that Q = 3.0 x 10^{-2} ft³/d per square meter of section perpendicular to the flow net.

Darcy's Law allows the amount of ground water flow through the area represented in figure 60 to be calculated using a flow net and the following equation:

$$Q = (mKH)/n,$$

where Q is the ground water flow rate, K is the hydraulic conductivity, H is the total change in hydraulic head across the flow net, m is the total number of flow tubes (the area between the flow lines), and n is the number of divisions of head in the flow net.

A standard flow net assumes that the aquifer is isotropic. When an aquifer is anisotropic (commonly the case in unconsolidated and sedimentary aquifers), the actual direction of ground water flow will not be perpendicular to the equipotential contours. Instead, the direction of flow will deviate from the perpendicular at an angle that depends on the ratio of the horizontal to the vertical hydraulic conductivity. Figure 61 illustrates how anisotropy in a fractured rock aquifer alters the direction of ground water flow compared to that expected in an isotropic aquifer.

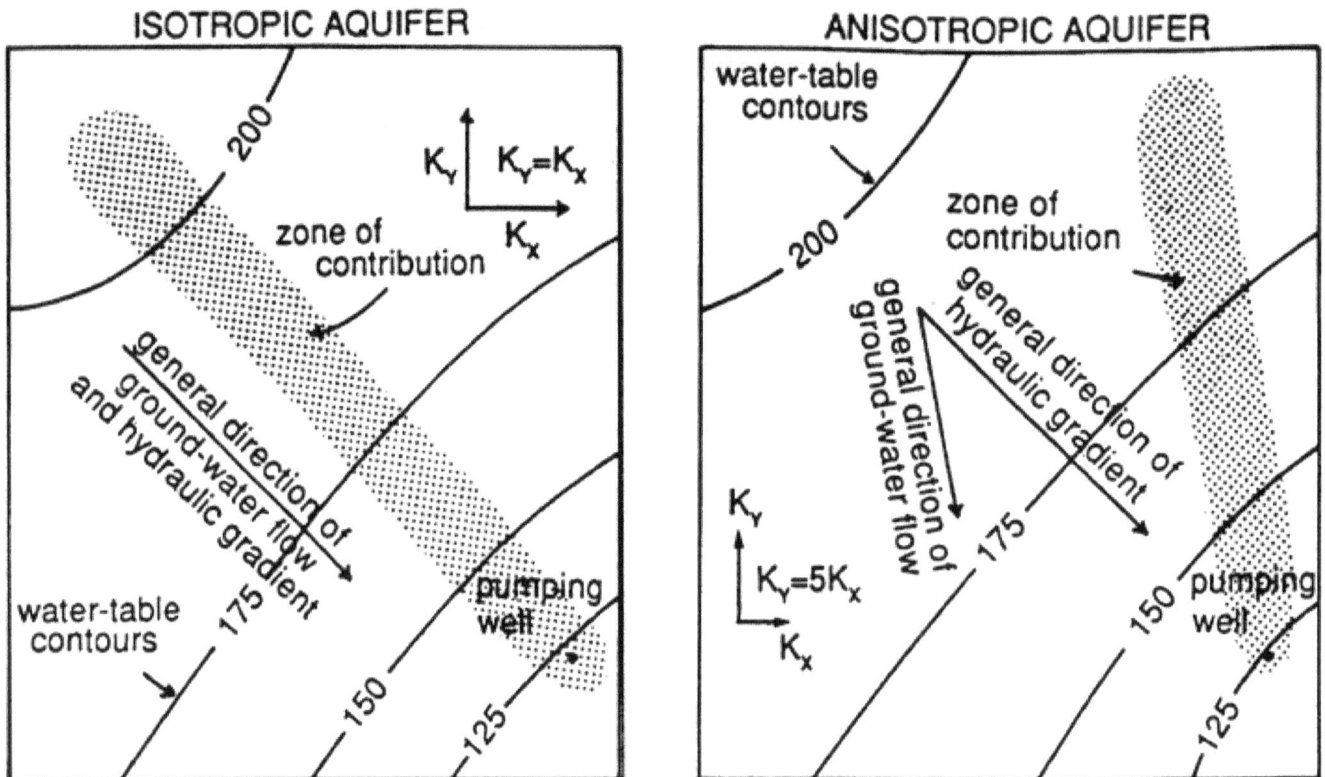

Figure 61. Effect of fracture anisotropy on the orientation of the zone of contribution to a pumping well (Bradbury and others 1991).

A potentiometric surface map can be developed into a flow net by constructing flow lines that intersect the equipotential lines or contour lines at right angles. Flow lines are imaginary paths that trace the flow of water particles through the aquifer. Although the number of both equipotential and flow lines is infinite, the former are constructed with uniform differences in elevation between them, while the latter are constructed so that they form, in combination with equipotential lines, a series of squares. A flow net carefully prepared in conjunction with Darcy's Law allows estimation of the quantity of water flowing through an area, and of the variability of transmissivity and hydraulic conductivity. Plan and cross-section views of flow nets drawn for a losing stream are shown in figure 62 and a gaining stream in figure 63. Plan view flow nets are valuable for delineating the zone of contribution to a well, or for boundary conditions for pumping wells.

192

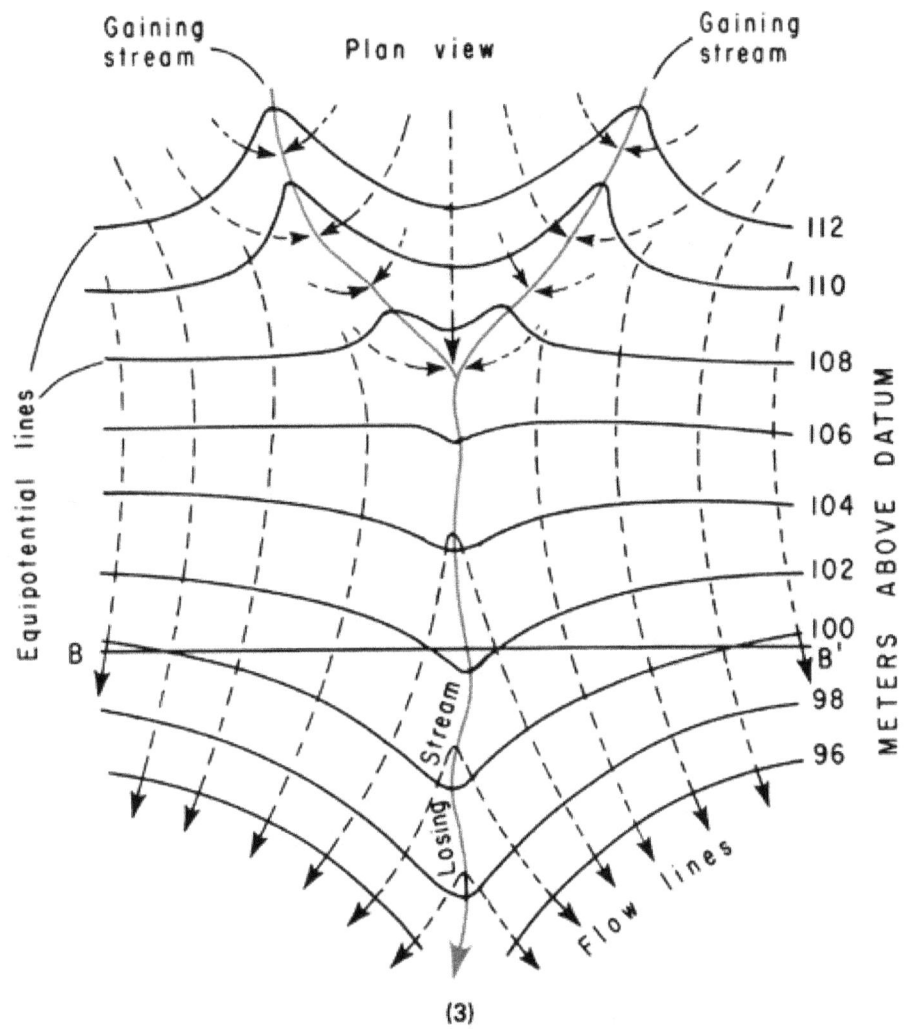

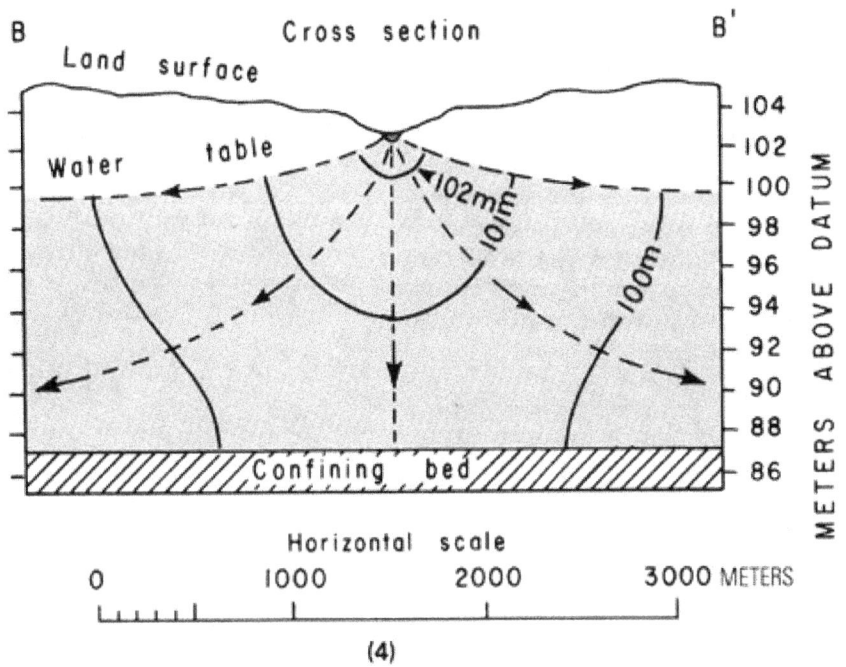

Figure 62. Plan view and cross section of flow net through losing stream segment (Heath 1983).

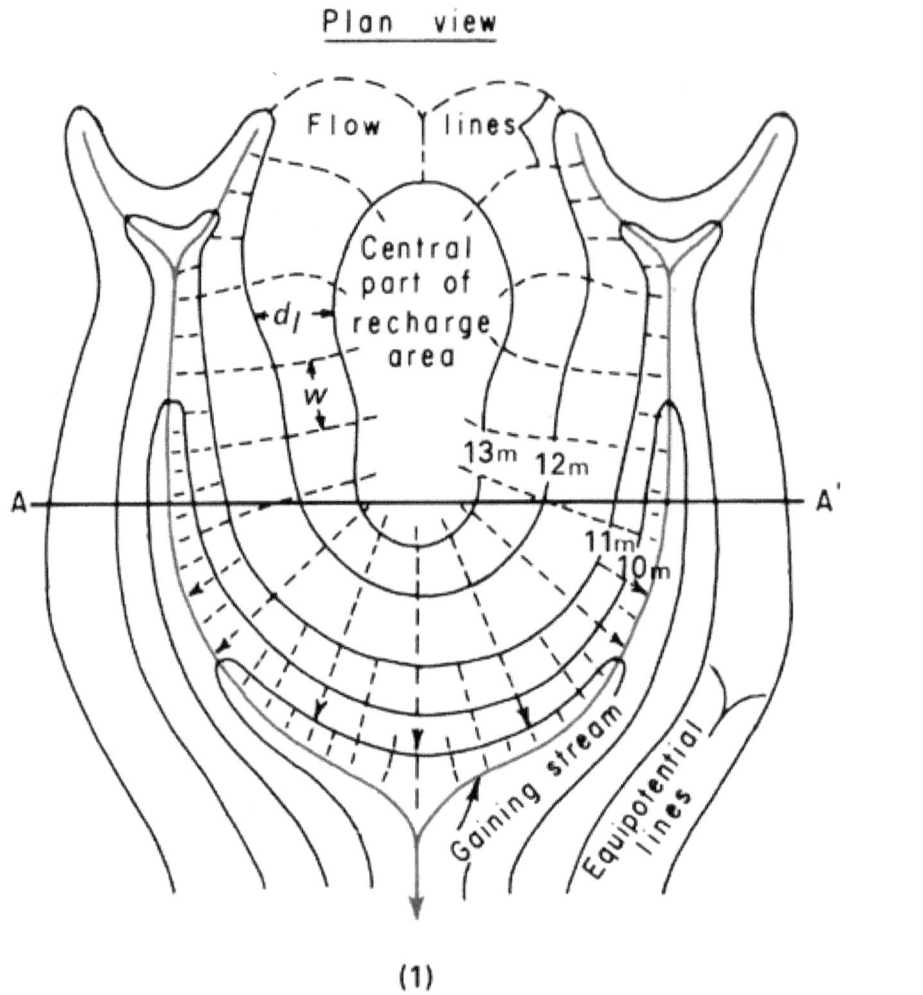

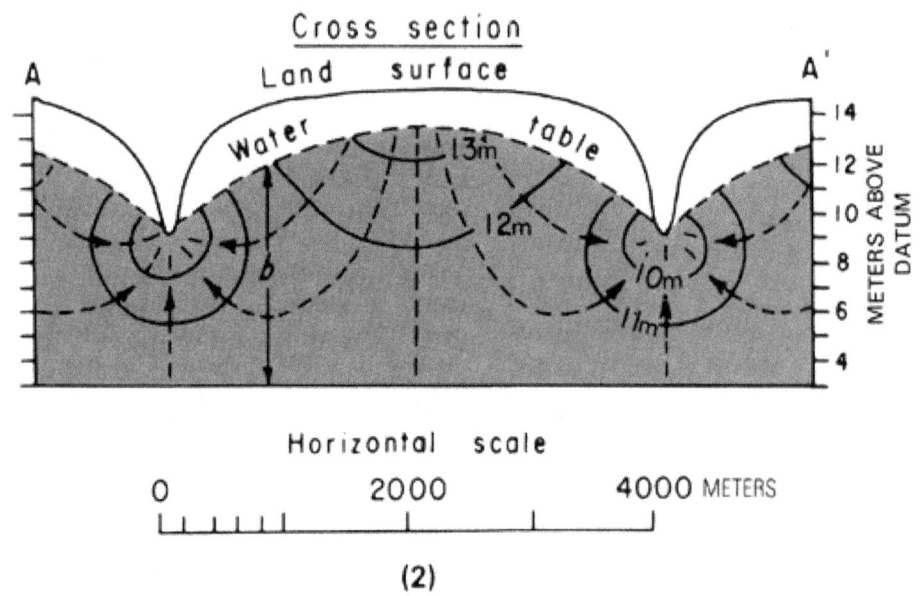

Figure 63. Plan view and cross section of flow net for gaining stream (Heath 1983).

ANALYSIS OF AQUIFER TEST DATA

Many different methods have been developed to analyze aquifer-test data, for both single-well and multiple-well tests. The correct analysis method to be used depends on the hydrogeological conditions at the test site, the type of data collected for the test, and how well the hydrogeological conditions match the assumptions inherent to each approach. For the test to be successful, it must be planned and conducted in a manner consistent with the site hydrogeology and the analysis method(s) to be used.

One of the simplest, and often the most cost-effective, aquifer test procedures is the specific-capacity test. This test, which is often conducted after well development by a driller, calculates the well yield per unit of drawdown in the well after a specified time (commonly 24 hours). The well is pumped at a constant, predetermined rate for the specified time, and the drawdown in the well is measured at the end of that time. The discharge divided by the drawdown is the specific capacity, usually reported in units of gallons per minute per foot of drawdown. The specific capacity value can change with the length of time that the well is pumped; for example, a short-duration test (1 hour or less) can result in a large value for specific capacity because of well-bore storage effects. For longer tests in unconsolidated aquifers, the specific capacity can decrease with time because of dewatering of the aquifer. Aquifer transmissivity (in gallons per day per foot) can be approximated from a specific-capacity value using the following equations (Driscoll 1986):

$$T = \text{Specific capacity x 2000, for a confined aquifer.}$$
$$T = \text{Specific capacity x 1500, for an unconfined aquifer.}$$

A more rigorous method for estimating transmissivity and hydraulic conductivity from specific capacity tests is described by Theis and others (1963). Bradbury and Rothschild (1985) describe a computer program to estimate hydraulic conductivity from specific-capacity tests.

Most aquifer-test data are analyzed using graphical procedures (many of which are now performed with the use of computer programs). One procedure involves analysis of the shape of a time-drawdown or distance-drawdown graph. Another involves curve-matching methods (Fig. 64). Detailed descriptions of each of these methods is beyond the scope of this section, and the reader is referred to one of the many textbooks or reports that can provide that level of detail, including Dawson and Istok (1991), Lohman (1972), Kruseman and deRidder (1991), or Driscoll (1986). Ground water flow models, which are discussed in a subsequent section, are also used to analyze aquifer-test data. Interpretation of aquifer-test data is often nonunique, however; for example, the time-drawdown responses are similar for leaky confined, unconfined, and bounded aquifer systems. Because a theoretical response curve can be matched to aquifer-test data does not prove that the aquifer fits the assumptions on which the curve is based (Freeze and Cherry 1979). Therefore, the experience and judgment of the analyst is critical to the proper interpretation of aquifer test data.

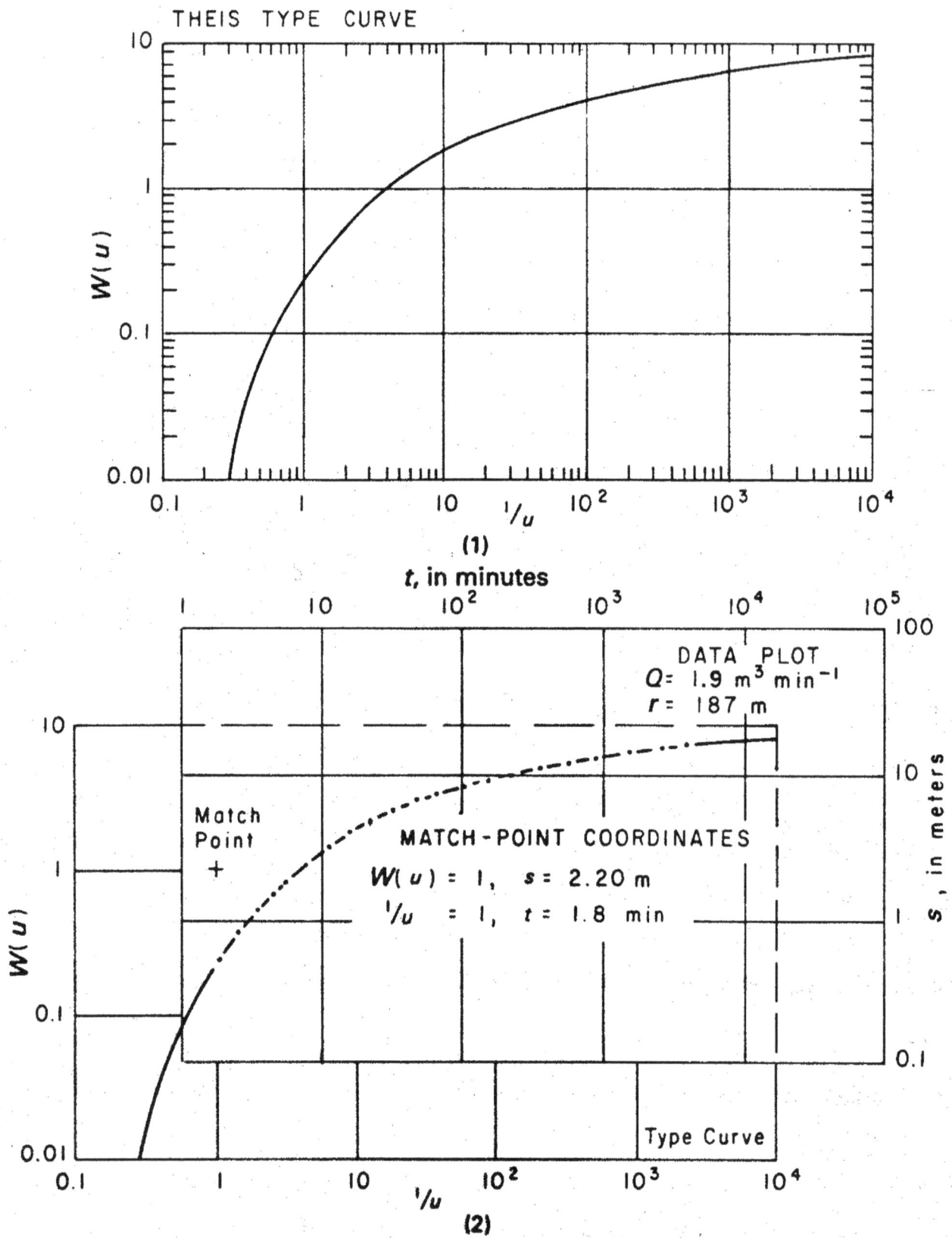

Figure 64. Example of a Theis type curve and a curve-matching plot for analysis of aquifer-test data (Heath 1983).

196

The use of aquifer tests to obtain hydraulic data in fractured-rock aquifers requires careful thought about the purpose and design of the test, the type of data to be collected, and the analyses of the data. Conventional slug tests and constant discharge/variable discharge pumping tests were designed for porous media flow and are difficult to apply to fractured rocks unless the fractures are highly connected. Identification and testing of water-bearing fractures are critical for the success of aquifer tests in fractured rocks. For single fractures or fracture zones, in situ measurements of average hydraulic conductivity can be made with a standard Lugen packer test (Singhal and Gupta 1999). Directional hydraulic conductivity can be measured with a modified Lugen packer test and/or a tracer injection test, and three-dimensional values of hydraulic conductivity can be measured with cross-hole hydraulic tests. For fractured-rock aquifers that have significant matrix porosity (with low matrix hydraulic conductivity) and regularly spaced fractures (high hydraulic conductivity), pumping test data can be used to estimate hydraulic characteristics of the fractures and the matrix blocks. Dual porosity models assume that porous media flow occurs within the matrix block and within the fractures.

Numerical Models

All of the analysis methods described so far contain assumptions or limitations that make them unsuitable for large-scale problems in complex hydrogeological settings. Numerical methods implemented through computer programs (computer models), however, can be well suited to these types of problems. See appendix V for more detailed information on numerical modeling. Numerical models can be much less burdened by the simplifying assumptions used in analytical models; therefore, they are inherently capable of addressing more complicated problems. They require significantly more input, however, and their solutions are inexact (numerical approximations); for example, in many models the assumptions of homogeneity and isotropy are unnecessary because the model can assign point (nodal) values of transmissivity and storage to hundreds or thousands of nodes. Likewise, the capacity to incorporate complex boundary conditions provides greater flexibility, and computer models can be used for both small-scale, site-specific problems and for large-scale (basin or multiple basins), complex problems. The user faces difficult choices, however, about model selection, boundary conditions, grid discretization, time steps, and ways to avoid truncation errors and numerical oscillations (Remson and others 1971, Javendel and others 1984). Improper choices may result in errors, such as mass imbalances, incorrect velocity distributions, and grid-orienting effects that are unlikely to occur with analytical approaches. Reilly and Harbaugh (2004) provide some guidelines and discussion of how to evaluate complex ground water flow models used in the investigation of ground water systems. Listed in table 13 are the relative advantages and disadvantages of analytical and numerical models.

A fundamental requirement of most numerical approaches is the creation of a discretized grid or mesh that represents the flow system being simulated. This discretization usually consists of rectangular- or triangular-shaped cells covering the lateral dimensions of the area of interest for which ground water parameters must be specified. The grid also extends vertically to represent one or more

Table 13. Relative advantages and disadvantages of analytical and numerical models.

Analytical Models	
Advantages	**Disadvantages**
Efficient when data on the system are sparse or uncertain	Limited to certain idealized conditions with simple geometry
Economical	May not be applicable to field problems with complex boundary conditions
Good for initial estimation of magnitude of contamination or drawdown	May not be able to readily handle spatial or temporal variations in system
Rough estimates of input data often possible from existing data sources	
Input data for computer codes usually simple	

Numerical Models	
Advantages	**Disadvantages**
Easily handle spatial and temporal variations of hydrogeologic system	Achieving familiarity with complex numerical programs can be time consuming and expensive
Easily handle complex boundary conditions	Errors because of numerical dispersion (artifacts of the computational process) may be substantial for transport models
Three-dimensional transient problems can be treated without much difficulty	May not be able to readily handle spatial or temporal variations in system
Rough estimates of input data often possible from existing data sources	More data input required, and can be time consuming
Input data for computer codes usually simple	

aquifers and/or confining units. The grid or mesh forms the basis for a matrix of equations to be solved. A new grid or mesh must be designed for each area to be modeled, based on the data collected during site characterization and on the conceptual model developed for the physical system. The size of the grid cells (or mesh elements) can vary from project to project, with smaller spacing (cell or element size) usually used in an area of the model where more detail is required (such as near well fields or sources of contaminants); however, this fine grid (or mesh) resolution also increases the requirements for data and the computational time necessary to reach a solution. Grid (or mesh) design is one of the most critical elements in the accuracy of computational results (EPA 1991).

Finite-difference and finite-element methods are the most frequently used numerical solution techniques. The finite-difference method approximates the solution of partial differential equations by using finite-difference equivalents. The finite-element method approximates differential equations by an integral approach. Perhaps the most frequently used finite-difference ground water model is MODFLOW. This model was originally developed by the USGS (McDonald and Harbaugh 1988). The computer code has been modified though the years, adding modules and refining the code for various situations. A recent version, MODFLOW-2000, includes options for parameter estimation and statistical evaluation of model results (Harbaugh and others 2000). Hill and others 2000) (see appendix V). Graphical interfaces have been written by several companies to ease the process of data input and to visualize model input and output.

Conceptualization, model design, and data input can take several hundred hours, but graphical interfaces and use of GIS techniques can substantially reduce that time. The time required to run the model is usually minimal, except for very large flow models with several hundred thousand nodes or contaminant transport models. The model is calibrated by adjusting model-input data until an acceptable match between simulated heads (concentrations for transport) and water-budget components and measured and estimated values are obtained. This process can take many months of effort. Model calibration can often result in a revised conceptualization of the ground water system and an identification of gaps in knowledge of the system and additional data needed to fill those gaps.

Additional numerical modeling tools may be necessary for particular investigations. If the purpose of the study is to predict the fate of a contaminant from a spill of hazardous chemicals, a solute-transport code may be required. Perhaps the most frequently used finite-difference solute-transport model is MT3D, which links easily with MODFLOW (Zheng 1990, Zheng and Wang 1998). See appendix III for a discussion of contaminant fate and transport mechanisms in ground water. Geochemical models, in which ground water quality is altered by water-rock interactions in an aquifer, are also valuable in conceptualizing and evaluating flow systems. Examples of geochemical models include MINTEQA2 (Allison and others 1991), PHREEQC (Parkhurst and Appelo 1999) and NETPATH (Plummer and others 1994). Model codes developed by the USGS and documentation of these codes can be obtained at no charge at http://water.usgs.gov/software/.

It is important to distinguish between the software, or computer code, used in a model and the model itself. The software is simply the analytical equation(s) to be solved and the algorithms for reading input data and for outputting simulation results. MODFLOW is an example of a simulation code. The model is the set of input data, simulation software, and output from the software. The code is generic. A model, however, includes a set of boundary and initial conditions as well as a site-specific grid, parameter values, and hydrologic stresses.

According to Anderson and Woessner (1992), the following are two prevalent and opposing opinions about models:

1. "Models are worthless because they require too many data; therefore, they are too expensive to assemble and run. Furthermore, they can never be proved to be correct and suffer from lack of scientific certainty."
2. "Models are essential in performing complex analyses and in making informed predictions."

Models do require extensive field, and sometimes laboratory, information for input data and calibration, and model solutions may be nonunique so that results may be uncertain; however, good modeling practices and an adequate amount of good-quality data will increase confidence in modeling results (Hill 1998). A ground water model is often the best way to make an informed analysis or prediction about consequences of a proposed action on a ground water flow system. Anderson and Woessner (1992) also state, "Models provide a framework for synthesizing field information and for testing ideas about how the system works. They can alert the modeler to phenomena not previously considered. They may identify areas where more field information is required." Much of the following discussion is taken directly from Anderson and Woessner (1992), and the reader is referred to that text for more detailed information.

Modeling is an excellent way to help organize and synthesize field data, but it is important to recognize that modeling is only one component of a hydrogeologic assessment and not an end in itself. In fact, the process of assembling and understanding the field data required for model input may provide the modeler with the answer to the problem before ever running the model. Conversely, a model that is based on inadequate field data can produce erroneous results that may not be obvious in the colorful graphical output from modern modeling software. The modeler must have some basic understanding of the geology and hydrology of the area being modeled, or should work in close collaboration with others who do have that understanding. In this way, model results that are hydrogeologically unreasonable, or that are based on unrealistic or erroneous data, can be recognized and addressed.

The adaptation of numerical ground water flow models to fractured-rock hydrogeological settings has progressed somewhat, but is still constrained in settings that exhibit significant anisotropy and heterogeneity (Forster and Smith 1988a). Ground water flow in these settings is often simulated as flow through porous media using MODFLOW or similar programs. This simplification is often adequate for large-scale flow systems, but may not be appropriate for small-scale (well-field) systems or contaminant transport problems. Watershed-scale models that distribute and attempt to balance elements of the water budget can be used to evaluate ground water and surface water development. Fracture network models, which utilize outcrop data on fracture geometry (for example, FRACMAN, Golder Associates), can be used to evaluate flow in discrete fracture networks (discrete volumes of rock), but they are constrained by the difficulties of obtaining sufficient data and by a poor correlation with depth.

Synthesis and Interpretation

A conceptual framework for a hydrological system is the final result of a hydrogeological study and pulls together all information gathered on the geological setting, the surface water and ground water system, and dependent ecosystems to provide a coherent, unified picture of the system and the important processes active within that system. Stone (1999) provides an excellent discussion on developing a conceptual framework (also known as a "conceptual model," though a much more refined version than the conceptual model step in the development of a numerical ground water model). Such a framework is the starting place for additional studies of water supply, waste disposal, inventory, and remediation. A conceptual framework varies with the scale of the study area. The study area can be conceptualized on a regional, aquifer, or project scale, or alternatively, in terms of flow systems. An ideal conceptual framework will include four components: geology, surface water, and saturated and unsaturated ground water. Formulating a conceptual framework involves describing the geological setting from a hydrological point of view and the interactions of surface, soil, and ground water within this setting.

An important subcomponent of a hydrological conceptual framework is the hydrochemistry. The framework should address the relationship between the hydrogeologic setting and its hydrochemistry including the concentrations of chemical constituents, contamination, geochemical transformations taking place along a flow path, trends in water quality, and comparisons to water-quality standards. Appendix VI contains a discussion of analysis and statistical methods for evaluating water-quality data.

Ongoing Data Analysis Costs

Costs of ongoing data analysis for ground water studies include those associated with periodic evaluations and report writing on the status or changes in the hydrogeologic system. Water-quality studies have long-term costs associated with data interpretation during the study to detect trends and provide a means to modify the study if the data indicate a need to change the strategy. The data retrieved from a ground water monitoring program also must be managed. A database must be developed, data collection and input forms must be prepared for field personnel, and data must be entered into the database and evaluated on a frequent and routine basis.

References

Aller, L.; Bennett, T.; Lehr, J.H.; Petty, R.J. 1985. DRASTIC—A standardized system for evaluating ground water pollution potential using hydrogeologic settings. Report EPA/600/2-85/018. Washington, DC: U.S. Environmental Protection Agency. 163 p.

Aller, L.T.; Bennett, T.; Lehr, J.H.; Petty, R.J. 1987. DRASTIC—A standardized system for evaluating ground water pollution potential using hydrogeologic settings. Report EPA/600/2-87/035. Ada, OK: R.S. Kerr Environmental Research Laboratory, U.S. Environmental Protection Agency. 455 p.

Aller, L.; Bennett, T.W.; Hackett, G.; Petty, R.J.; Lehr, J.H. Handbook of Suggested Practices for the Design and Installation of Ground water Monitoring Wells (Revised). Report EPA/600/4-89/034-REV. Las Vegas, NV: Environmental Monitoring Systems Laboratory, U.S. Environmental Protection Agency, 231 p.

Alley, W.M.; Reilly, T.E .; Franke, O.L. 1999. Sustainability of ground water resources. Circular 1186. Washington, DC: U.S. Geological Survey. 79 p. http://water.usgs.gov/pubs/circ/circ1186/.

American Petroleum Institute. 1993. Environmental guidance document: well abandonment and inactive well practices for U.S. exploration operations. Bulletin E352. Washington DC: American Petroleum Institute.

Allison, J.D.; Brown, D.S.; Novo-Gradac, K.J. 1991. MINTEQA2/PRODEFA2, A geochemical assessment model for environmental systems: version 3.0 user's manual. Report EPA/600/3-91/021. Athens, GA: Environmental Research Laboratory, U.S. Environmental Protection Agency, 107 p. http://www.epa.gov/ceampubl/mmedia/minteq/index.htm/.

American Society for Testing and Materials (ASTM). 1990. Standard practice for decontamination of field equipment used at nonradioactive waste sites. Standard D5088-02. West Conshohocken, PA: American Society for Testing and Materials.

American Society for Testing and Materials (ASTM). 1996. Standard guide for design of ground water monitoring systems in karst and fractured-rock aquifers. Standard D5717-95e1. West Conshohocken, PA: American Society for Testing and Materials.

Anderson, M.P.; Woessner, W.W. 1992. Applied groundwater modeling, simulation of flow and advective transport. San Diego, CA: Academic Press, Inc. 381 p.

Anderson, H.W.; Hoover, M.D.; Reinhart, K.G. 1976. Forests and water: effects of forest management on floods, sedimentation, and water supply. General Technical Report PSW-18. Berkeley, CA: U.S. Department of Agriculture, Forest Service, Pacific Southwest Forest and Range Experiment Station. 115 p.

Anderson, T.W.; Freethey, G.W.; Tucci, P. 1992. Geohydrology and water resources of alluvial basins in south-central Arizona and parts of adjacent States. Professional Paper 1406-B. Washington, DC: U.S. Geological Survey. 67 p.

Baird, A.J.; Wilby, R.L. (eds.). 1999. Eco-hydrology: plants and water in terrestrial and aquatic environments. London: Routledge Press. 402 p. Barcelona, M.J. 1984. TOC determinations in ground water. Ground Water. 22(1): 18–24.

Barcelona, M.J.; Gibb, J.P.; Miller, R.A. 1983. A guide to the selection of materials for monitoring well construction and ground water sampling. Illinois State Water Survey Contract Report, Report EPA 600/S2-84/024. Washington, DC: U.S. Environmental Protection Agency. 78 p.

Barcelona, M.J.; Gibb, J.P.; Helfrich, J.A.; Garske, E.E. 1985. Practical guide for ground water sampling. Illinois State Water Survey Contract Report 374. Las Vegas, NV: U.S. Environmental Protection Agency, Environmental Monitoring and Support Laboratory. 94 p.

Bossong, C.R.; Caine, J.S.; Stannard, D.I., Flynn, J.L.; Stevens, M.R.; Heiny-Dash, J.S. 2003. Hydrologic conditions and assessment of water resources in the Turkey Creek watershed, Jefferson County, Colorado, 1998–2001. Water-Resources Investigation Report 03-4034. Washington, DC: U.S. Geological Survey. 140 p. http://water.usgs.gov/pubs/wri/wri03-4034.

Bedinger, M.S.; Reed, J.E. 1988. Practical guide to aquifer test analysis. Las Vegas, NV: U.S. Environmental Protection Agency, Environmental Monitoring Systems Laboratory. 81 p.

Berger, A.R.; Iams, W.J. 1996. Geoindicators: assessing rapid environmental changes in earth systems. Rotterdam: A.A. Balkema. 466 pp.

Bradbury, K.R.; Rothschild, E.R. 1985. A computerized technique for estimating the hydraulic conductivity of aquifers from specific capacity data. Ground Water. 23(2): 240–246.

Bradbury, K.R.; Muldoon, M.A.; A Zaporozec, A.; Levy, J. 1991. Delineation of wellhead protection areas in fractured rocks. Report EPA 570/9-91/009. Office of Groundwater and Drinking Water. Washington, DC: U.S. Environmental Protection Agency. 144 p.

Brahana, J.V.; Mesko, T.O. 1988. Hydrogeology and preliminary assessment of regional flow in the upper Cretaceous and adjacent aquifers in the northern Mississippi Embayment. Water-Resources Investigations Report 87-4000. Washington, DC: U.S. Geological Survey. 65 p.

Bredehoeft, J.D.; Papadopulos, S.S.; Cooper, H.H. Jr. 1982. Groundwater—the water-budget myth. In: Scientific basis of water-resource management. Washington, DC: National Academy Press: 51–57.

Brobst, R.B. 1984. Effects of two selected drilling fluids on ground water sample chemistry. In: Monitoring wells: their place in the water well industry. National Well Water Association Meeting and Exposition. Educational Session. Las Vegas, NV: National Well Water Association.

Caine, J.S.; Evans, J.P.; Forster, C.B. 1996. Fault zone architecture and permeability structure. Geology. 24: 1025–1028.

Campbell, C.J. 1970. The ecological implications of riparian vegetation management. Journal of Soil and Water Conservation. 25: 49–52.

Carr, M.R.; Winter, T.C. 1980. An annotated bibliography of devices developed for direct measurement of seepage. Open-File Report 80-344. Washington, DC: U.S. Geological Survey.

Carter, V. 1996. Wetland hydrology, water quality, and associated functions. In: National water summary—wetland resources. Water-Supply Paper 2425. Washington, DC: U.S. Geological Survey. 431 p.

Clark, I.D.; Fritz, P. 1997. Environmental isotopes in hydrogeology. Boca Raton, FL: CRC Press. 328 p.

Cruden, D.M.; Varnes, D.J. 1996. Landslide types and processes. In: Turner, A.K.; Schuster, R.L. (eds.) Landslides—investigation and mitigation. Transportation Research Board Special Report 247. Washington, DC: National Academy Press: 36–75.

Culler, R.C.; Hanson, R.C.; Murick, R.M.; Turner, R.M.; Kipple, F.P. 1982. Evapotranspiration before and after clearing phreatophytes: Gila River floodplain, Graham County, Arizona. Professional Paper 655-P. Washington, DC: U.S. Geological Survey.

Danielopol, D.L. 1989. Ground water fauna associated with riverine aquifers. Journal of the North American Benthological Society. 8: 18–35.

Darcy, H. 1856. Les fontaines publiques de la ville de Dijon. Paris: Victor Dalmont. 674 p.

Davis, S.N.; DeWiest, R.J.M. 1966. Hydrogeology. New York: John Wiley & Sons. 463 p.

Dawson, K.J.; Istok, J.D. 1991. Aquifer testing: design and analysis of pumping and slug tests. Boca Raton, FL: Lewis Publishers, Inc. 368 p.

DeBruin, R.H.; Lyman, R.M.; Jones, R.W.; Cook, L. 2001. Coalbed methane in Wyoming. Wyoming State Geological Survey Information Pamphlet 7. Laramie, WY: Wyoming State Geological Survey.

De Vries, J.J.; Simmers, I. 2002. Groundwater recharge: An overview of process and challenges. Hydrogeology Journal. 10: 5–17.

Dissmeyer, G.E. (ed.). 2000. Drinking water from forests and grasslands: a synthesis of the scientific literature. Gen. Tech. Rep. SRS-39. Asheville, NC: U.S. Department of Agriculture, Forest Service, Southern Research Station. 246 p. http://www.srs.fs.usda.gov/pubs/gtr/gtr_srs039/index.htm.

Dosskey, M.G. 2001. Toward quantifying water pollution abatement in response to installing buffers on crop land. Environmental Management. 8: 577–598.

Drever, J.I. 1997. The geochemistry of natural waters. Upper Saddle River, NJ: Prentice Hall. 436 p.

Driscoll, F.G. 1986. Groundwater and wells. St. Paul, MN: Johnson Division. 1089 p.

Dudley, W.W.; Larson, J.D. 1976. Effect of irrigation pumping on desert pupfish habitats in Ash Meadows, Nye County, Nevada. Professional Paper 927. Washington, DC: U.S. Geological Survey. 52 p.

Dunne, T. 1990. Hydrology, mechanics, and geomorphic implications of erosion by subsurface flow. In: Higgins, C.G.; Coates, D.R. (eds.). Groundwater geomorphology: the role of subsurface water in earth-surface processes and landforms. Special Paper 252. Boulder, CO: Geological Society of America: 1–28.

Durnford, D.S.; Thompson, K.R.; Ellerbrook, D.A.; Loftis, J.C.; Davies, G.S. 1990. Screening methods for ground water pollution potential from pesticide use in Colorado agriculture. Completion Report 157. Fort Collins, CO: Colorado Water Resources Research Institute. 165 p.

Edmunds, W.M. 1996. Indicators in the groundwater environment of rapid environmental change. In: Berger, A.R.; Iams, W.J. (eds.). Geoindicators: assessing rapid environmental changes in earth systems. Rotterdam: A.A. Balkema: 135–153.

Ellis, A.J. 1938. The divining rod—a history of water witching. Water Supply Paper 416. Washington, DC: U.S. Geological Survey. 59 p.

Ellyett, C.D.; Pratt, D.A. 1975. A review of potential applications of remote sensing techniques to hydrogeological studies in Australia. Technical Paper 13. Canberra: Australian Water Resources Council.

Erman, N.A.; Erman, D.C. 1995. Spring permanence, Trichoptera species richness, and the role of drought. Journal of the Kansas Entomological Society. 68: 50–64.

Evans, R.B.; Schweitzer, G.E. 1984. Assessing hazardous waste problems. Environmental Science and Technology. 18(11): 330A–339A.

Farrell, S.O. 1985. Evaluation of color infrared aerial surveys of wastewater soil absorption systems. Report EPA 600/2-85/039. Washington, DC: U.S. Environmental Protection Agency.

Fenneman, N.M. 1938. Physiography of Western United States. New York: McGraw Hill. 534 p.

Fetter, C.W. 1999. Contaminant hydrogeology. Upper Saddle River, NJ: Prentice Hall. 500 p.

Fetter, C.W. 2000. Applied hydrogeology. 4th Edition. Upper Saddle River, NJ: Prentice Hall. 598 p.

Field, M.S.; Wilhelm, R.G.; Quinlan, J.F.; Aley, T.J. 1995. An assessment of the potential adverse properties of fluorescent tracer dyes used for groundwater tracing. Environmental Monitoring and Assessment. 38: 75–96.

Forster, C.; Smith, L. 1988a. Groundwater flow systems in mountainous terrain: 1. numerical model technique. Water Resources Research. 42(7): 999–1010.

Forster, C.; Smith, L. 1988b. Groundwater flow systems in mountainous terrain: 2. controlling factors. Water Resources Research. 42(7): 1011–1023.

Franke, O.L. 1997. Conceptual frameworks for ground water quality monitoring. Denver, CO: Intergovernmental Task Force on Monitoring Water Quality. http://water.usgs.gov/wicp/gwfocus.pdf.

Franke, O.L.; Reilly, T.E.; Pollock, D.W.; LaBaugh, J.W. 1998. Estimating areas contributing recharge to wells: lessons learned from previous studies. Circular 1174. Washington, DC: U.S. Geological Survey. 14 p. http://water.usgs.gov/ogw/pubs/Circ1174/.

Freeze, R.A.; Cherry, J.A. 1979. Groundwater. Upper Saddle River, NJ: Prentice-Hall. 604 p.

Galloway, D.L.; Jones, D.R.; Ingebritsen, S.E. 1999. Land subsidence in the United States. Circular 1182. Washington, DC: U.S. Geological Survey. 175 p. http://water.usgs.gov/pubs/circ/circ1182/.

Galloway, D.L.; Alley, W.M; Barlow, P.M; Reilly, T.E.; Tucci, P. 2003. Evolving issues and practices in managing ground water resources: case studies on the role of science. Circular 1247. Washington, DC: U.S. Geological Survey. 73 p. http://water.usgs.gov/pubs/circ/2003/circ1247/.

Garber, M.S.; Koopman, F.C. 1968. Methods of measuring water levels in deep wells. In: U.S. Geological Survey techniques of water-resources investigations. Washington, DC: U.S. Geological Survey: Book 8, Chapter A1. http://water.usgs.gov/pubs/twri/twri8a1/.

Gerhart, J.M. 1984. A model of regional ground water flow in secondary-permeability terrain. Ground Water. 22: 168–175.

Gilbert, J. 1996. Do ground water ecosystems really matter? In: Barber, C.; Davis, G. (eds.). Proceedings, ground water and land-use planning conference. Perth: Centre for Ground water Studies, CSIRO Division of Water Resources.

Gilbert, J.; Danielopol, D.L.; Stanford. J.A., eds. 1998. Ground water ecology. San Diego, CA: Academic Press, Inc. 571 p.

Gillham, R.W.; Robin, M.J.L.; Barker, J.F.; Cherry, J.A. 1983. Ground water monitoring and sample bias. Publication 4367. Washington, DC: American Petroleum Institute.

Gorelick, S.M.; Evans, B.; Remsan, I. 1983. Identifying sources of ground water pollution: an optimization approach. Water Resources Research. 19(3): 779–780.

Greenlee, J.T. 1998. Ecologically significant wetlands in the Flathead, Stillwater, and Swan River Valleys. Helena, MT: Montana Department of Environmental Quality. Montana Natural Heritage Program Report. 192 p.

Griffiths, R.P.; Entry, J.A.; Ingham, E.R.; Emmingham, W.H. 1997. Chemistry and microbial activity of forest and pasture riparian-zone soils along three Pacific Northwest streams. Plant and Soil. 190: 169–178.

Gurrieri, J.T.; Furniss, G. 2004. Estimation of groundwater exchange in alpine lakes using non-steady mass-balance methods. Journal of Hydrology. 297: 187–208.

Haijtema, H.M. 1985. Modeling three-dimensional flow in confined aquifers by superposition of both two- and three-dimensional analytic functions. Water Resources Research. 21(10): 1557–1566.

Halford, J.; Meyer, G.C. 2000. Problems associated with estimating groundwater discharge and recharge from stream-discharge records. Ground Water. 38(3): 331–342.

Hamerlinck, J.D.; Ameson, C.S. 1998. Wyoming ground water vulnerability assessment technical guide. Spatial Data Visualization Center Report SDVC 98-01. Laramie, WY: University of Wyoming. 2 vol. http://www.sdvc.uwyo. edu/groundwater/report.html.

Harbaugh, A.W.; Banta, E.R.; Hill, M.C.; McDonald, M.C. 2000. MODFLOW-2000, the U.S. Geological Survey modular ground water model: user guide to modularization concepts and the ground water flow process. Open-File Report 00-92. Washington, DC: U.S. Geological Survey. 121 p. http://water.usgs.gov/ nrp/gwsoftware/modflow2000/ofr00-92.pdf.

Harvey, J.W.; Bencala, K.E. 1993. The effect of streambed topography on surface-subsurface water exchange in mountain catchments. Water Resources Research. 29: 89–98.

Hatton, T.; Evans, R. 1998. Dependence of ecosystems on ground water and its significance to Australia. Occasional Paper No. 12/98. Australia: Land and Water Resources Research and Development Corp., CSIRO. 77 p.

Hayashi, M.; Rosenberry, D.O. 2002. Effects of ground water exchange on the hydrology and ecology of surface water. Ground Water. 40: 309–316.

Heath, R.C. 1983. Basic ground water hydrology. Water-supply Paper 2220. Washington, DC: U.S. Geological Survey. 84 p. http://water.usgs.gov/pubs/wsp/ wsp2220.

Heath, R.C. 1984. Ground water regions of the United States. Water-supply Paper 2242. Washington, DC: U.S. Geological Survey. 78 p.

Heath, R.C. 1988. Hydrogeologic settings of regions. In: Back, W.; Rosenhein, P.R.; Seaber, P.R. (eds.). The decade of North American geology. Boulder, CO: Geological Society of America. Vol. 2. Plate 2.

Helsel, D.R.; Hirsch, R.M. 1992. Statistical methods in water resources. New York: Elsevier Publishers. 529 p.

Hem, J.D. 1989. Study and interpretation of the chemical characteristics of natural water. 3rd ed. Water Supply Paper 2254. Washington, DC: U.S. Geological Survey. 263 p. http://pubs.usgs.gov/wsp/wsp2254/.

Hendrickson, D.A.; Minckley, W.L. 1984. Cienegas: vanishing climax communities of the American Southwest. Desert Plants. 6: 131–175.

Hibbert, A.R. 1983. Water yield improvement potential by vegetation management on western rangelands. Water Resources Bulletin. 19: 375–381.

Higgins, C.G.; Coates, D.R. (eds.). 1990. Groundwater geomorphology: the role of subsurface water in earth-surface processes and landforms. Special Paper 252. Boulder, CO: Geological Society of America. 28 p.

Hill, M.C. 1998. Methods and guidelines for effective model calibration. Water-Resources Investigations Report 98-4005. Washington, DC: U.S. Geological Survey. 90 p. http://water.usgs.gov/nrp/gwsoftware/modflow2000/WRIR98-4005.pdf.

Hill, M.C.; Banta, E.R.; Harbaugh, A.W.; Anderman, E.R. 2000. MODFLOW-2000, the U.S. Geological Survey modular ground water model. User guide to the observation, sensitivity, and parameter-estimation processes and three post-processing programs. Open-File Report 00-184. Washington, DC: U.S. Geological Survey. 209 p. http://water.usgs.gov/nrp/gwsoftware/modflow2000/ofr00-184.pdf.

Hornbeck, J.W.; Leak, W.B. 1992. Ecology and management of northern hardwood forests in New England. General Technical Report NE-159. Radnor, PA: U.S. Department of Agriculture, Forest Service, Northeastern Forest Experiment Station. 44 p.

Horton, R.E. 1933. The role of infiltration in the hydrologic cycle. EOS Transactions of the American Geophysical Union. 14: 446–460.

Humphreys, W.F. 1999. Relict stygofaunas living in sea salt, karst, and calcrete habitats in arid northwestern Australia contain many ancient lineages. In: Ponder, W.; Lunney, D. (eds.). The other 99%. The conservation and biodiversity of invertebrates. Transactions of the Royal Zoological Society of New South Wales. 219–227.

Hunt, R.J.; Walker, J.F.; Krabbenhoft, D.P. 1999. Characterizing hydrology and the importance of ground water discharge in natural and constructed wetlands. Wetlands. 19(2): 458–472.

Hunt, R.J.; Haitjema, H.M.; Krohelski, J.T.; and Feinstein, D.T. 2003. Simulating ground water-lake interactions: approaches and insights. Invited contribution for special MODFLOW 2001 issue, Ground Water 41(2), p. 227-237.

Hutson, S.S.; Barber, N.L; Kenny, J.F. [and others]. 2004. Estimated use of water in the United States in 2000. Circular 1268, Washington, DC: U.S. Geological Survey. http://water.usgs.gov/pubs/circ/2004/circ1268/.

Hynes, H.B.N. 1970. The ecology of running waters. Toronto: University of Toronto Press. 555 p.

Jackson, T.J. 1986. Passive microwave remote sensing of soil moisture. Advances in Hydroscience. 14: 123–159.

Jackson, T.J.; Schmugge, T.J.; O'Neill, P. 1982. Passive microwave sensing of soil under vegetative canopies. Water Resources Research. 18(4): 1137–1142.

Javendel. I.; Doughty, C.; Tsang, C.F. 1984. Ground water transport: technical guide of mathematical models. Water Resources Monograph 10. Washington, DC: American Geophysical Union.

Jones, J.B.; Mulholland, P.J. 2000. Streams and ground waters. San Diego, CA: Academic Press.

Kabata-Pendias, A.; Pendias, H. 2000. Trace elements in soil and plants. 3rd ed. London: CRC Press. 432 p.

Kellogg, R.L.; Wallace, S.; Alt, K.; Goss, D.W. 1997. Potential priority watersheds for protection of water quality from nonpoint sources related to agriculture [Poster]. 52nd Annual Soil and Water Conservation Society Conference; 1997 July; Toronto, Canada. http://www.nrcs.usda.gov/technical/land/pubsl.

Kendall, C.; McDonnell, J.J. 1998. Isotope tracers in catchment hydrology. Amsterdam: Elsevier. 839 p.

Kenney, C. 1984. Properties and behaviours of soils relevant to slope instability. In: Brunsden, D.; Prior, D.B. (eds.). Slope instability. New York: John Wiley & Sons: 27–65.

Kenoyer, G.J.; Anderson, M.P. 1989. Groundwater's dynamic role in regulating acidity and chemistry in a precipitation-dominated lake. Journal of Hydrology. 109: 287–306.

Keppeler, E.; Brown, D. 1998. Subsurface drainage processes and management impacts. General Technical Report PSW-GTR-168. Berkeley, CA: U.S. Department of Agriculture, Forest Service, Pacific Southwest Research Station.

Keys, W.S. 1990. Borehole geophysics applied to ground water investigations. In: Techniques of Water-resources Investigations Book 2, Chapter E2. Washington, DC: U.S. Geological Survey. http://water.usgs.gov/pubs/twri/twri2-e2.

Kimball, B.A. 1997. Use of tracer injections and synoptic sampling to measure metal loading from acid mine drainage. Fact Sheet FS-245-96. Washington, DC: U.S Geological Survey. 8 p.

Krabbenhoft, D.P.; Bowser, C.J.; Anderson, M.P.; Valley, J.W. 1990a. Estimating groundwater exchange with lakes: 1. the stable isotope mass-balance method. Water Resources Research. 26: 2445–2453.

Krabbenhoft, D.P.; Bowser, C.J.; Anderson, M.P.; Valley, J.W. 1990b. Estimating groundwater exchange with lakes: 2. calibration of a three-dimensional, solute transport model to a stable isotope plume. Water Resources Research. 26: 2455–2462.

Kruseman, G.P.; deRidder, N.A. 1991. Analysis and evaluation of pumping test data. 2nd edition. Publication 47. Wageningen, the Netherlands: International Institute for Land Reclamation and Improvement. 377 p.

Lane, J.W.; Haeni, F.P.; Watson, W.M. 1995. Use of a square-array direct-current resistivity method to detect fractures in crystalline bedrock in New Hampshire. Ground Water. 33(3): 476–485.

Lapham, W.W.; Wilde, F.D.; Franceska, D.; Koterba, M.T. 1997. Guidelines and standard procedures for studies of ground water quality: selection and installation of wells and supporting documentation. Water-Resources Investigations Report 96-4233. Washington, DC: U.S. Geological Survey. 110 p. http://water.usgs.gov/owq/pubs/wri/wri964233/.

Lerman, A.; Imboden, D.; Gat, J. 1995. Physics and chemistry of lakes. Berlin: Springer-Verlag, 334 p.

Lohman, S.W. 1972. Ground water hydraulics. Professional Paper 708. Washington, DC: U.S. Geological Survey. 70 p.

Longley, G. 1992. The subterranean aquatic ecosystem of the Balcones Fault Zone Edwards Aquifer in Texas: threats from overpumping. In: Stanford, J.A.; Simons, R., Tampa, FL. April 26-29, 1992eds. Proceedings, 1st international conference on ground water ecology. Bethesda, MD: American Water Resources Association: 291–300.

Lyon, J.S.; Hilliard, T.J; Bethell, T.N. 1993. Burden of gilt. Washington, DC: Mineral Policy Center. 68 p.

Lynch, S.D.; Reynders, A.G.; Schulze, R.E. 1994. Preparing input data for a national-scale ground water vulnerability map of Southern Africa. Water in South Africa. 20(3): 239–246.

Maxwell, J.R.; Edwards, C.J.; Jensen, M.E. [and others]. 1995. A hierarchical framework of aquatic ecological units in North America (neartic zone). General Technical Report NC-176. St. Paul, MN: U.S. Department of Agriculture, Forest Service, North Central Forest Experiment Station.

McBride, M.S., and H.O. Pfannkuch. 1975. The distribution of seepage within lakebeds. Journal of Research of the U.S. Geological Survey, 3(5): 505-512.

McCuen, R.H. 1998. Hydrologic analysis and design. Upper Saddle River, NJ: Prentice Hall. 814 p.

McDonald, M.G.; Harbaugh, A.W. 1988. A modular three-dimensional finite-difference ground water flow model. Techniques of Water-resources Investigations, Book 6, Chapter A1. Washington, DC: U.S. Geological Survey. 586 p.

McNeill, J.D. 1986. Geonics EM39 borehole conductivity meter: theory and operation. Technical Note TN-20. Mississaugua, Ontario: Geonics Ltd. 11 p.

Meinzer, O.E. 1923. Outline of ground water hydrology. Water Supply Paper 494. Washington, DC: U.S. Geological Survey. 71 p.

Melby, J.T. 1989. A comparative study of hydraulic conductivity determinations for a fine-grained aquifer. M.S. thesis. Stillwater, OK: School of Geology, Oklahoma State University. 171 p.

Mifflin, M.D. 1988. Region 5, Great Basin. In: Geology of North America, Volume O-2, Hydrogeology. Boulder, CO: Geological Society of America: 69–78.

Miller, J. (comp.). 1998. Principal aquifers [1:5,000,000]. In: National atlas of the United States. Washington, DC: U.S. Geological Survey. 1 sheet.

Milligan, J. H.; Marshall, R.E.; Bagley, J.M. 1966. Thermal springs of Utah and their effect on manageable water resources. Report WG23-6. Logan, UT: Utah Water Resources Research Laboratory, Utah State University.

Mills, W.B.; Procella, D.B.; Ungs, M.J. [and others].1985. Water quality assessment: a screening procedure for toxic and conventional pollutants, part II. Report EPA 600/6-85/002b. Washington, DC: U.S. Environmental Protection Agency.

Montana Department of Environmental Quality and U.S. Department of Agriculture Forest Service. 2001. Final environmental impact statement, Rock Creek project.

Muckel, D.C. 1966. Phreatophytes: water use and potential savings. The Journal of the Irrigation and Drainage Division, American Society of Civil Engineers. 92: 27–34.

National Ground Water Association. 2002. Fractured-rock aquifers. In: Proceedings, National Ground Water Association. Westerville, OH: National Ground Water Association. 222 p.

National Mining Association. 1999. Facts about coal: 1999–2000. Washington, DC: National Mining Association.

National Research Council. 1979. Redistribution of accessory elements in mining and mineral processing, part 1: coal and oil shale. Washington, DC: National Academy of Science, National Academy Press. 180 p.

National Research Council. 1981. Effect of environment on nutrient requirements of domestic animals. Washington, DC: National Research Council Subcommittee in Environmental Stress, National Academy Press. 168 p.

National Research Council. 1991. Mitigating losses from land subsidence in the United States. Washington, DC: National Academy of Science, National Academy Press. 58 p.

National Research Council. 1993. Ground water vulnerability assessment: contamination potential under conditions of uncertainty. Washington, DC: National Academy of Science, National Academy Press. 210 p. http://books.nap.edu/books/0309047994/html.

National Research Council. 1996. Rock fractures and fluid flow: contemporary understanding and applications. Washington, DC: National Academy of Science, National Academy Press. 551 p.

National Research Council. 2004. Groundwater fluxes across interfaces. Washington, DC: National Academy of Science, National Academy Press. 85 p.

Neary, D.G.; Swift, L.W. Jr. 1987. Rainfall thresholds for triggering a debris avalanching event in the Southern Appalachian Mountains. In: Costa, J.E.; Wieczorek, G.F. (eds.). Debris flows/avalanches: process, recognition, and mitigation. Reviews in Engineering, Volume 7, Geology. Boulder, CO: Geological Society of America: 81–92.

Newton, J.G. 1986. Development of sinkholes resulting from man's activities in the Eastern United States. Circular 968. Washington, DC: U.S. Geological Survey. 54 p.

Nilsson, S.I.; Miller, H.G.; Miller, J.D. 1982. Forest growth as a possible cause of soil and water acidification: an examination of the concepts. Oikos. 39: 40–49.

Nimick, D.A.; Cleasby, T.E. 2001. Quantification of metal loads by tracer injection and synoptic sampling in Daisy Creek and the Stillwater River, Park County, Montana, August 1999. Water-Resources Investigations Report 00-4261. Washington, DC: U.S. Geological Survey. 51 p.

Nimick, D. A.; von Guerard, P. (eds.). 1998. Science for watershed decisions on abandoned mine lands. OF Report 98-297. Washington, DC: U.S. Geological Survey. 71 p.

O'Brien, C.; Blinn, D.W. 1999. The endemic spring snail Pyrgulopsis Montezumenis in a high CO_2 environment: importance of extreme chemical habitats as refugia. Freshwater Biology. 42: 225–234.

Office of Technology Assessment. 1984. Protecting the Nation's Groundwater from Contamination, U.S. Congress, OTA-O-276, Volumes I and II, Washington, DC.

Paillet, F.L. 1994. Application of borehole geophysics in the characterization of flow in fractured rocks. Water-Resources Investigations Report 93-4214. Washington, DC: U.S. Geological Survey. 36 p.

Paillet, F.L. 2000. A field technique for estimating aquifer parameters using flow log data. Ground Water. 38(4): 510–521.

Parizek, R.R. 1976. On the nature and significance of fracture traces and lineaments in carbonate and other terrains. In Yevjevich, V. (ed.). Karst hydrology and water resources. Fort Collins, CO: Water Resources Publications: 3-1 to 3-62. Vol. 1.

Parkhurst, D.L.; Appelo, C.A.J. 1999. User's guide to PHREEQC (version 2): a computer program for speciation, batch-reaction, one-dimensional transport, and inverse geochemical calculations. Water-Resources Investigations Report 99-4259. Washington, DC: U.S. Geological Survey. 312 p.

Paulsen, R.J.; Smith, C.F.; O'Rourke, D.; Wong, T.F. 2001. Development and evaluation of an ultrasonic ground water seepage meter. Ground Water. 39(6): 904–911.

Peterken, G.F., ed. 1957. Guide to the check sheet of IBP areas. Technical Guide 4. Oxford, UK: International Biological Programme. 133 p.

Pilliod, D.S.; Bury, R.B.; Hyde, E.J.; Pearl, C.A.; Corn, P.S. 2003. Fire and amphibians in North America. Forest Ecology and Management. 178: 163–181.

Plummer, L.N.; Friedman, L.C. 1999. Tracing and dating young ground water. Fact Sheet 134-99. Washington, DC: U.S. Geological Survey. 4 p.

Plummer, L.N.; Michel, R.L.; Thurman, E.M.; Glynn, P.D. 1993. Environmental tracers for age-dating young ground water. In: Alley, W.M. (ed.). Regional ground water quality. New York: Van Nostrand Reinhold: 255–294.

Plummer, L.N.; Prestemon, E.C.; Parkhurst, D.L. 1994. An interactive code (NETPATH) for modeling NET geochemical reactions along a flow PATH (version 2.0). Washington, DC: Water-Resources Investigations Report 94-4169. U.S. Geological Survey. 130 p.

Poland, J.F.; Lofgren, B.E.; Ireland, R.L.; Pugh, R.G. 1975. Land subsidence in the San Joaquin Valley, California, as of 1972. Professional Paper 437-H. Washington, DC: U.S. Geological Survey. 78 p.

Potter, K.W.; Bowser, C.; Amann, M.A.; and Bradbury, K.R. 1995. Estimating the spatial distribution of groundwater recharge rates using hydrologic, hydrogeologic, and geochemical methods. WRC GRR 95-07. Water Resources Center, University of Wisconsin-Madison. 25 pp. http://digicoll.library.wisc.edu/cgi-bin/EcoNatRes/EcoNatRes-idx?id=EcoNatRes.WRCGRR95-07.

Prellwitz, R.W.; Koler, T.E.; and Steward, J.E. 1994. Slope stability reference for the national forests in the United States. EM-7170-13. Moscow, ID: U.S. Department of Agriculture, Forest Service, Rocky Mountain Research Station.

Price, J.C. 1980. The potential of remotely sensed thermal infrared data to infer surface soils moisture and evaporation. Water Resources Research. 16(4): 787–795.

Regional Groundwater Center. 1995. Groundwater vulnerability study for Tyrone Township, Livingston County, Michigan. Flint, MI: University of Michigan, Flint, Regional Groundwater Center.

Reid, M.E.; LaHusen, R.G. 1998. Real-time monitoring of active landslides along Highway 50, El Dorado County. California Geology. 51(3): 17–20.

Reilly, T.E.; Harbaugh, A.W. 2004. Guidelines for evaluating ground water flow models. Scientific Investigations Report 5038. Washington, DC: U.S. Geological Survey. 30 p. http://water.usgs.gov/pubs/sir/2004/5038.

Reilly, T.E.; Pollock, D.W. 1993. Factors affecting areas contributing recharge to wells in shallow aquifers. Water-Supply Paper 2412. Washington, DC: U.S. Geological Survey. 21 p. http://water.usgs.gov/pubs/wsp/wsp_2412.

Remson, I.; Hormberger, G.M.; Molz, F.J. 1971. Numerical methods in subsurface hydrology. New York: John Wiley & Sons. 389 p.

Robinson, T.W. 1967. The effect of desert vegetation on the water supply of arid regions. In: Proceedings, International Conference on Water for Peace, Washington, DC, May 23-31, 1967. Washington, DC: US Government. 622–630. Vol. 3.

Rorabaugh, M.I. 1964. Estimating changes in bank storage and ground water contribution to streamflow. IUGG Volume XIII, Publication 63. Surface Waters Symposium, Gentbrugge, Belgium, August 19-31, 1963. [Place of publication unknown]. International Association of Scientific Hydrology (now International Association of Hydrological Sciences).

Rosenberry, D.O.; Striegl, R.G.; Hudson, D.C. 2000. Plants as indicators of focused ground water discharge to a northern Minnesota lake. Ground Water. 38(2): 296–303.

Rubin, J.; Steinhardt, R. 1963. Soil water relations during rain infiltration: I. theory. Proceedings, Soil Science Society of America. 27: 246–251.
Rubin, J.; Steinhardt, R. 1964. Soil water relations during rain infiltration: II. moisture content profiles during rains of low intensities. Proceedings, Soil Science Society of America. 28: 1–5.

Rutledge, A. T. 1993. Computer programs for describing the recession of ground water discharge and for estimating mean ground water recharge and discharge from streamflow records. Water Resources Investigation Report 93-4121. Washington, DC: U.S. Geological Survey. 45 p.

Rutledge, A.T. 1998. Computer programs for describing the recession of ground water discharge and for estimating mean ground water recharge and discharge from streamflow records—update. Water-Resources Investigations Report 98-4148. Washington, DC: U.S. Geological Survey. 43 p. http://water.usgs.gov/pubs/wri/wri984148.

Rutledge, A.T. 2000. Considerations for use of the RORA program to estimate ground water recharge from streamflow records. Open-File Report 00-156. Washington, DC: U.S. Geological Survey. 44 pp. http://pubs.usgs.gov/of/2000/ofr00-156.

Rutledge, A.T.; Mesko, T.O. 1996. Estimated hydrologic characteristics of shallow aquifer systems in the Valley and Ridge, and the Piedmont physiographic provinces based on analysis of streamflow recession and baseflow. Professional Paper 1422-B. Washington, DC: U.S. Geological Survey. 58 p.

Sacks, L.A. 2002. Estimating ground water inflow to lakes in central Florida using the isotope mass-balance approach. Water-Resources Investigations Report 02-4192. Washington, DC: U.S. Geological Survey. 59 p. http://fl.water.usgs.gov/PDF_files/wri02_4192_sacks.pdf.

Sada, D.W. 2002. Distribution of spring snails (Family Hydrobiidae), Inland Feeder Project Area, San Bernardino County, California. Publication 41175. Orange, CA: Desert Research Institute, Division of Hyrologic Sciences, University and Community College System of Nevada. 10 p.

Salama, R.B.; Tapley, I.; Ishii, T.; Hawkes, G. 1994. Identification of areas of recharge and discharge using Landsat-TM satellite imagery and aerial photography mapping techniques. Journal of Hydrology. 162: 119–141.

Sanders, T. G.; Ward, R.C.; Loftis, J.C., Steele, T.D.; Adrian, D.D.; Yevjevich, V. 2000. Design of networks for monitoring water quality. Highlands Ranch, CO: Water Resources Publications. 336 p.

Scalf, M.R.; McNabb, J.F.; Dunlap, W.J.; Cosby, R.L.; Fryberger, J. 1981. Manual of ground water quality sampling procedures. Report EPA 600/2-1/160. Washington, DC: U.S. Environmental Protection Agency. 269 p.

Searcy, J.K. 1959. Flow-duration curves. In: Manual of hydrology: part 2, low-flow techniques. Water-supply Paper 1542-A. Washington, DC: U.S. Geological Survey. 33 p.

Seelig, B. 1994. An assessment system for potential groundwater contamination from agricultural pesticide use in North Dakota. Extension Bulletin 63. Fargo, ND: North Dakota State University Extension Service. http://www.ext.nodak.edu/extpubs/h2oqual/watgrnd/eb63w.htm.

Sharp, J.M. Jr. 1988. Alluvial aquifers along major rivers. In: Back, W.; Rosenshein, J.S.; Seaber, P.R. (eds.). The geology of North America. Volume 0-2, Hydrogeology. Boulder, CO: Geological Society of America: 273-282.

Shukla, S.; Mostaghimi, S.; Shanholt, V.O.; Collins, M.C.; and Ross, B.B. 2000. A county-level assessment of ground water contamination by pesticides. Ground Water Monitoring and Review. 20(1): 104–119.

Sinclair Knight Merz Pty Ltd. 2001. Environmental water requirements of ground water dependent ecosystems. Technical Report 2. Canberra, Australia: Environmental Flows Initiative.

Singhal, B.B.S.; Gupta, R.P. 1999. Applied hydrogeology of fractured rocks. The Netherlands: Kluwer Academic Press. 400 p.

Slater, L.; Sandberg, S.; Jankowski, M. 1998. Improvement in the azimuthal EM method: the value of signal processing. Chicago, IL. March 22-26, 1978 In: Proceedings, conference of the environmental and engineering geophysical society symposium on the application of geophysics to environmental and engineering problems, 1998. Wheat Ridge, CO: Environmental and Engineering Geophysical Society: 177–186.

Smart, P.L.; Laidlaw, I.M.S. 1977. An evaluation of some fluorescent dyes for water tracing. Water Resources Research. 13: 15–33.

Smith, S.D.; Devitt, D.A.; Sala, A.; Cleverly, J.R.; Busch, D.E. 1998. Water relations of riparian plants from warm desert regions. Wetlands. 18: 687–696.

Sonderegger, J.L. 1970. Hydrogeology of limestone terrains: photogeologic investigations. Bulletin 94C. Tuscaloosa, AL: Geological Survey of Alabama.

Stanford, J.A.; Gonser, T. (eds.). 1998. Rivers in the landscape: special issue on riparian and ground water ecology. Freshwater Biology. 40(3): 402–585.

Stanford, J.A.; Ward, J.V. 1993. An ecosystem perspective of alluvial rivers: connectivity and the hyporheic corridor. Journal of the North American Benthological Society. 12(1): 48–60.

Stanford, J.A.; Ward, J.V. 1988. The hyporheic habitat of river ecosystems. Nature. 335: 64–66.

Stauffer, R.E. 1985. Use of tracers released by weathering to estimate ground water inflow to seepage lakes. Environmental Science and Technology. 19: 405–411.

Stephenson, D.A.; Fleming, A.H.; Mickelson, D.M. 1988. Glacial deposits. In: Back, W.; Rosenshein, J.S.; Seaber, P.R. (eds.). The geology of North America, Volume O-2, Hydrogeology. Boulder, CO: Geological Society of America. 301–314.

Stone, W.J. 1999. Hydrogeology in practice: a guide to characterizing ground water systems. Upper Saddle River, NJ: Prentice Hall. 248 p.

Strack, O.D.L. 1989. Groundwater mechanics. Engelwood Cliffs, NJ: Prentice Hall. 732 p.

Sun, R.J. (ed.). 1986. Regional aquifer-system analysis program of the U.S. Geological Survey: summary of projects, 1978–1984. Circular 1002. Washington, DC: U.S. Geological Survey. 264 p.

Taylor, C.J.; Alley, W.M. 2001. Ground water-level monitoring and the importance of long-term water-level data. Circular 1217. Washington, DC: U.S. Geological Survey. 68 p.http://water.usgs.gov/pubs/circ/circ1217/.

Taylor, R.W.; Fleming, A.H. 1988. Characterizing jointed systems by azimuthal resistivity surveys. Ground Water. 26 (4): 464-474.

Terzaghi, K. 1950. Mechanism of landslides. In: Application of geology to engineering practice. New York: Geological Society of America: 83–123.

Theis, C.V.; Brown, R.H.; Meyer, R.R. 1963. Estimating the transmissibility of aquifers from the specific capacity of wells. In: Bentall, R. (ed.). Methods of estimating permeability, transmissibility, and drawdown. Water-supply Paper 1536-I. Washington, DC: U.S. Geological Survey: 331–341.

Thomas, H.E. 1952. Ground water regions of the United States: their storage facilities. Washington, DC: Interior and Insular Affairs Committee, U.S. House of Representatives. 76 p.

Todd, D.K. 1980. Groundwater hydrology. New York: John Wiley & Sons. 533 p.

Triska, F.J.; Kennedy, V.C.; Avanzio, R.J.; Zellweger, G.W.; Bencala, K.E. 1989. Retention and transport of nutrients in a third-order stream in northwestern California: hyporheic processes. Ecology 70: 1893–1905.

Tucci, P.; Martinez, M.I. 1995. Hydrology and simulation of ground water flow in the Aguadilla to Rio Camuy area, Puerto Rico. Water-resources Investigations Report 95-4028. Washington, DC: U.S. Geological Survey. 39 p.

Turner, A.K.; Schuster, R.L. 1996. Landslides: investigation and mitigation. Special Report 247. Washington, DC: Transportation Research Board, National Academy Press. 673 p.

U.S. Army Corps of Engineers. 2003. Lake Tahoe Basin framework study groundwater evaluation. Sacramento, CA: U.S. Army Corps of Engineers.

U.S. Bureau of Land Management (U.S BLM). 2001. Riparian area management: a guide to managing restoring, and conserving springs in the Western United States. Technical Reference 1737-17. Denver, CO: Bureau of Land Management. 70 p.

U.S. Bureau of Reclamation. 1963. Lower Colorado River water salvage phreatophyte control: Arizona-California-Nevada, reconnaissance report. Boulder City, NV: U.S. Bureau of Reclamation Region 3.

U.S. Environmental Protection Agency (EPA). 1987. Report to Congress: management of wastes from the exploration, development and production of crude oil, natural gas, and geothermal energy. Washington, DC: U.S. Environmental Protection Agency, Office of Solid Waste and Emergency Response.

U.S. Environmental Protection Agency (EPA). 1989. Report on minimum criteria to ensure data quality. Report EPA 530/SW-90/021. Washington, DC: U.S. Environmental Protection Agency. 36 p.

U.S. Environmental Protection Agency (EPA). 1991. Technical guide: ground water, volume II: methodology. Report EPA 625/6-90/016b. Washington, DC: U.S. Environmental Protection Agency. 141 pp.

U.S. Environmental Protection Agency (EPA). 1993a. Subsurface characterization and monitoring techniques. Report EPA 625/R-93/003. Cincinnati, OH: U.S. Environmental Protection Agency Center for Environmental Research Information.

U.S. Environmental Protection Agency (EPA). 1993b. Use of airborne, surface, and borehole geophysical techniques at contaminated sites. Report EPA 625/R-92/007. Washington, DC: U.S. Environmental Protection Agency Office of Research and Development.

U.S. Environmental Protection Agency (EPA). 1994. Ground water and wellhead protection. Report EPA 625/R-94/001. Washington, DC: U.S. Environmental Protection Agency Office of Ground Water and Drinking Water. 269 p.

U.S. Environmental Protection Agency (EPA). 1997. State source water assessment and protection programs guidance: final guidance. Report EPA 816-R-97-009. Washington, DC: U.S. Environmental Protection Agency. 127 p.

U.S. Geological Survey (USGS). 1982. Evaporation and transpiration. In: National technical guide of recommended methods for water data acquisition. Reston, VA: U.S. Geological Survey Office of Water Data Coordination.

U.S. Geological Survey (USGS). 1993. Water dowsing. 1993-337-936. Washington, DC: U.S. Geological Survey. 14 p.

U.S. Geological Survey (USGS). 1996. U.S. Geological Survey programs in Montana. Fact Sheet FS-026-96. Washington, DC: U.S. Geological Survey.

U.S. Geological Survey (USGS). 1997 to present. National field manual for the collection of water-quality data. In: *Techniques of Water-resources Investigations*, Book 9, Chapters A1–A9.. Washington, DC: U.S. Geological Survey: variously paged. 2 vol. http://pubs.water.usgs.gov/twri9A. Updates and revisions are ongoing and are summarized at http://water.usgs.gov/owq/FieldManual/index.html.

U.S. Water Resources Council. 1980. Essentials of ground water hydrology pertinent to water-resources planning. Rev. Hydrology Committee Bulletin 16. Washington, DC: U.S. Water Resources Council Hydrology 38 p.

van der Heijde, P.K.M.; El-Kadi, A.I.; Williams, S.A. 1988. Groundwater modeling: an overview and status report. Indianapolis, IN: International Ground Water Modeling Center, Butler University.

van Everdingen, R.O. 1991. Physical, chemical, and distributional aspects of Canadian springs. In: Williams, D.D.; Danks, H.V. (eds.). Arthropods of springs with particular reference to Canada. Memoirs of the Entomological Society of Canada No.155. Ottawa, Ontario, Canada: Entomological Society of Canada: 7–28.

Walton, W.C. 1970. Ground water resource evaluation. New York: McGraw-Hill Book Co. 535 p.

Walton, W. C. 1984. Technical guide of analytical ground water models. Report 84-06. Indianapolis, IN: International Groundwater Modeling Center, Holcomb Research Institute, Butler University.

Wehrmann, H.A. 1983. Monitoring well design and construction. Ground Water Age. 4: 35–38.

Weight, W.D.; Sondregger, J.L. 2001. Manual of applied field hydrogeology. New York: McGraw Hill. 608 p.

Welder, G.E. 1988. Hydrologic effects of phreatophyte control, Acme-Artesia reach of the Pecos River, New Mexico, 1967–82. Water-resources Investigations Report 87-4148. Albuquerque, NM: U.S. Geological Survey. 46 p.

Western Governors' Association. 2004. Coal bed methane best management practices: a handbook. Denver, CO: Western Governors' Association. http://www.westgov.org/wga/initiatives/coalbed/index.htm.

Winter, T.C. 1981. Effects of water-table configuration on seepage through lakebeds. Limnology and Oceanography. 26: 925–934.

Winter, T.C. 1995. Hydrological processes and the water budget of lakes. In: Lerman, A.; Imboden, D.; Gat, J. (eds.). Physics and chemistry of lakes. Berlin: Springer-Verlag: 37-62.

Winter, T.C.; Harvey, J.W.; Franke, O.L.; Alley, W.M. 1998. Ground water and surface water: a single resource. Circular 1139. Washington, DC: U.S. Geological Survey. 79 p. http://water.usgs.gov/pubs/circ/circ1139

Wireman, M. 2002. Recharge to the subsurface via infiltration as a method for disposal of produced coal-bed methane water. Technical Memorandum. Washington, DC: U.S. Environmental Protection Agency. 6 p.

Wood, W.W. 1976. Guidelines for collection and field analysis of groundwater samples for selected unstable constituents. In: \Techniques of Water-resources Investigations, Book 1, Chapter D2. Washington, DC: U.S. Geological Survey. http://water.usgs.gov/pubs/twri/twri1-d2.

Wood, W.W.; Fernandez, L.A. 1988. Volcanic rocks. In: Geology of North America, Volume. O-2,Hydrogeology. Boulder, CO: Geological Society of America: 353–365.

Wood, P.J.; Dykes, A.P. 2002. The use of salt dilution gauging techniques: ecological considerations and insights. Water Research. 36: 3054–3062.

Woods, L.G. 1966. Increasing watershed yield through management. Journal of Soil and Water Conservation. 21: 95–97.

Young, M.K.; Gresswell, R.E.; Luce, C., eds. 2003. The effect of wildland fire on aquatic ecosystems in the Western USA. Forest Ecology and Management. 178,(1 and 2): 1–229.

Zheng, C. 1990. MT3D, A modular three-dimensional transport model for simulation of advection, dispersion, and chemical reactions of contaminants in groundwater systems, Report to the Kerr Environmental Research Laboratory, US Environmental Protection Agency, Ada, OK.

Zheng, C.; Wang, P.P. 1999. MT3DMS: A modular three-dimensional multispecies model for simulation of advection, dispersion and chemical reactions of contaminants in groundwater systems; Documentation and User's Guide, Contract Report SERDP-99-1, U.S. Army Engineer Research and Development Center, Vicksburg, MS.

Ziemer, R.R. 1992. Effect of logging on subsurface pipeflow and erosion: coastal northern California, USA. In: Walling, D.E.; Davies, T.R.; Hasholt, B. (eds.). Erosion, debris flows and environment in mountain regions. Proceedings of the Chengdu Symposium, July 5-9, 1992; Chengdu, China. IUGG Volume XXI, Publication No. 209. Wallingford, UK: International Association of Scientific Hydrology: 187-197.

Zohdy, A.A.R.; Eaton, G.P.; Mabey, D.R. 1974. Application of surface geophysics to ground water investigations. In: U.S. Geological Survey techniques of water-resource investigations. Washington, DC: U.S. Geological Survey: Book 2, Chapter D1. http://water.usgs.gov/pubs/twri/twri2-d1.

Zwahlen, F. (ed.). 2003. COST action 620: vulnerability and risk mapping for the protection of carbonate (karst) aquifers. Final evaluation report submitted to commission COST secretariat. Luxemborg: European Communities. 297 p.

Resources

U.S. Geological Survey Ground Water Contacts

National Issues

Office of Ground Water
USGS National Center, MS-411
12201 Sunrise Valley Drive
Reston, VA 20192
Phone: 703–648–5035

Regional Issues

Northeast Region Ground water Specialist
USGS National Center, MS-433
12201 Sunrise Valley Drive
Reston, VA 20192
Phone: 703–648–5814

Southeast Region Ground water Specialist
USGS
Spalding Woods Office Park, Suite 160
3850 Holcomb Bridge Road
Norcross, GA 30092
Phone: 770–409–7716

Central Region Ground water Specialist
USGS, MS-406
Box 25046
Denver Federal Center, Bldg. 53, Room F-1200
Lakewood, CO 80225
Phone: 303–236–5950, ext. 213

Western Region Ground water Specialist
USGS
7801 Folsom Blvd., Suite 325
Sacramento, CA 95826
Phone: 916–379–3737

State or Local Issues

Contact the State ground water specialist, through the USGS State representative on the Internet at http://interactive2.usgs.gov/contact_us/index.asp.

USGS Online Resources

The USGS home page, with links to many earth-science-related topics and information on USGS water programs, technical resources (such as computer programs), publications, and water data, can be obtained on the Internet at http://water.usgs.gov/.

Many USGS reports are now available online. These reports can be accessed at http://water.usgs.gov/pubs/.

You can search for a publication by report series (for example, Water-supply Paper, Open-file Report, Techniques of Water-resources Investigations) andS number, or by keyword. A listing of the most recently published USGS reports can be obtained at http://pubs.usgs.gov/publications/index.shtml.

The U.S. Geological Survey series of print publications *The Ground Water Atlas of the United States* describes the location, the extent, and the geologic and hydrologic characteristics of the important aquifers of the Nation. This series can be accessed online at http://capp.water.usgs.gov/gwa/gwa.html.

Current drought information can be obtained at http://water.usgs.gov/.

Information on various ground water issues being addressed by USGS can be obtained at http://water.usgs.gov/ogw/issues.html.

The home page for the USGS Office of Ground Water Branch of Geophysics, which specializes in the application of geophysical methods to ground water investigations, is at http://water.usgs.gov/ogw/bgas/.

Other Online Resources

The National Park Service National Cave and Karst Research Institute's home page is http://www2.nature.nps.gov/nckri/.

The EPA ground water research lab in Ada, OK (USEPA GWERD Library) has published numerous ground water issue papers focused on contaminant hydrology. They can be found at http://www.epa.gov/ada/publications.html

The Association of American State Geologists (AASG) is an organization of the chief executives of the State geological surveys in 50 States and Puerto Rico. The responsibilities of the various State surveys differ from State to State, depending on the enabling legislation and the traditions under which the particular survey evolved. Some have regulatory responsibilities for water, oil and gas, land reclamation, and so on. http://stategeologists.org

Appendix I.
Legal Framework for Ground Water Use in the United States

This document provides an overview of doctrine governing ground water in the 43 States in which National Forest System land is located. **Information here is not a substitute for legal advice from the USDA Office of the General Counsel.**

Rights to use ground water are regulated by States through application of common law, State statutes and regulations, and/or judicial precedent. The ownership and allocation rules applicable to ground water are usually different from those applying to surface water. A brief overview of ground water law in the United States is given below. While ground water schemes can be divided into a few general categories, there are variations in every State. The USDA Office of the General Counsel should be consulted as specific questions about ground water laws arise. States generally follow one of four basic systems of ground water allocation: (1) the "English" rule of absolute ownership, (2) the "American" rule of reasonable use, (3) the prior appropriation rule, and (4) the correlative rights rule.[1]

The Reserved Rights Doctrine

While the central focus of this document is an overview of state laws and regulations regarding ground water, Federal law may have limited application when managing ground water resources. This doctrine is known as reserved rights, and it applies to land reserved from the public domain. The U.S. Supreme Court has decided that when the Federal Government reserves land from the public domain, it also implicitly, or sometimes explicitly, reserves the water needed to fulfill the reservation's primary legislative purposes.[2] As part of the creation of national forests, water rights were reserved for the purposes of securing favorable conditions of water flows and to furnish a continuous supply of timber.[3] The U.S. Supreme Court rejected the United States' claim of reserved water rights for maintenance of in-stream flows, recreation, stock watering, and wildlife within the Gila National Forest.[4]

The amount of water reserved is "only that amount of water necessary to fulfill the purpose of the reservation, no more."[5] However, the reservation encompasses an amount of water "sufficient for the future requirements of the area reserved."[6] The date of the reservation establishes the priority right and the water right applies only to previously unappropriated waters.[7] In *Cappaert*,

---

[1] Malone, Linda A., The Necessary Interrelationship between Land Use and Preservation of Ground water Resources, 9 UCLA J. Environmental Law & Policy 1, 5 (1990).
[2] *Winters v. United States*, 207 U.S. 564 (1908).
[3] 16 U.S.C. § 475; *United States v. New Mexico*, 438 U.S. 696, 707-08, 718 (1978)
[4] *Id.* at 708, 716-17
[5] *Cappaert v. United States*, 426 U.S. 128, 141 (1976).
[6] *Arizona v. California*, 373 U.S. 546, 601 (1963).
[7] *Cappaert v. United States*, 426 U.S. at 139

the Supreme Court held that the reservation of land withdrawn under the American Antiquities Preservation Act, reserved subterranean water necessary for the maintenance of the pupfish at Devil's Hole National Monument, and the United States did not have to perfect its water rights according to State law. However, in doing so the Supreme Court did not define the subsurface waters where the pupfish lived as "ground water."[8] The Supreme Court and Circuit Courts of Appeal have never made a determination as to whether the reserved rights doctrine applies to water lying beneath federal lands.

The Federal Courts have left the question of whether reserved rights in ground water exist for a later day. Wyoming and Arizona have addressed whether there are federally reserved rights in ground water. Arizona came to the conclusion that the Federal Government did have reserved rights in stationary ground water and that those reserved rights entitle the federal government to greater protection than permittees with only State law rights. For additional discussion see section on Arizona Water Law.[9]

Should Federal Courts establish that the Federal Government has reserved rights in ground water, Federal Agencies will likely face similar difficulties to those encountered in the *New Mexico* decision; namely that the use of the ground water would be confined to the statutory purposes of the reservation of the land.

Absolute Ownership

The absolute ownership doctrine is based on the English precedent of a landowner owning the airspace above and the soil beneath one's property.[10] Under this doctrine, the landowner overlying an aquifer has an absolute right to extract all ground water from the aquifer beneath the landowner's property. The overlying landowner can pump as much water as needed without regard to the needs or effect on other overlying landowners. The doctrine worked well in areas where abundant water was available; however, the drawbacks of the doctrine became apparent in the arid environment of the Western States.[11] Most of the States that initially followed this rule abandoned it during the late nineteenth and early twentieth century in favor of the reasonable use or "American" rule.[12] States still following the absolute ownership rule include Connecticut, Georgia, Indiana, Louisiana, Maine, Massachusetts, Mississippi, Rhode Island, and Texas.[13]

Reasonable Use

The reasonable-use rule is a modified absolute ownership rule wherein ground water use by an overlying landowner must be "reasonable" and must be used for a beneficial purpose on the overlying land.[14] Use of ground water on

---

[8] *Cf. Cappaert v. United States*, 508 F.2d 313,317 (9th Cir., 1974) (the Ninth Circuit characterized the waters of Devil's Hole were ground water and found a reserved right).

[9] *In re General Adjudication of All Rights to use the Gila River System and Source III*, 195 Ariz. 411 (1999).

[10] *Acton v. Blundell*, 152 Eng. Rep. 1223 (Exch. 1843).

[11] Ashley, Jeffrey S. and Smith, Zachary A., *Ground water Management in the West*, University of Nebraska Press, 1999.

[12] A. Tarlock, *Law of Water Rights and Resources*, §4.04, Clark Boardman Callaghan, 1997.

[13] Malone at 5, fn. 25

[14] Malone at 6.

nonoverlying land is considered unreasonable. Reasonableness is based on such factors as well location, amount of water, and the proposed use and placement of the water.[15] Waste of water is not a reasonable use if it interferes with the right of adjacent landowners to use the water for the beneficial use of their overlying lands.[16] If the requirements of the rule are met, a landowner may withdraw ground water even if doing so deprives another landowner of the reasonable use of the ground water.[17] States applying the reasonable use rule include Alabama, Florida, Illinois, Kentucky, Maryland, New York, Ohio, North Carolina, and Tennessee.[18]

Prior Appropriation

The prior appropriation doctrine gives priority to ground water users who put ground water to beneficial uses that are first in time. During water shortages, first in time appropriators have priority over later appropriators.[19] Many States have statutory systems requiring permits to establish priority use. Idaho, Kansas, Montana, Nevada, New Mexico, North Dakota, Oregon, South Dakota, Utah, and Wyoming apply the doctrine of prior appropriation to ground water.[20] California applies it where surplus water exists above the needs of overlying owners. Arizona, once an absolute ownership State, now has a statutory scheme that creates Active Ground Water Management Areas, grandfathers pre-1980 water rights in these areas, and sets up a permit administration system.[21] The States of Colorado, Kansas, Montana, Nebraska, Nevada, New Mexico, Washington, and Oregon have combined prior appropriation with critical area legislation to designate areas where new pumping may be prohibited and existing pumping may be restricted to preserve ground water.[22] Courts in Idaho have upheld laws limiting water extraction to the annual recharge rate and have issued injunctions against junior wells that exceed reasonably anticipated future rate of recharge.[23] Arizona, Colorado and New Mexico further limit ground water mining and extraction to a rate that will restore the aquifer to the level necessary for economically feasible extraction.[24] Some States exempt ground water that is a by-product of secondary oil and gas recovery (Wyoming), geothermal resources (California), or water from mine dewatering (New Mexico).[25]

---

[16] *Id.*[16] Ashley at 9.

[17] Tarlock at §4.05(1).

[18] Malone at 6.

[19] Malone at 8.

[20] Turlock at §6.03(1).

[21] Patrick, Kevin L and Archer, Kelly E., *A Comparison of State Ground water Laws*, 30 Tulsa L.J. 123, 132-33.

[22] L. Malone at 9-10.

[23] Malone at 10, fn. 48.

[24] Malone at 10.

[25] Tarlock at 6.03(3).

Correlative

The correlative rights doctrine gives each overlying property owner a common right to the reasonable, beneficial use of the basin supply on the overlying land. This is similar to the doctrine of riparian rights to surface water. All overlaying landowners have equal rights to percolating ground water and all must share in any water shortages;[26] however, overlying landowners do not have a right to maintenance of the natural water table.[27] The States that have adopted the correlative rights doctrine include Arkansas, California, Delaware, Minnesota, Missouri, Nebraska, and New Jersey.[28]

Subject to future requirements on overlying lands, ground water that is surplus to the needs of overlying owners is available for appropriation for uses on non-overlying land. The burden of proof is on the appropriator to prove that a surplus exists beyond prior vested-right uses of overlying landowners. In the event of a shortage, overlying landowners have first priority.[29]

Some uses of ground water on land overlying a basin have been held to constitute appropriative uses. For example, the public use of ground water is typically not an overlying use. Municipalities or public water agencies generally have appropriative rights, not overlying rights, to the water pumped from a ground water basin to supply their customers. They do not exercise the overlying rights of their inhabitants.[30]

Most States have a permit system for ground water extraction. Permit requirements differ in each state. Some States require a permit for all extractions. Others require permits where water is proposed to be withdrawn from certain designated areas. Some States have a common permit system for surface and ground water.[31]

The definition of "beneficial use" is a critical issue in analyzing ground water law in any State. Some uses are universally considered to be beneficial. They include the use of water for domestic, irrigation, manufacturing or stock-watering purposes;[32] however, the States differ on whether protection of fish, recreation, aesthetic, or scenic uses are beneficial uses of water.[33]

---

[26] *Tehachapi-Cummings County Water District v. Armstrong*, 49 Cal. App. 3d 992, 1001 (1975).
[27] *Katz v.Walkinshaw* 141 Cal. 116 (1903) [74 P. 766].
[28] A. Tarlock at §4.06(2).
[29] Montecito Valley Water Co. v. Santa Barbara, 144 Cal. 578, 584-85 (1904).
[30] Hutchins, *The California Law of Water Rights*, 1956, p. 458; *San Bernardino v. Riverside*, 186 Cal. 7, 25 (1921) [198 P. 784].
[31] Malone at 12.
[32] Ashley at 10.
[33] *Id.*

Appendix II.
Common Ground Water Terms and Definitions

The following section defines, in relatively simple terms, terminology and properties commonly associated with ground water. More thorough discussions of each term or concept can be found in ground water hydrology textbooks (for example, Fetter 2001, Freeze and Cherry 1979). A list of definitions of common hydrological terms is provided online by the U.S. Geological Survey at http://capp.water.usgs.gov/GIP/h2o_gloss/

Hydraulic Head and Hydraulic Gradient

Hydraulic head (often simply referred to as "head") can be considered simply as the elevation of the water surface in a well, although the actual definition of hydraulic head is more complex. A water-level measurement made under static conditions is a measurement of the hydraulic head in the aquifer at the depth of the screened or open interval of a well (fig. 1). Because hydraulic head represents the energy of water, ground water flows from locations of higher head to locations of lower head. The change in hydraulic head over a specified distance in a given direction is called the "hydraulic gradient."

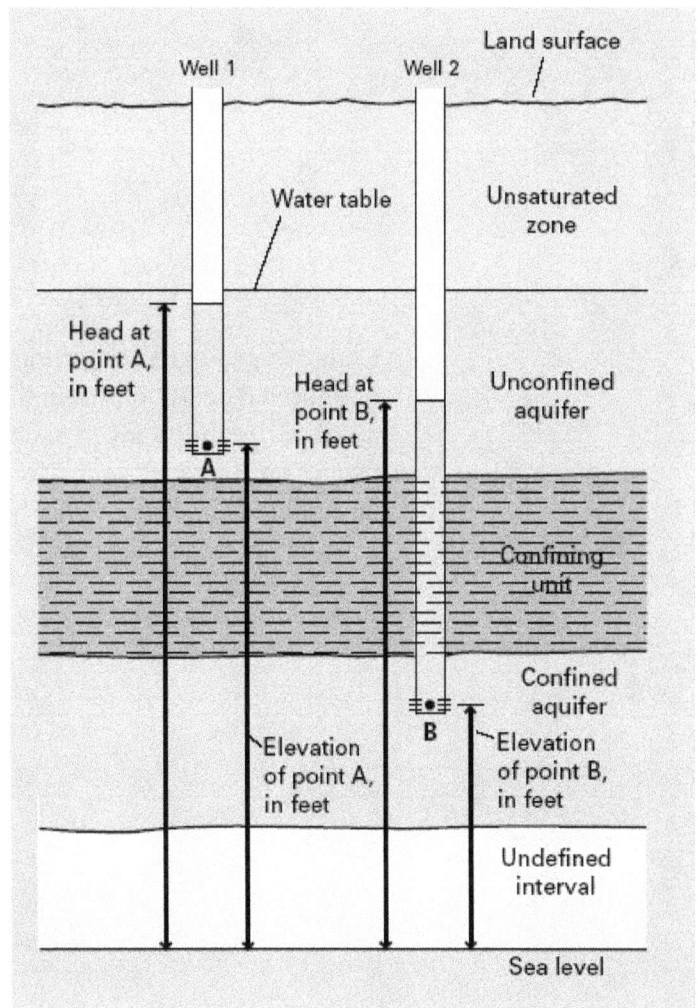

Figure 1. The relation between hydraulic head and water level in two observation wells. Well 1 is screened in an unconfined aquifer, and Well 2 is screened in a confined aquifer (Taylor and Alley 2001).

229

Saturated and Unsaturated Zones

When rain falls or snow melts, some of the water evaporates, some is transpired by plants, some flows overland and collects in streams, and some infiltrates into the pores or cracks of the soil and rocks. The first water that enters the soil replaces water that has been evaporated or used by plants during a preceding dry period. Between the land surface and the aquifer water is the unsaturated zone. In the unsaturated zone, there usually is at least a little water, mostly in smaller openings of the soil and rock; the larger openings usually contain air instead of water. After a significant rain, the zone may become almost saturated; after a long dry spell, it may become almost dry. However, some water is always held in the unsaturated zone by molecular attraction.

After the water requirements for plant and soil are satisfied, any excess water will infiltrate to the water table —the top of the zone below which the openings in rocks are fully saturated (the saturated zone). The water table is often considered the boundary between the saturated and unsaturated zones, but in reality a capillary fringe often exists between the two zones (fig. 2). At the water table the fluid pressure within pore spaces is exactly equal to atmospheric pressure, but within the capillary fringe the fluid pressure is less than atmospheric.

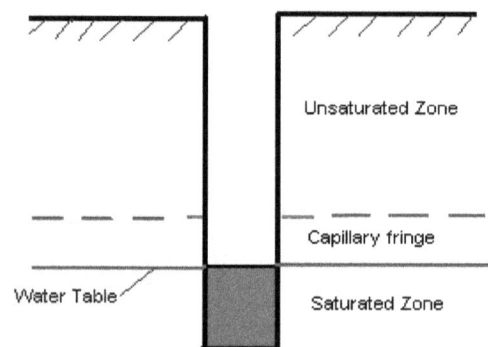

Figure 2. The relationships among the unsaturated zone, capillary fringe, saturated zone, and water table.

Complex geological environments can lead to more complex saturated-unsaturated conditions than those previously discussed. The presence of a low permeability layer, such as a clay layer, within a highly permeable formation can result in the formation of a discontinuous saturated lens in which unsaturated conditions exist both above and below the lens. Such a lens is called a "perched water body" (fig. 3).

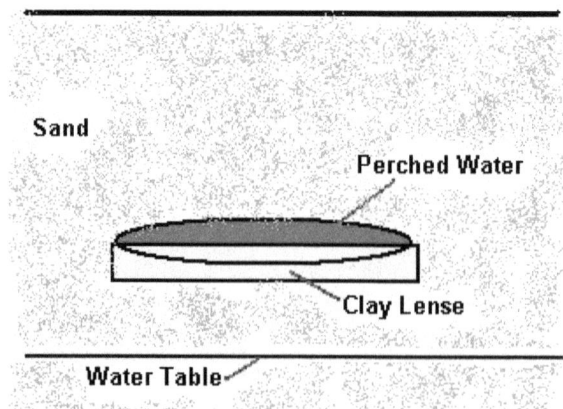

Figure 3. An example of a perched water body.

Aquifers and Confining Units

An aquifer is a geological formation, group of formations, or part of a formation that contains sufficient saturated, permeable material to yield significant quantities of water to wells and springs (Taylor and Alley 2001). Examples of aquifer materials include sand and gravel, cavernous or fractured limestone, sandstone, and fractured crystalline rock.

Two general classes of aquifers — unconfined and confined — are recognized (fig. 4). In unconfined aquifers (sometimes referred to as "water-table aquifers"), hydraulic heads fluctuate freely in response to changes in recharge, discharge, and barometer pressure. Water levels measured in the upper part of an unconfined aquifer help define the elevation of the water table. In confined aquifers, water in the aquifer is confined by an overlying geologic formation that is much less permeable than the aquifer. Water levels in tightly cased wells completed in confined aquifers may rise above the elevation of the top of the aquifer (fig. 4), and may even flow at land surface. These aquifers are considered to be "artesian". These water levels define an imaginary surface, referred to as the potentiometric surface, which represents the potential height to which water will rise in wells completed in the confined aquifer. Many aquifers are intermediate between being completely confined or unconfined, and in some cases an aquifer can be both confined and unconfined at different locations (fig. 5).

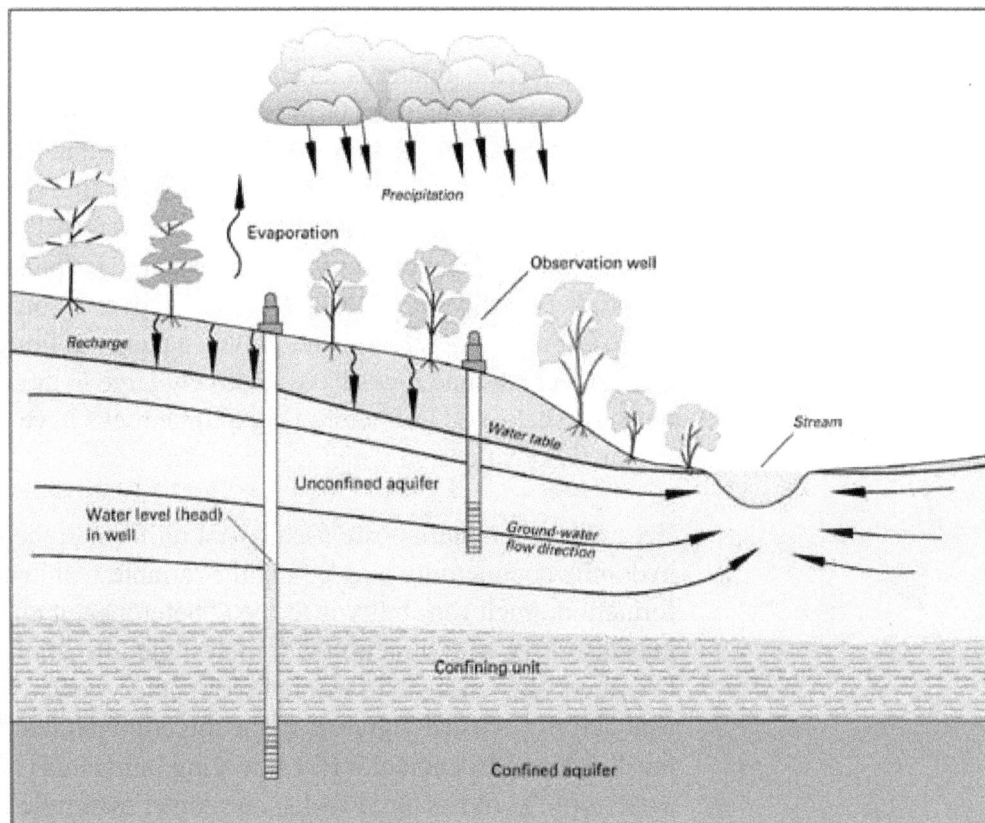

Figure 4. A typical ground water flow system showing the relation between an unconfined and a confined aquifer, a water table, and other hydrologic elements (Taylor and Alley 2001).

231

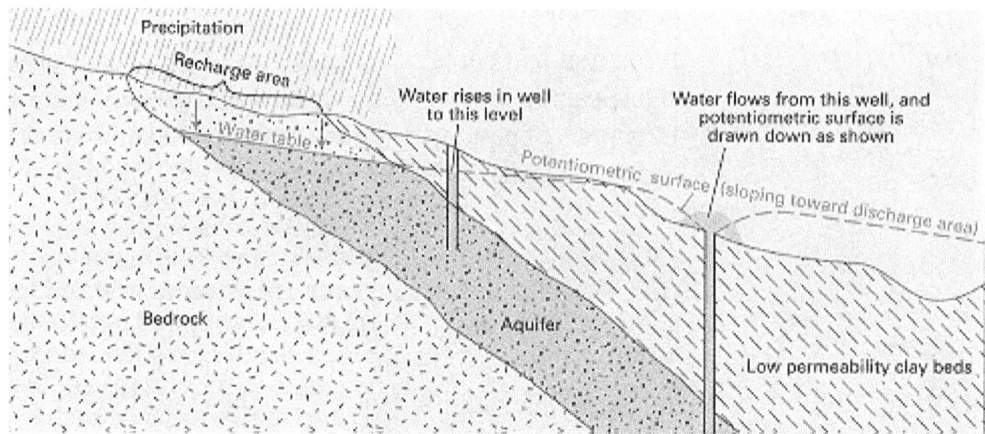

Figure 5. This aquifer is unconfined, in the area beneath the recharge area on the left, and confined on the right side of this illustration.

The geological unit that isolates a confined aquifer and restricts the movement of water between aquifers is called a confining unit (sometimes referred to as an "aquitard" or "aquiclude"). A confining unit is composed of geological materials that are significantly less permeable than the adjacent aquifer(s). Examples of confining unit materials include clay, shale, glacial till, and unfractured crystalline rock.

Hydraulic Conductivity

Hydraulic conductivity (K) is a measure of the capacity of an aquifer to transmit water, and it is expressed in units of velocity, such as feet per day or centimeters per second. In general, the greater the hydraulic conductivity of an aquifer, the greater is its ability to provide water to a well. Hydraulic conductivity is sometimes used interchangeably with the term "permeability," but these terms are technically somewhat different. Permeability is an intrinsic property of the aquifer material, whereas hydraulic conductivity captures not only the size and interconnectedness of the water-filled openings in the aquifer but also the physical properties of the water. Hydraulic conductivity of earth materials is highly variable, and can range over 12 orders of magnitude (fig. 6). For example, sand and gravel, cavernous limestone, and highly fractured crystalline rocks have relatively large hydraulic conductivity values, but clay, shale, and unfractured crystalline rocks have relatively small hydraulic conductivity values.

Because earth materials are usually not uniform in their physical properties, hydraulic conductivity may be highly variable within a single geologic formation. Such variability is termed "heterogeneity." Hydraulic conductivity may also vary with direction within a single formation, and this variability is termed "anisotropy." In bedded sedimentary rocks, for example, hydraulic conductivity is usually greater in the direction parallel to the bed than in the direction perpendicular (for flat-lying units, this is known as "vertical anisotropy"). In fractured rocks, horizontal hydraulic conductivity is often greater in the direction parallel to the fracture planes (termed "horizontal anisotropy).

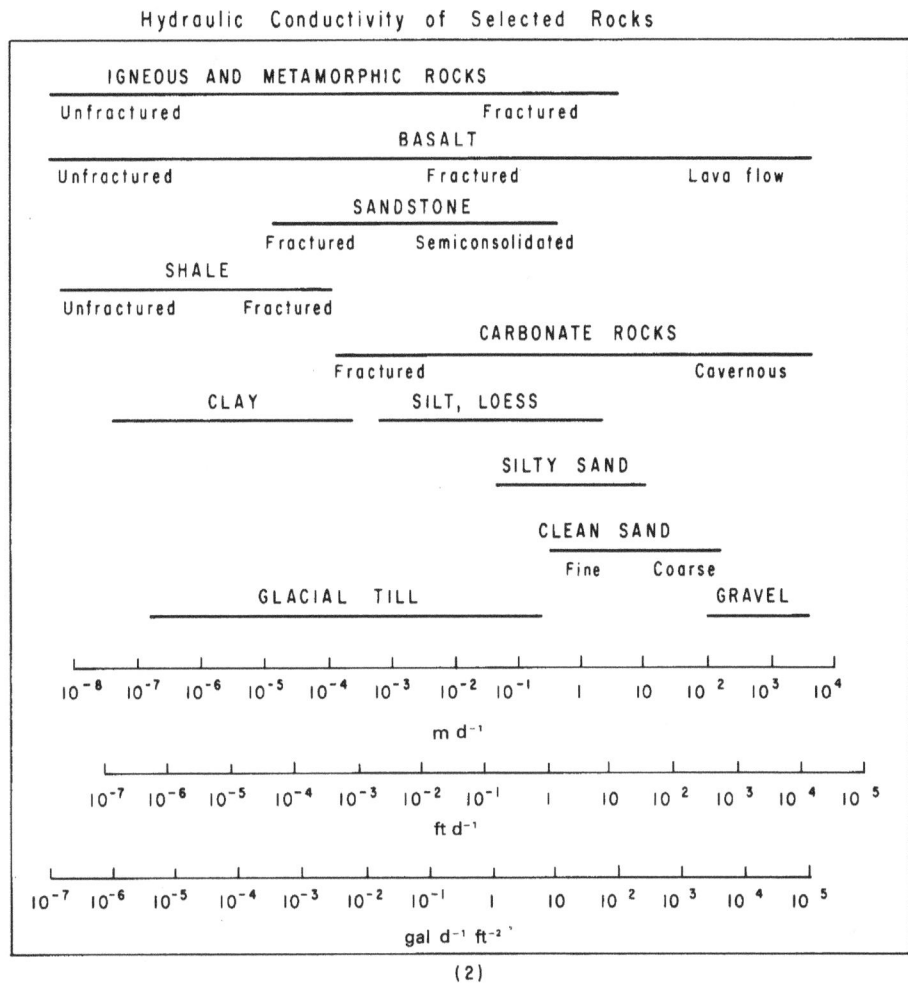

Hydraulic Conductivity of Selected Rocks

Figure 6. Range in hydraulic conductivity for selected earth materials (Heath 1983).

Transmissivity

Transmissivity is another measure of an aquifer's ability to transmit water. It is the product of the aquifer's hydraulic conductivity (K) and the saturated thickness of the aquifer (b), such that:

$$T = Kb.$$

Transmissivity, commonly expressed in ft^2/day or cm^2/s, is usually the aquifer property that is solved for when analyzing an aquifer (pumping) test.

Porosity and Effective Porosity

The ratio of openings (voids) to the total volume of a soil or rock is referred to as porosity, which is unitless and usually expressed as a percentage or a decimal fraction. Porosity depends on the range in grain size (sorting) and on the shape of the void spaces, but not necessarily on the size of the grains. For example a gravel deposit may be less porous than a clay deposit, because the clay is composed of a more uniform grain size and has a very open internal structure (often described as a "house of cards" structure). The individual pore spaces in the clay are smaller than those of the gravel, but the overall volume of pores in a clay will be tend to be greater than those of an equal volume of gravel.

233

Porosity also varies widely for different earth materials. Unfractured crystalline rocks can have almost no porosity, but clays and some modern carbonate rocks (fig. 7) may have porosities of 40 percent or more.

Effective porosity is that portion of the porosity that is interconnected and able to transmit fluids. The effective porosity is the ratio of the volume of interconnected voids to the total volume and is also unitless. Because it leaves out the dead-end void spaces and those void spaces that are too small to admit water molecules, it is typically less than the total porosity.

Specific Yield and Storage Coefficient

Porosity is important in ground water hydrology because it tells us the maximum amount of water that a rock or soil can contain when it is saturated (Heath 1983). It is equally important, however, to know that only a part of this water is available to supply a well. Water in storage in the ground (total saturated porosity) is divided into the part that will drain under the influence of gravity (specific yield) and the part that is retained as a film on rock surfaces and in very small openings because of capillary forces (specific retention). Specific yield is the measure of how much water is available for use, and specific retention tells us how much water will remain in the rock after it is drained by gravity. Specific yield generally ranges between 10 and 30 percent in unconsolidated deposits, and is generally 10 percent or less in consolidated rocks. In confined aquifers, the term "storage coefficient" is usually used in place of specific yield. Storage coefficient values tend to be much smaller than values of specific yield, commonly on the order of less than 1 percent.

Figure 7. A core sample, approximately four inches in length, of highly porous Miami oolite (limestone) obtained near Miami, FL. (USGS 2002)

Water Budget

Budgets or balances of the amounts of precipitation, consumption, transpiration, evaporation, runoff, streamflow, and ground water flow within a basin may be performed to infer how much ground water is discharged to streams and becomes baseflow. A water budget is simply a statement of mass balance for hydrology (fig. 8). The following is the governing equation:

$$\text{Inflow} - \text{Outflow} = \text{Change in Storage}$$

Watershed modeling and ground water modeling alike rely on the water balance approach. Although straightforward in concept, water budgets are difficult to determine in practice. The primary obstacle is obtaining the requisite data in sufficient detail, spatially and temporally. Many of the difficulties in making projections with watershed models attend attempts to perform complete ground water and surface water balances on basins and sub-basins. Despite the difficulties, water budgets are useful and desirable.

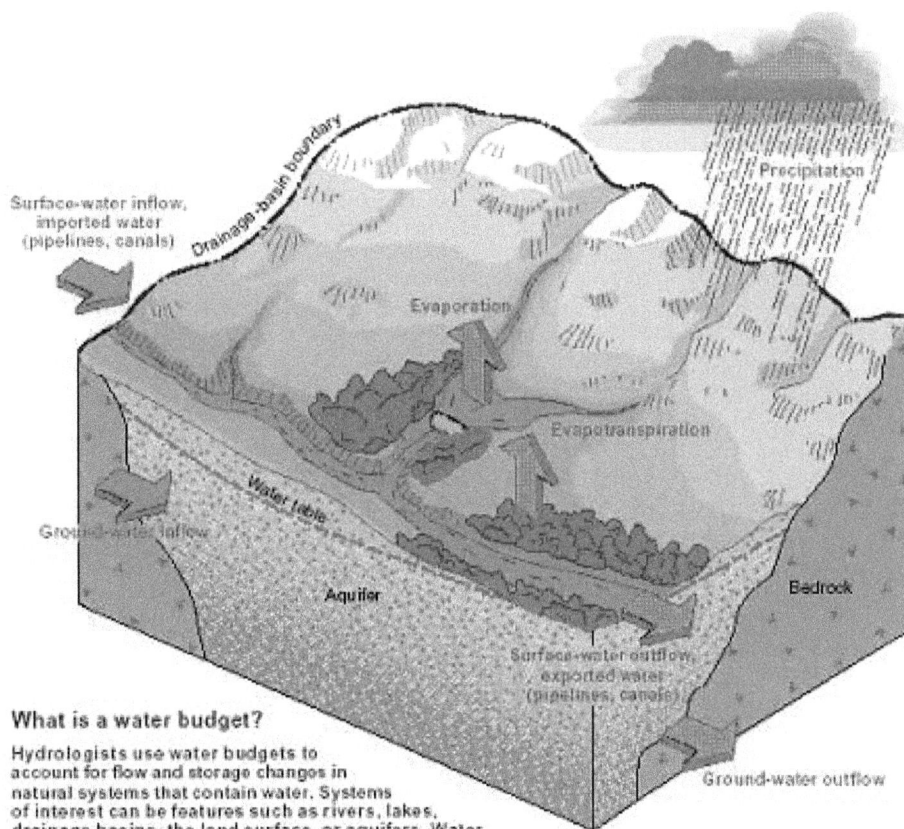

What is a water budget?

Hydrologists use water budgets to account for flow and storage changes in natural systems that contain water. Systems of interest can be features such as rivers, lakes, drainage basins, the land surface, or aquifers. Water budgets for each of these systems use the relation:

(WATER INFLOW) – (WATER OUTFLOW) = (CHANGE IN WATER STORAGE)

Typical water budget components

WATER INFLOW	WATER OUTFLOW	CHANGE IN WATER STORAGE, increased/decreased water in:
—Precipitation	—Evaporation	—Snowpack
—Surface-water flow into basin	—Transpiration by vegetation (evapotranspiration)	—Unsaturated soil zone
—Imported water	—Surface-water outflow	—Streams, rivers, reservoirs
—Ground-water inflow	—Exported water	—Aquifers
	—Ground-water outflow	

Figure 8. Water budget components for a typical watershed.

References

Fetter, C.W. 2001. Applied hydrogeology. Englewood Cliffs, NJ: Prentice-Hall, 598 p.

Freeze, R.A.; Cherry, J.A. 1979. Groundwater. Englewood Cliffs, NJ: Prentice-Hall, 604 p.

Heath, R.C. 1983. Basic ground water hydrology. Water-Supply Paper 2220. Washington, DC: U.S. Geological Survey, 84 p.

Taylor, C.J.; Alley, W.M. 2001. Ground-water-level monitoring and the importance of long-term water-level data. Circular 1217. Washington, DC: U.S. Geological Survey. 68 p. [http://water.usgs.gov/pubs/circ/circ1217/]

U.S. Geological Survey (USGS). 2002. Report to Congress. Concepts for national assessment of water availability and use. Circular 1223. Reston, VA.: U.S. Geological Survey. 34 p. [http://water.usgs.gov/pubs/circ/circ1223/]

Appendix III.
Contaminant Fate and Transport

Several mechanisms influence the spread of a contaminant in a ground water flow field. Dispersion and differences in density and viscosity may accelerate contaminant movement, while various retardation processes slow the rate of movement compared to that predicted by simple advective transport. Fetter (1999) presents a comprehensive discussion of contaminant hydrogeology. The major mechanisms of contaminant fate and transport in the subsurface are summarized below.

Ground Water Advection

In its natural state, ground water moves very slowly, but continuously. Advection is the process by which dissolved solutes are carried along with the flowing groundwater. Advecting solutes are traveling at the same rate as the average linear velocity of the ground water if the solutes are not subject to any sort of reactions with the porous media. These movement patterns are generally governed by the space occupied by the mass of the liquid flowing through the media and the rate(s) of flow encountered within these spaces. The hydraulic conductivity of a geological formation depends on a variety of physical factors within the formation, such as effective porosity; particle size, arrangement, distribution, and shape; and secondary features, such as fracturing and dissolution. Generally, hydraulic conductivity values for unconsolidated porous materials vary with particle size. Fine-grained clayey materials exhibit lower values than those of coarse-grained, sandy materials.

Effective Porosity

Effective porosity is basically an estimated parameter, because the actual measurement of the volume of interconnected pore spaces in most porous media is not known. Therefore, effective porosity is usually estimated as being somewhat less than total porosity, which is calculated from the ratios of saturated and dry porous materials. In coarse-grained materials that drain freely, effective porosity is essentially equal to total porosity and is generally defined as the ratio of the volume of water that drains by gravity to the total volume of saturated porous material.

Diffusion

Diffusion is the process by which a solute moves from areas of higher chemical potential (high concentration) to areas of lower chemical potentials (low concentration). This process is also known as molecular diffusion. Diffusion occurs in the absence of any bulk hydraulic movement of the solution; that is, solutes diffuse (spread) regardless of whether the bulk mass of liquid is static or moving through the hydrogeological medium.

Hydrodynamic Dispersion

Ground water molecules move at different rates depending on position within the aquifer and within the interconnected pores in the aquifer; some are faster than the average linear velocity while some are slower (Mills and others 1985). There are three causes for this phenomenon: friction on pore

237

walls, variations in pore sizes, and variations in path length. As ground water moves through the pores, it will move faster at the center of the pore than along the walls because of friction. In cases where different size pores exist, ground water will move through larger pores faster. Ground water molecules have tortuous flow paths and some will travel longer pathways than others. Because the invading solute-containing water is not all moving at the same rate, mixing occurs along the flow path. This mixing is termed mechanical dispersion. The mixing that occurs along the direction of fluid flow is termed longitudinal dispersion, whereas the mixing that occurs normal to the direction of fluid flow is termed transverse dispersion. Because molecular diffusion cannot be readily separated from mechanical dispersion in flowing ground water, the two are combined into a parameter called hydrodynamic dispersion. Because of hydrodynamic dispersion, the concentration of a solute will decrease over distance along the flow path. Generally speaking, the solute will spread more in the direction of ground water flow than in the direction normal to the ground water flow because longitudinal dispersivity is typically substantially higher than transverse dispersivity. Because dispersion anisotropy is often difficult to measure, a default value of a factor of 10 higher for longitudinal relative to transverse is often used. In fact, most solute plumes are long and thin.

Quantifying dispersion may be important in fate assessment, because contaminants can move more rapidly through an aquifer by this process than by simple plug flow (uniform movement of water through an aquifer with a vertical front). In other words, physical conditions, such as the presence of more permeable zones where water can move more quickly, and chemical processes, such as movement by molecular diffusion of dissolved species at greater velocities than the water, result in more rapid contaminant movement than would be predicted by ground water equations for physical flow, which assume average values for hydraulic conductivity.

Chemical Reactions

There are many types of chemical reactions that can be important in ground water systems. These include oxidation-reduction, acid-base, dissolution-precipitation, sorption, complexation, and ion exchange. A detailed discussion of the chemistry of natural waters is beyond the scope of this document. Some additional information is provided below on two of the most common reactions in contaminant transport, ion exchange and sorption. More information on aquatic chemistry is available in Chapelle (2000), Drever (1997), Langmuir (1997), Stumm and Morgan (1996), and Morel and Hering (1993).

Ion Exchange

Ion exchange processes exert an important influence on retarding the movement of chemical constituents in ground water. In ground water systems, ion exchange occurs when ions in solution displace ions associated with geological materials. This process removes constituents from the ground water and releases others to the flow system. One major consideration in ion exchange is that the exchange capacity of a given geological material is limited. A measure of this capacity is quantified in a term called "ion exchange capacity" and is defined as the amount of exchangeable ions in milliequivalents per 100 grams of solids at pH 7. Typically, clay materials such as montmorillonite exhibit greater cation (negatively charged ions) exchange capacities than other minerals such as quartz, which is the primary component of sand. This difference is attributable to the often much greater surface area of clays than other minerals.

Anionic (positively charged ions) exchange in aquifer systems is not as well understood as cationic exchange. Anions such as sulfate, chloride, and nitrate would not be expected to be retarded significantly by anion exchange because most mineral surfaces in natural water systems are negatively charged. Chloride ions may be regarded as conservative or noninteracting ions, which move largely unretarded with the advective velocity of the ground water mass.

It is important to recognize that the ion-exchange capacity of a geological material may retard contaminant movement from a waste or other source for years or even decades. However, if the source continues to supply a strongly ionic leachate, it is possible to exceed the exchange capacity of the geological material, eventually allowing unretarded transport of the contaminant. Changes in environmental conditions or ground water solution composition can also cause the release of constituents formerly bound to the geological materials.

Sorption

Sorption involves the surface interaction of a dissolved constituent with a solid material. More specifically, the term encompasses both adsorption-desorption reactions and absorption. The former refers to a buildup or a release of a constituent on the surface of a solid as a result of molecular-level interactions, while the latter implies a more or less uniform penetration of the solid by a contaminant. In many environmental settings, this distinction may serve little purpose as there is seldom information about the specific nature of the interaction. A number of factors control the interaction of a contaminant and the surfaces of soil or aquifer materials. These include chemical and physical characteristics of the constituent, composition of the surface of the solid, and the fluid media encompassing both. By gaining an understanding of these factors, logical conclusions can often be drawn about the impact of sorption on the movement and distribution of constituents in the subsurface. The failure to take sorption into account can result in a significant underestimation of the amount of a contaminant at a site, the time required for it to move from one point to another, and the cost and time involved for remediation. The properties of a contaminant that have a profound effect on its sorptive behavior include water solubility, polar/ionic character, octanol/water partition coefficient, acid/base chemistry, and oxidation/reduction chemistry.

Biotrans-formation and Biodegradation

The transformation of both organic and inorganic chemicals by microorganisms readily occurs in many subsurface environments, including landfills and septic systems. Microbial processes may be a major factor in the transformation of both natural and anthropogenic organic materials present in ground water. These transformations usually result in the formation of CO_2, CH_4, H_2, H_2S, N_2, NH_3, and NO gases, among other compounds. Under the appropriate circumstances, pollutants can be completely degraded to harmless products; whereas, under other circumstances, they can be transformed to new substances that are more mobile or more toxic than the original contaminant. Quantitative predictions of the fate of biologically reactive substances are primitive in comparison with predictions for other processes that affect pollutant transport and fate.

Biotransformations in ground water were previously thought to mimic those known to occur in surface water bodies, but detailed fieldwork has demonstrated the fallacy of this assumption. With the relatively long residence times and stable environments in ground water systems, water-table aquifers are now known to harbor appreciable numbers of metabolically active microorganisms distinctly different from those in surface waters. These ground water organisms frequently can effectively degrade organic contaminants in the subsurface that would not be effectively degraded on the surface. Thus, it is necessary to consider biotransformation as a process that affects pollutant transport and fate.

Contaminant residence time in ground water is usually long, at least measured in weeks or months, and frequently in years or even decades. Further, contaminant concentrations that are high enough to be of environmental concern are often high enough to elicit adaptation of the microbial community. For example, the U.S. Environmental Protection Agency maximum contaminant limit (MCL) for benzene is 5 ug/L. This is very close to the concentration of alkylbenzenes required to elicit adaptation to this class of organic compounds in soils. As a result, the biotransformation rate of a contaminant in the subsurface environment is not a constant, but increases after exposure to the contaminant in an unpredictable way. Careful fieldwork has shown that the transformation rate in aquifers of typical organic contaminants such as alkylbenzenes can vary as much as two orders of magnitude over a meter vertically and a few meters horizontally. This surprising variability in transformation rate is not related in any simple way to system geology or hydrology. Biological activity may promote or catalyze chemical reactions as well, and stimulation of the native microbial population and the addition of contaminant-specific "seed" microorganisms for the restoration of contaminated aquifers by in situ biological treatment has been explored vigorously.

Radionuclides in Ground Water

Most ground water sources have very low levels of radioactive contaminants ("radionuclides"). The most natural radionuclides in ground water are referred to as primordial radionuclides and have exceptionally long half-lives. These very low levels are not considered to be a public health concern. Of the small percentage of drinking water systems with radioactive contaminant levels high enough to be of concern, most of the radioactivity is naturally occurring. Certain rock types have naturally occurring trace amounts of "mildly radioactive" elements (radioactive elements with very long half-lives) that serve as the "parent" of other radioactive contaminants ("daughter products"). These radioactive contaminants, depending on their chemical properties, may accumulate in drinking water sources at levels of concern. The "parent radionuclide" often behaves very differently from the "daughter radionuclide" in the environment. Because of this, parent and daughter radionuclides may have very different drinking water occurrence patterns. For example, ground water with high radium levels tends to have low uranium levels and vice versa, even though uranium-238 is the parent of radium-226.

Most parts of the United States have very low "average radionuclide occurrence" in ground water sources; however, some parts of the country have, on average, elevated levels of particular radionuclides compared to the national average. For example, some parts of the Midwest have significantly higher average combined radium-226/radium-228 levels. On the other hand, some Western States have elevated average uranium levels compared to the national average. In general, however, average uranium levels are very low compared to the EPA Maximum Contaminant Level for drinking water throughout the United States. While there are other radionuclides that have been known to occur in a small number of drinking water supplies, their occurrence is thought to be rare compared to radium-226, radium-228, and uranium. Uranium is present in ground water in amounts ranging from 0.05 parts per billion (ppb) to 10 ppb (the median is about 1.5 ppb).

Radon-222, a naturally occurring radionuclide of concern in ground water, has a half-life of 3.8 days and is produced continuously in aquifers by the disintegration of the parent nuclide radium-226. Radioactivity in ground water is normally measured in the units of microcuries per milliliter (μCi/ml). Normal ground water contains from less than 1×10^{-7} μCi/ml to about 3×10^{-5} μCi/ml radon; the median is about 2×10^{-6} μCi/ml.

Ground water has been contaminated with radionuclides beyond background levels through the mining, refinement, and processing of uranium ore; initial production of nuclear fuels and explosives; reprocessing used reactor elements; discharge of cooling water that has been exposed to nuclear activation; escape of volatile material from evaporation and burning; dispersion of products of nuclear explosions; and the release of radionuclides used in science and medicine. The safe disposal of wastes from reactor operations and fuel reprocessing is one of the major problems in the widespread utilization of nuclear power. Disposal practices depend on the radioactivity of the waste, the general chemical character of the waste, the design of protective containment, and the physical environment of the disposal area. The radioactivity of liquid waste is broadly referred to as low level if it has fractions of a microcurie per gallon, intermediate level if it has a less than a few curies per gallon but greater than a microcurie per gallon, and high level if it has more than a few curies per gallon. Low-level wastes have been disposed into the subsurface at the National Reactor Testing Station in Idaho, at the Savanna River Plant, SC, and at Hanford, WA. Sorption on soil particles plus decay of the radionuclides with short half-lives has for the most part limited undesirable movement of contaminants into the ground water at these sites.

References

Chapelle, F. H. 2000. Ground water microbiology and geochemistry. New York: John Wiley and Sons, 496 p.

Drever, J. I. 1997. Geochemistry of natural waters, the surface and groundwater environments. Upper Saddle River, NJ: Prentice Hall, 436 p.

Fetter, C.W. 1999. Contaminant hydrogeology. Englewood Cliffs, NJ: Prentice Hall. 500 p.

Langmuir, D. 1997. Aqueous environmental geochemistry. Upper Saddle River, NJ: Prentice Hall, 600 p.

Mills, W.B.; Borcella, B.B.; Ungs, M.J.; Gherini, S.A.; Summers, K.V.; Lingsung, M.; Rupp, G.L.; Bowie, G.L.; and Haith, D.A. 1985. Water Quality Assessment: A screening procedure for toxic and conventional pollutants in surface and ground water, Parts 1 and 2. Report EPA 600/6-85/002a,b. Athens, GA: Environmental Research Laboratory, U.S. Environmental Protection Agency.

Morel, F. M. M.; Hering, J. G. 1993. Principles and applications of aquatic chemistry. New York: Wiley-Interscience, 608 p.

Stumm, W.; Morgan, J.J. 1996. Aquatic Chemistry, Chemical Equilibria and Rates in Natural Waters, 3rd ed. New York: John Wiley & Sons, Inc., 1022 p.

Appendix IV.
Ground Water Remediation

Once ground water is contaminated, it is difficult and typically very expensive to restore to natural or pre-contamination conditions. The broad range of chemical, physical, and biological characteristics of the thousands of potential ground water contaminants coupled with the complex heterogeneities of subsurface flow and contaminant transport make it very difficult to determine the exact nature and extent of ground water contamination in a given area or aquifer. If the value of the ground water that has been contaminated is great enough, it is very important to conduct an appropriate remedial investigation that is aimed at determining the nature and extent of the contaminant or contaminants. This remedial investigation is then used to scope and conduct a feasibility study that will focus on evaluating potential remedial options.

Strategies and technologies that are typically used to remediate contaminated ground water include the following general categories:

(1)   Those aimed at removing or controlling the source of contamination.
(2)   Those aimed at hydraulically controlling the contaminant plume(s) to isolate the contaminated ground water.
(3)   Those that include treatment of the contaminated ground water, either in situ or by collecting, treating and returning the ground water to the aquifer.

The decision regarding which remedial option is appropriate for a given situation depends largely on the following factors:

Source Removal

(1)   The compatibility of the remedial option to the hydrogeologic setting.
(2)   The ability to achieve the remedial goals.
(3)   The cost and time required to implement the remedy.

The objective of source removal is to reduce or eliminate the volume of waste (solid or liquid) or non-waste that is the source of the ground water contaminant(s). Removal should stop or minimize ongoing contamination; however, it is important to not transfer the problem from one location to another. To determine if source removal is a viable option, it is necessary to consider the following items: (1) problems associated with excavation and transport of the source material; (2) accessibility, distance, and road conditions between the origin and disposal sites; (3) cost; and (4) political, social, and legal factors.

Source Control

Over the last couple of decades, a multitude of measures have been developed to control contaminant sources; some of those have been successful under certain conditions, while others have not demonstrated much success. Two commonly utilized source control measures, surface runoff control and ground water barriers, are discussed below. Both of these options are aimed at preventing water from moving into and through the contaminant source, thus minimizing or stopping the leaching and subsequent transport of contaminants. A complete discussion of potential source control measures is beyond the scope of this document.

Surface-runoff controls. Surface-runoff control measures are used to minimize or prevent infiltration of precipitation and overland flow. Overland flow over an area of concern can be prevented by contouring the land using dikes, berms, ditches, terraces, benches, levees, and sedimentation basins. These features can be used to divert or collect the overland flow to prevent infiltration. If feasible, a contaminated site or buried waste (landfill, mine waste) can be capped to control or prevent infiltration of precipitation into the underlying waste. In areas with low annual precipitation, a water balance cap may be appropriate. A water balance cap is constructed in a way that allows for infiltration of precipitation at a rate close to the uptake rate of grasses, shrubs or trees that are planted on top of the cap. This infiltration allows for a revegetated cap and minimizes water management considerations. In areas where a water balance cap is not appropriate, a low-permeability cap can be installed to divert most precipitation off the cap to a collection or diversion system. The goal of this type of cap is to limit infiltration into the underlying waste. Caps can be constructed of native soils, clays, synthetic membranes, or a combination of these materials. Revegetation can also be a cost-effective method for helping to control overland flow and infiltration, especially if combined with contouring and/or capping. Vegetation reduces the impact of rainfall, decreases overland flow velocity, and strengthens soil structure.

Ground water barriers. Ground water barriers are designed to stop ground water flow into, through, or from a certain location, thus limiting the mixing of uncontaminated ground water with contaminated ground water or source materials. Ground water barriers are commonly used in combination with other treatment strategies and technologies, such as pump and treat. Common types of barriers include (1) slurry trench walls, (2) grout curtains and seals, and (3) cutoff walls.

Slurry trench walls are suitable for placing upgradient of a contaminated site to limit ground water flow through the site, downgradient to limit offsite flow of contaminated ground water, or completely around a site to contain contaminated ground water. Slurry walls can be constructed so that they extend well below the water table, if desirable. A slurry wall is constructed by excavating a trench to the desired depth (up to 100 feet, under appropriate conditions) and backfilling the trench with a slurry mixture that forms the final wall. The mixture can be composed of soil and bentonite, cement and bentonite, or concrete. Generally, a soil slurry should contain 5 to 7 percent by weight suspension of bentonite in water. The slurry will provide for trench wall stability and forms a low-permeability filter cake on the walls. Slurry walls are reported to have long service lives. Two separate slurry walls were constructed along the boundary of the Rocky Mountain Arsenal near Denver, Colorado in the early 1980s and are reported to still be in operation.

Grouting involves the pressure-injection of stabilizing materials into the subsurface to fill and seal voids, cracks, and fissures. Subsurface grouting has been used for decades in geotechnical applications related to subsurface and dam construction. Grouts are usually composed of mixtures of materials involving bentonite and cement with small amounts of additives to promote penetration and manage the set time. For application to ground water contamination problems, grout curtains are typically formed by injecting grout through tubes. Grout curtains can create a fairly effective barrier to ground water movement, depending on the degree of completeness of the curtain. The amount of grout needed is a function of the volume of void space, the density of the grout, and pressures required for injecting the grout. Grout can also be used to seal the bottom of an excavation or waste impoundment. Grout is injected through drill holes to form a curved or horizontal barrier to prevent downward migration of a leachate.

Hydrodynamic Controls

A properly located array of recharge and discharge wells can be used to prevent a ground water contamination plume from (1) moving into the zone of influence of a water supply well or well field, (2) moving into another aquifer or aquifer zone, or (3) connecting to surface water. By controlling the rate of ground water discharge and recharge at selected locations and vertical intervals, the magnitude and direction of hydraulic gradient of the water table or potentiometric surface can be controlled. It can allow the location, direction of movement and velocity of a plume to be controlled. Discharge wells can be used to create cones of depression with known diameter and depths. Recharge wells can be used to develop a hydraulic pressure ridge that will function as a hydraulic barrier.

The system design of these gradient-control techniques is very sensitive to the local subsurface geology. Optimizing system operations with regard to well construction and maintenance, pumping schedules, costs and time frames is very site specific. It is wise to utilize computer simulations to evaluate system design elements. It is important to recognize that establishing hydrodynamic control in a given situation can involve the management of large amounts of potentially contaminated water.

Ground Water Collection and Treatment

If it is legally required or desirable to treat contaminated ground water, and it cannot be achieved by in situ treatment, it will be necessary to collect the contaminated water, treat it by methods appropriate for the contaminant(s), and return it to the subsurface. The pump-and-treat system is the most common and the most successful collection and treatment technique. Depending on the site hydrogeology, the nature of the ground water contaminant(s), and the extent of the plume, an array of extraction wells is installed to remove contaminated ground water for treatment. Once treatment has been completed, the "clean" ground water is returned to the subsurface through infiltration ponds/trenches, spreading basins, or injection wells or is discharge to the surface. The location and operation of extraction wells is highly dependent on subsurface hydrogeological conditions. It is important to recognize that the operation of an array of extraction wells will result in the formation of a stagnant zone—an area downgradient of an extraction well where ground water flow is not affected by pumping. Contaminated ground water within these stagnation zones will not be collected for treatment; thus, it is important to construct and locate a sufficient number of wells to mitigate this effect.

Another significant constraint associated with pump-and-treat systems is the asymptotic decrease in concentrations of low-solubility contaminants over time as ground water flows along a geologic pathway. This slow decrease in concentrations is caused by (1) the slow release of contaminants from small pore spaces into the larger pores that comprise the primary flow paths and (2) the desorption of sorbed contaminants as the concentration in the pore waters is reduced. This phenomenon can significantly increase the time required to achieve water-quality goals. It is also important to recognize that pump-and-treat systems are effective primarily for dissolved contaminants. Contaminants that readily sorb to organic and mineral particles in the aquifer will not be readily collected by pumping. "Pulsed" pumping can be used to aid in the effectiveness of removal of some contaminants. This method involves intermittent pumping, which allows time for contaminant concentrations to come into equilibrium with regard to diffusion and partitioning. Alternating the pattern of pumping within an array of extraction wells can also modify active flow paths.

Subsurface drains are an alternative collection system that may be more appropriate in some situations than an array of extraction wells. A subsurface or "French drain" functions as an infinite line of extraction wells and creates a zone of influence in which ground water passively flows towards the drain. French drains usually have a perforated polyvinyl chloride (PVC) pipe installed within a bed of high-permeability gravel. For some applications, it can be necessary to line the bottom of the trench with a low-permeability liner.

Ground water pump-and-treat systems are commonly used in combination with other remedial methods. Examples include the use of a barrier wall or a subsurface funnel and gate to control the location of a plume and optimize collection by extraction wells and the use of surface ponds or enhanced irrigation to flush contaminants from the unsaturated zone prior to collection by extraction wells.

In Situ Remediation

In general, in situ remedial techniques have not been as effective for restoring large volumes of contaminated ground water as ground water collection and treatment systems. This is attributable to a number of factors, including:

(1) Inability to deliver the in situ treatment to all parts of a contaminated aquifer.

246

    (2)   Difficulty in maintaining the correct biological and chemical conditions for treatment optimization.

    (3)   Constraints presented by the considerable subsurface heterogeneity with respect to hydrogeological conditions.

In situ treatment techniques can be grouped into two categories: (1) physical/chemical treatment processes and (2) biological treatment processes. In situ treatment processes include the injection of a treatment medium into an aquifer. As contaminated ground water comes in contact with this medium, specific chemical/biological reactions are catalyzed, causing the contaminant to participate in reaction(s) that reduce its concentration and/or toxicity or break it down into nontoxic constituents.

Oxidation-reduction reactions can be effectively used to remediate selected metals (chrome, copper, zinc, manganese). These reactions are catalyzed by microorganisms and oxygen and can cause metals to change from toxic to less toxic or nontoxic species. For example, calcium polysulfide has proven to be effective in reducing Cr^{6+} to Cr^{3+} in a high-permeability valley-fill aquifer along the South Platte River in Denver. Cr^{6+} concentrations have been reduced by two to three orders of magnitude after 2 to 3 years of injecting a calcium polysulfide slurry into the aquifer. Contaminants that are held within the aquifer by sorption can be mobilized by the introduction of a solvent or surfactant which can enhance the solubility of the sorbed contaminant. Examples of physical processes that can immobilize or reduce dissolved concentration include precipitation, volatilization, and polymerization.

Biological treatment below ground involves the injection of nutrients and oxygen into the contaminated aquifer to enhance the activity of microorganisms that utilize the contaminant in their metabolic processes. This treatment facilitates the breakdown of toxic organic compounds into nontoxic constituents and results in the destruction of the contaminant. To effectively utilize subsurface biological treatment, it is necessary to (1) control the anaerobic/aerobic conditions, (2) provide the correct amount and timing of nutrients and (typically) oxygen to the consortium of microorganisms, (3) understand what the likely degradation products will be, and (4) maintain optimal conditions for the microorganisms for the period of time required to meet the water-quality goals. Experience at experimental sites as well as regulated ground water contamination sites has shown that it is difficult to maintain ideal chemical and biological conditions for the microorganisms to be effective over long periods of time. Delivery of the treatment medium to all of the contaminated parts of the aquifer has also proven to be difficult.

Performance Monitoring

A performance monitoring program must be developed and implemented at ground water contamination sites that are being, or have been, remediated. The performance monitoring plan should be designed to provide data that can be used to determine whether the remedies that were utilized have achieved the established water-quality goals. Ground water-quality goals for any particular ground water contamination site are established based on legal and political requirements, use requirements, and/or the constraints of remediation technology. When developing a performance monitoring program, the following factors should be considered:

(1) The extent of the ground water contamination.
(2) The potential receptors of the contaminated ground water, such as a stream or water-supply well.
(3) The applicable regulatory requirements.
(4) The hydrogeological setting.
(5) The sampling frequency and methodology.
(6) An appropriate parameter list.
(7) Sample collection, transport, and analysis.
(8) Sound quality assurance and quality control procedures.

In general, performance monitoring programs should include the following four features:

(1) Clearly established compliance locations.
(2) Clearly established compliance limits and schedules.
(3) Early warning and trigger-level limits and locations.
(4) Appropriate contingency measures to be implemented in the event compliance cannot be achieved.

When developing plans for managing a ground water contamination site, it is important to allocate an appropriate budget, staff time, field time, and lab time.

Appendix V.
Ground Water Modeling

The Modeling Process

A generalized process for developing a ground water model includes the following components: model conceptualization, code selection, model design, model calibration, sensitivity analysis, and prediction (fig. 1). Few modeling studies will incorporate all of the steps in the process shown in figure 1; however, all should include the steps through calibration and sensitivity analysis, and the model should be completely documented in a written report.

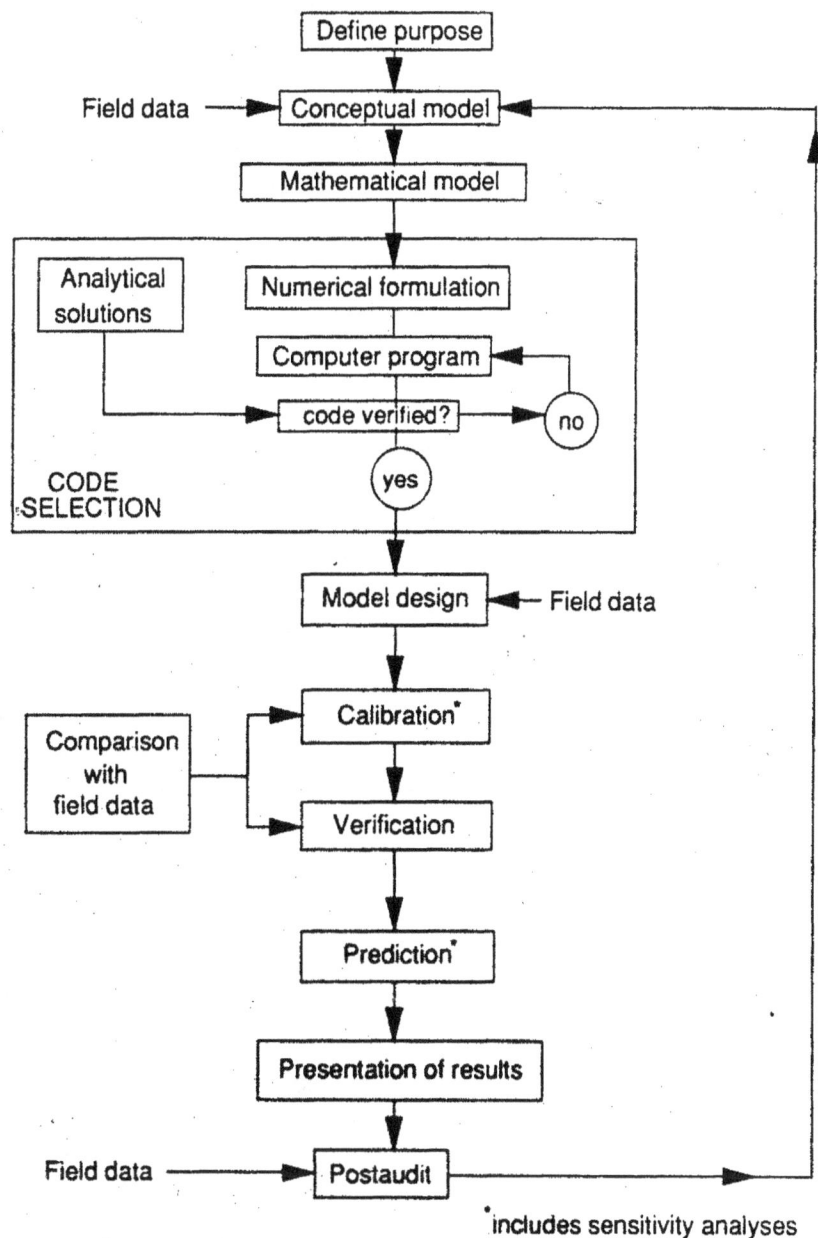

Figure 1. A generic process for modeling (Anderson and Woessner 1992).

The ideal modeling process would include the following steps (Anderson and Woessner 1992):

1. Establish the purpose of the model. The purpose will determine what type of model is needed and which codes should be considered.

2. Develop a conceptual model of the hydrologic system of interest. Hydrostratigraphic units and system boundaries are identified. Field data are assembled, including information on the water balance and data needed to assign values to aquifer parameters and hydrological stresses. A visit to the field site is essential during this step, to provide the modeler with information that cannot be adequately conveyed on a map or in a report. At this stage, a hydrogeological framework for the study area is developed by means of hydrogeologic maps and sections.

3. Select the appropriate computer code or codes for simulation. There are many codes commercially available for aquifer simulation, and the code(s) selected should be able to adequately simulate the field conditions of importance for the study. For example, if the study area contains multiple aquifers critical to the study that are separated by confining units, then code(s) that can simulate a fully three-dimensional system should be used. Whichever code is selected, it should be one that is verified (either by the developer or someone else) and fully documented.

4. Design the model. The conceptual model is put into a form suitable for modeling. This step includes design of the model grid, setting boundary and initial conditions, and selection of values for aquifer parameters and hydrologic stresses. Figure 2 shows an example of the relation between geologic units, hydrogeologic units, and model layers that is typical of the conversion of the conceptual model to a computer model.

5. Calibrate the model. The purpose of calibration is to establish that the model can reproduce observed heads and flows, such as spring flow or measured ground water contribution to streams. Values for aquifer parameters and stresses are systematically adjusted through a reasonable range of values until the differences between simulated and observed heads and flows are minimized. Calibration can be done by a trial-and-error process or automatically by using a parameter-estimation code such as UCODE (Poeter and Hill 1998) or PEST (Doherty 1994). This step is sometimes called solving the inverse problem. The calibration process may result in identification of areas of data deficiencies that need to be filled before the model can be adequately calibrated, or the process may result in a redefinition of the conceptual model. In the latter cases, the modeler may have to go back to Step 2, collect new data, and possibly redefine the conceptual model before proceeding to Step 6. Hill (1998) provides detailed guidance on model calibration.

6. Perform a sensitivity analysis for the calibrated model. The calibrated model can contain a large degree of uncertainty and nonuniqueness, because of the inability to exactly define the spatial and temporal distribution of aquifer parameters and hydrological stresses. During a sensitivity analysis, the calibrated parameters and stresses are varied over the range of uncertainty or over a range of hydrogeologically reasonable values to establish how calibrated results may vary because of this uncertainty.

7. Conduct a model verification. During model verification the calibrated model is used to reproduce a second set of observed heads and flows. For example, the model may have been calibrated to hydrological conditions for the period 1960–70, and the model is then used to simulate conditions observed for 1970–90 as a verification. If these later conditions are simulated within accepted criteria, then greater confidence can be placed in the model as a representation of the real world.

8. Use the model to predict the response to a new set of anticipated or proposed hydrological stresses. The model is run with calibrated values for parameters and stresses, except for those stresses that are expected to change in the future.

9. Perform a sensitivity analysis for the predictive model. Uncertainty in the predictive simulation results is the result of uncertainties inherent in the calibrated model and the inability to accurately predict future hydrological conditions and stresses. The sensitivity analysis helps to bracket the range of possible predicted responses to the simulated stresses and reduce the uncertainty.

10. Document the model design, process, and results to effectively communicate the modeling effort for other potential users of the model or model results. It is important that this documentation include a discussion of model uncertainty and any limitations on future use of the model. For example, if the model was calibrated only to steady-state conditions, it may not be appropriate to use it to simulate transient conditions without additional calibration.

11. Conduct a "postaudit" of the model predictions. A postaudit is conducted several years after the modeling study is completed. New field data are collected to determine if the model predictions were accurate. Postaudits often show that model predictions were not accurate, primarily because simulated stresses did not accurately duplicate those that actually occurred. For example, a new well field may have come on line or a new surface-water source was developed that reduced the amount of pumping from an existing well field, and these conditions were not included in the predictive simulation.

12. Redesign and recalibrate the model, either as a result of the postaudit or because new data have been collected in areas where data were previously lacking. The redesigned model should also be documented.

The degree to which the modeling study incorporates all or some of these steps is dependent on the objectives of the study and the amount of resources (time, personnel, and funding) available for the study. Additional details for each of these steps are provided in Anderson and Woessner (1992).

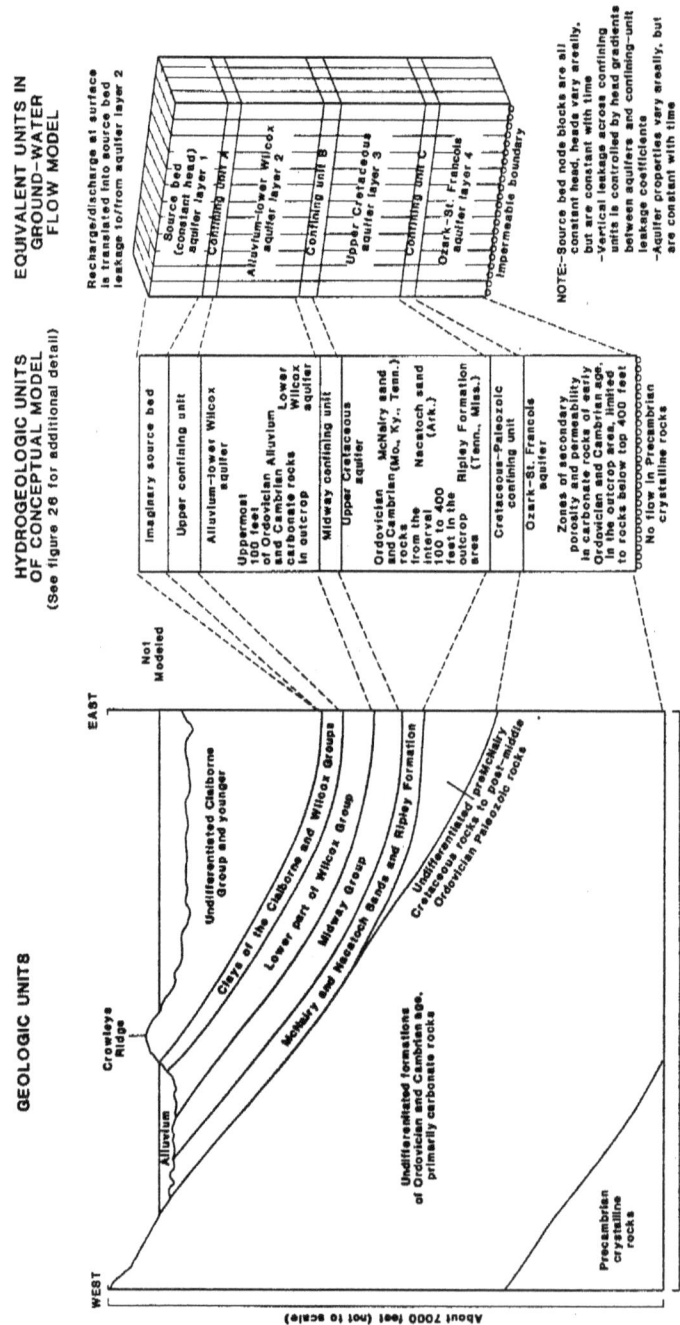

Figure 2. The relationship between geologic units, hydrogeologic units, and model layers for the northern Mississippi Embayment area (Brahana and Mesko 1988).

252

Types of Computer Models

The purpose of the study will determine the type of model and the computer code that are needed to achieve the study objectives. If the purpose of the study is simply to predict the effect of a new well field on the ground water system, a flow-model code such as MODFLOW (McDonald and Harbaugh 1988) or GFLOW (Haitjema Consulting, Inc.) may be all that is required. If the purpose of the study is to predict the fate of a contaminant from a spill of hazardous chemicals, then a solute-transport code such as MOC3D (Konikow and others 1996) will also be required. If processes in the unsaturated zone are important to the study, then a code such as TOUGH2 (Pruess 1991) may be required. Geochemical models that account for changes in ground water quality as a result of water-rock interactions in an aquifer are also valuable in conceptualizing and evaluating flow systems. Examples of geochemical models include MINTEQA2 (Allison and others 1991), PHREEQC (Parkhurst and Appelo 1999), and NETPATH (Plummer and others 1994). Each code has its own capabilities and limitations, and the modeler must carefully define what is required of the model when selecting the code to use.

There are two basic numerical methods used to incorporate the ground water flow equations into a model code: finite-difference methods and finite-element methods. Each of these methods requires discretizing the real world into distinct blocks for which the ground water flow equations are approximated. A detailed discussion of these methods is beyond the scope of this section. It is sufficient here to say that finite-difference model codes discretize the flow system into rectangular blocks and the flow equations are approximated for a point at either the center of the block or at the corners of the block. Figure 3 shows an example of a finite-difference grid. Finite-element codes discretize the flow system into triangular or other polygonal-shaped blocks (that can also be rectangular), and approximate the flow equations at the corners of the blocks. Finite-element codes allow the modeler to more closely approximate the shape of highly irregularly shaped aquifer systems and better represent seepage faces than a finite-difference code. Figure 4 shows an example of a finite-element grid.

Several numerical codes exist for parameter-estimation-based calibration of a model (see modeling procedure Step 5). Codes such as PEST (Doherty 1994) and UCODE (Poeter and Hill 1998) are independent of the flow model used. MODFLOW2000 (Hill and others 2000) includes a parameter-estimation process that runs inside of the MODFLOW code. In parameter estimation, the hydrological parameters to be estimated such as hydraulic conductivity and recharge are automatically adjusted within a preset range to minimize the difference between simulated and observed heads and fluxes. Benefits of parameter estimation include the quantification of (1) the quality of the calibration, (2) data needs, and (3) confidence in estimates and predictions (Poeter and Hill 1997).

A more recent type of model code uses the "analytic-element" method, which does not require the use of a model grid or specification of boundary conditions at the grid perimeter (Strack 1989; Haitjema 1995). An analytic-element

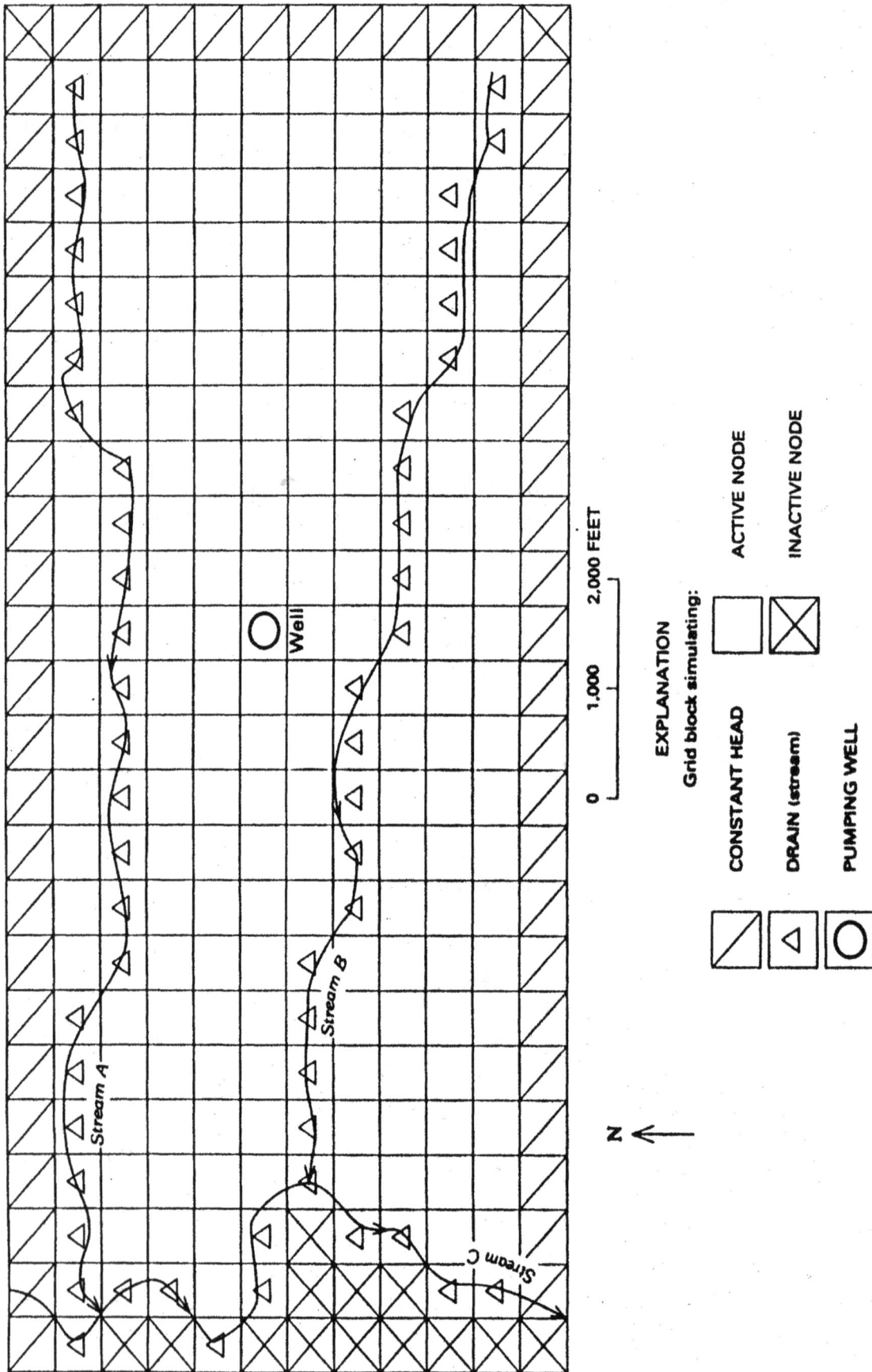

Figure 3. Example of a finite-difference model grid and boundary conditions (Tucci 1994).

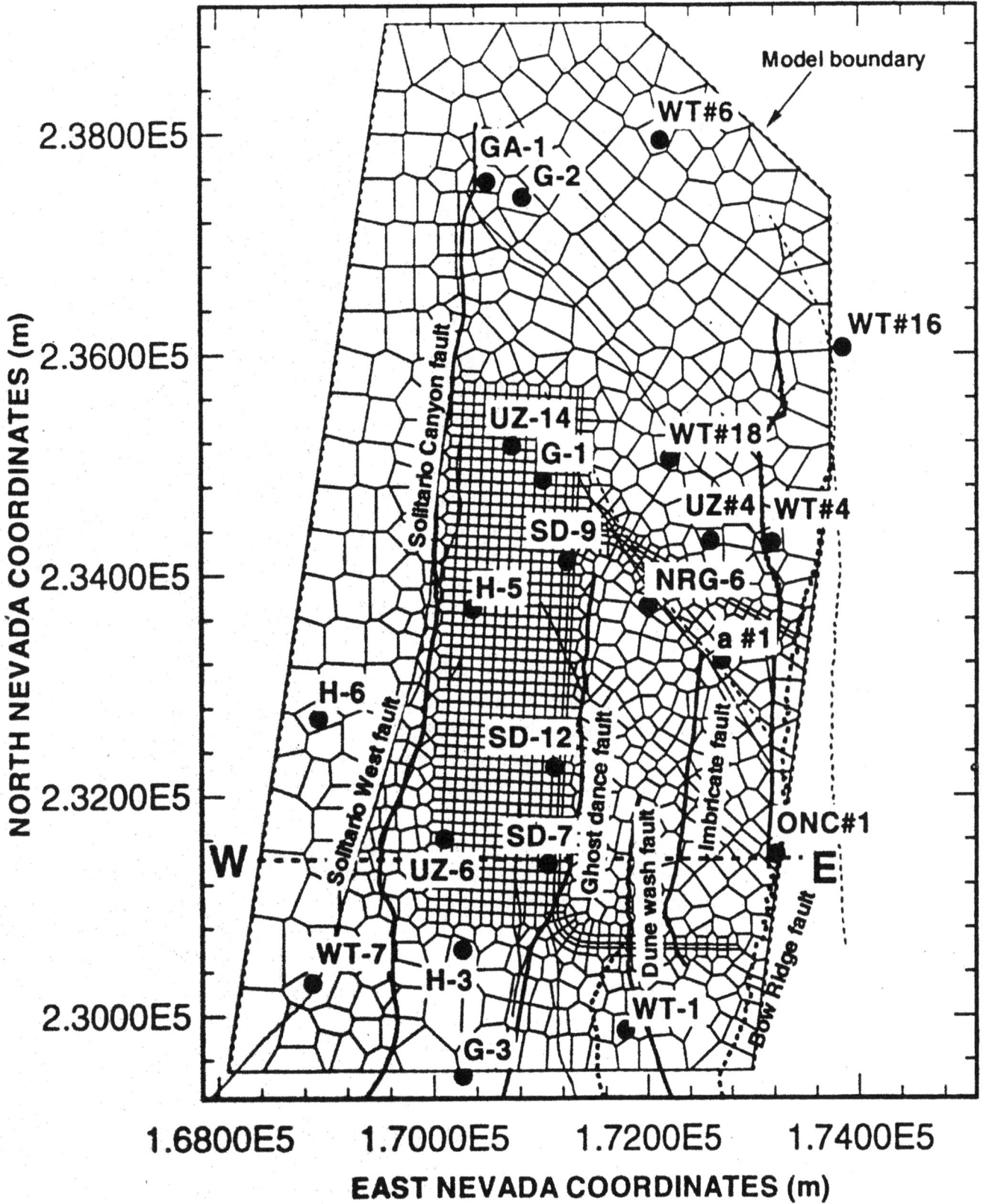

Figure 4. Example of a finite-element model grid (Wu and others 1999).

model uses superposition of closed-form analytical solutions to the differential equation to approximate both local and regional flow. These models allow for representation of large domains that include many hydrologic features outside the immediate area of interest, and easy modification of the regional flow field by adding analytic elements representing regional hydrologic features (Hunt and others 1998).

Data Needed for Models

Data needed for ground water flow models can be considered to fall into three general categories:

1. *Data needed to define the physical and hydrogeological framework.* Topographic maps, geological maps, cross sections, well and boring logs (driller's, geological, geophysical), well-construction information, maps showing the areal extent and thickness of aquifers and confining units, hydraulic properties of aquifers and confining units, and streambed and lakebed characteristics.
2. *Data needed to define the water budget.* Precipitation, evapotranspiration, streamflow, springflow, and pumping.
3. *Data needed to define the flow system.* Ground water levels, potentiometric-surface maps, stream stages, lake stages, and spring discharge elevations. Information about how these data vary with time is also needed.

If simulation of solute transport is needed, additional data are needed on water-quality characteristics of the aquifer and/or contaminant plumes over time. Of course, if the model is being developed solely to test hydrological concepts that are generic in nature, no real data are required. The types and amount of data required, therefore, are directly related to the objective(s) of the modeling study.

A Word of Caution

According to Anderson and Woessner (1992, 6), "Modeling is an excellent way to help organize and synthesize field data, but it is important to recognize that modeling is only one component of a hydrogeologic assessment and not an end in itself." In fact, the process of assembling and understanding the field data required for model input may provide the modeler with the answer to the problem before ever running the model. Conversely, a model that is based on inadequate field data can produce erroneous results that may not be obvious in the colorful graphical output from modern modeling software. The modeler must have some basic understanding of the geology and hydrology of the area being modeled, or should work in close collaboration with others who do have that understanding. In this way, model results that are hydrogeologically unreasonable, or that are based on unrealistic or erroneous data, can be recognized and addressed.

Private consultants, university researchers or other government agencies (e.g., USGS, state geological surveys) are often contracted to develop ground water flow models. Careful review and evaluation of such models on the part of the user is important to insure that the modeling was done correctly and fulfills the contract obligations. Reilly and Harbaugh (2004) provides guidelines for evaluating models, and is also useful in planning a modeling study.

References

Allison, J. D.; Brown, D. S.; Novo-Gradac, K. J. 1991. MINTEQA2/PRODEFA2, A geochemical assessment model for environmental systems: version 3.0 user's manual. Report EPA/600/3-91/021. Athens, GA: Environmental Research Laboratory, U.S. Environmental Protection Agency, 107 p.[http://www.epa.gov/ceampubl/mmedia/minteq/index.htm].

Anderson, M.P.; Woessner, W.W. 1992. Applied groundwater modeling, simulation of flow and advective transport. San Diego, CA: Academic Press, Inc., 381 p.

Brahana, J.V.; Mesko T.O. 1988. Hydrogeology and preliminary assessment of regional flow in the upper Cretaceous and adjacent aquifers in the northern Mississippi Embayment. Water-resources Investigations Report 87-4000. Washington, DC: U.S. Geological Survey. 65 p.

Doherty, J. 1994. Manual for PEST package – Fifth Edition. Watermark Numerical Computing, Brisbane, Australia, 2002.

Haitjema, H. M. 1995. Analytic element modeling of groundwater flow. San Diego: Academic Press, 394 p.

Hill, M.C. 1998. Methods and guidelines for effective model calibration. Water-resources Investigations Report 98-4005. Washington, DC: U.S. Geological Survey. 90 p. [http://water.usgs.gov/nrp/gwsoftware/modflow2000/WRIR98-4005.pdf].

Hill, M.C.; Banta, E.R.; Harbaugh, A.W.; Anderman, E.R. 2000. MODFLOW-2000, the U.S. Geological Survey modular ground water model: User guide to the observation, sensitivity, and parameter-estimation processes and three post-processing programs. Open-file Report 00-184. Washington, DC: U. S. Geological Survey, 209 p. [http://water.usgs.gov/nrp/gwsoftware/modflow2000/ofr00-184.pdf].

Hunt, R.J.; Anderson, M.P.; Kelson, V.A. 1998. Improving a complex finite-difference ground water flow model through the use of an analytical element screening model. Ground Water. 36 (6): 1011-1017.

Konikow, L.F.; Goode, D.J.; Hornberger, G.Z. 1996. A three-dimensional method-of-characteristics solute-transport model (MOC3D). Water-resources Investigations Report 96-4267. Washington, DC: U.S. Geological Survey, 87 p.

McDonald, M.G.; Harbaugh, A.W. 1988. A modular three-dimensional finite-difference ground water flow model. *In* Techniques of Water-resources Investigations. Book 6, Chapter A1. Washington, DC: U.S. Geological Survey, 586 p. [http://water.usgs.gov/pubs/twri/twri6a1].

Parkhurst, D.L.; Appelo, C.A.J. 1999. User's guide to PHREEQC (Version 2): A computer program for speciation, batch-reaction, one-dimensional transport, and inverse geochemical calculations. Water-resources Investigations Report 99-4259. Washington, DC: U.S. Geological Survey. 312 p.

Plummer, L.N.; Prestemon, E.C.; Parkhurst, D.L. 1994. An interactive code (NETPATH) for modeling NET geochemical reactions along a flow PATH-- Version 2.0. Water-resources Investigations Report 94-4169. Washington, DC: U.S. Geological Survey. 130 p.[http://water.usgs.gov/software/netpath.html].

Poeter, E.P.; Hill, M.C. 1997. Inverse models: A necessary next step in ground water modeling. Ground Water. 35 (2): 250-260.

Poeter, E.P.; Hill, M.C. 1998. Documentation of UCODE, a computer code for universal inverse modeling. Water-resources Investigations Report 98-4080. Washington, DC: U.S. Geological Survey. 116 p. [http://water.usgs.gov/software/ucode.html].

Pruess, K. 1991. TOUGH2: A general-purpose numerical simulator for multiphase flow and heat flow. Berkeley, CA: Lawrence Berkeley Laboratory Report LBL-29400, 102 p.

Reilly, T.E.; Harbaugh, A.W. 2004. Guidelines for evaluating ground water flow models. Scientific Investigations Report 2004-5038. Washington, DC: U.S. Geological Survey. 30 p. [http://water.usgs.gov/pubs/sir/2004/5038].

Strack, O.D.L. 1989. Groundwater mechanics. Engelwood Cliffs, NJ: Prentice Hall. 732 p.

Tucci, P. 1994. Simulated effects of horizontal anisotropy on ground- water flow paths and discharge to streams. *In* Marston, R.A., and Hasfurther, V.R. (eds.) Effects of human- induced changes on hydrologic systems. New York, NY: American Water Resources Association Technical Publications Series TPS-94-3, pp. 851-860.

Wu, Y.S.; Ritcey, A.C.; Bodvarsson, G.S. 1999. A modeling study of perched water phenomena in the unsaturated zone at Yucca Mountain. Journal of Contaminant Hydrology. 38: 157-184.

Appendix VI.
Water-quality Data: Statistics, Analysis, and Plotting

This section describes some statistical procedures recommended for analysis of water-quality data. Most of these statistical procedures are explained in Gilbert (1987), U.S. Environmental Protection Agency (1992), or Helsel and Hirsch (1992), Sanders and others (2000).

Statistics for Water-quality Data

Water-quality data possess unique characteristics that may require specialized approaches to statistical analysis. Data sets generally have a base limit of zero because only positive values are possible for most parameters, and can contain censored (less than) values, outliers, multiple detection limits, missing values, and serial correlation. These characteristics commonly present problems in the use of conventional parametric statistics based on an assumption of normally distributed data sets. The presence of censored data, non-negative values, and outliers may lead to an asymmetric or non-normal distribution instead of a normal, symmetric, or bell-shaped (Gaussian) distribution, which is common for many data sets. These skewed data sets may require use of specific nonparametric statistical procedures for their analysis. The use of nonparametric statistical procedures is also preferred when determining trends of many constituents at multiple stations. Additionally, nonparametric statistical tests are more powerful when applied to non-normally distributed data, and almost as powerful (under certain conditions) as parametric tests when applied to normally distributed data (Helsel and Hirsch 1992).

Seasonal Variation

A major cause of variation in water-quality data is the effect of seasonality, which needs to be compensated for to discern specific anthropogenic or natural processes that affect water quality over time. Seasonal variation may be the result of a variety of conditions, including specific land-use practices, biological activity, or changes in sources and volumes of water. As an example, precipitation-induced stream discharge may predominate during specific months of the year, whereas baseflow (driven by ground water seepage) may be dominant at other times of the year. Another example is the increase in biological activity that occurs in surface waters during summer because of warmer temperatures. The result may be seasonal variation in nutrient concentrations.

Trend Analysis

A trend in water quality is defined as a monotonic change in a particular constituent with time. Investigators must employ parametric and nonparametric tests that are designed to deal with characteristics unique to water-quality data. Because all data may not have been collected at the same frequency for the duration of a project or monitoring program, specific seasonal definitions are needed to prevent bias in the trend results.

The statistical approach that is used to compensate for seasonal variability in water-quality data is the distribution-free, nonparametric seasonal Kendall trend test. This test, modified from the Mann-Kendall test (Helsel and Hirsch 1992), compares relative ranks of data values from the same season. For example, January values are compared to January values, February values are compared to February values, and so forth. No comparisons are made across seasonal boundaries. A plus value is recorded if the subsequent value in time is higher, and a minus is recorded if the subsequent value in time is lower. If pluses predominate, a positive trend exists; if minuses predominate, a negative trend results. No trend is the result of pluses and minuses being equal. The null hypothesis is that the concentration of the water-quality constituent is independent of time (Smith and others 1982). The test assumes that the data are independent and from the same statistical distribution. The seasonal Kendall test statistic is the summation of the Mann-Kendall test results from all the seasons. The attained significance level (or p-value) is the probability of incorrectly rejecting the null hypothesis of no trend when actually there is a trend. The seasonal Kendall slope estimator is computed according to the method of Sen (1968); it is the median slope of all the pairwise comparisons from all of the seasons expressed as rate of change per year in original units (usually in milligrams per liter depending on the constituent) and in percent per year.

Recommended Statistical Procedures

Summary Statistics

Summary statistics are simple procedures that allow an investigator to quickly analyze a data set.

Time-Series Plot. Displays the variability in concentration levels over time for constituents and can be used to examine possible outliers. More than one station can be compared on the same plot to look for differences between stations. They can be used to examine the data for trends.

Histogram. Displays the frequency distribution for constituents. More than one station can be compared on the same plot to look for differences between stations.

Box-and-Whiskers Plot. An efficient way to visually display the distribution of data for constituents at a given station. This plot can also be utilized by year or by season. The plot locates the median and the 25th and 75th quartiles and the whiskers extend to the minimum and maximum values of the data set.

Wilcox Diagram. This plot can be used to quickly determine the viability of water for irrigation purposes (also known as the USDA diagram).

Kruskal-Wallis Test. Tests for seasonality. If seasonality is found to exist in a time series of concentrations, then data can be deseasonalized prior to running further statistical analyses.

Rank Von Neumann. This procedure tests for serial correlation at a station and also reflects the presence of trends or cycles such as seasonality.

Statistical Outlier Test (Dixon's Test). The outlier test identifies data points that do not appear to fit the distribution of the rest of the data set and determines if they differ significantly from the rest of the data set.

Shapiro-Wilk Test, Shapiro-Francia Test, Chi-Squared-Goodness-Fit Test. These tests, called normality statistics, evaluate the distribution of the data.

Censored Data Substitution Functions

Detection Limit Substitution. All censored data are usually corrected to half of the detection limit or to the detection limit of the least sensitive analytical procedure prior to running statistical analyses. This procedure provides a closer comparison between samples and time periods, but results in a large loss of information. This procedure should not be used if the percentage of censored data exceeds 50 percent of the total number of records.

Cohen's Adjustment, Aitchison's Adjustment. As alternatives to replacement of the detection limit with arbitrary constant values, these procedures calculate a corrected sample mean that accounts for data below the detection limit. The methods use probability theory to estimate the shape of the tail of the population probability density function that was censored, thus preserving the sample variability and mean that would have been estimated had the detection limit been zero and had no values been censored.

Mean/Median Analysis

Analysis of Variance (ANOVA) (Interstation or Intrastation; Parametric or Nonparametric). Compares the means or median values of different groups of observations to determine if a statistical difference exists among groups.

Mann-Whitney Test (Interstation or Intrastation). Tests whether the measurements from one population are significantly higher or lower than those of another population.

Trend Analysis

Sen's Slope, Mann-Kendall Trend Tests. Used to detect a general increase or decrease in observed values over time and determine the significance and magnitude of the trend.

Seasonal Kendall Test. This test is an extension of the Mann-Kendall test that removes seasonal cycles and tests for trend.

Shewhart-CUSUM Control Chart. These charts monitor the statistical variation of data collected at a station and flag anomalous results. If a result falls outside the predetermined control limits, then the process is considered "out of control."

261

Excursion Analysis

Proportion Estimate. This test computes the proportion of observations in the record which exceed a stated excursion limit and computes a confidence limit.

Tolerance Limit (Interstation or Intrastation). Tolerance limits define an interval that contains a specified fraction of the population with specified probability. They are used to compare concentrations from compliance stations to the upper limit of the tolerance interval.

Prediction Limit (Interstation or Intrastation). Used to determine whether a single observation is statistically representative of a group of observations.

Confidence Interval. A confidence interval is constructed from sample data and is designed to contain the mean concentration of a station analyte, with a designated level of confidence.

Graphical Display of Chemical Data

The following graphical methods are used for displaying water-quality data on maps or for analyzing different water types and compositions.

Bar Chart. The most widely used graphical procedure for displaying ion concentrations is the vertical bar system developed by Collins (1923). This method uses a vertical bar whose weight is proportional to the total concentration of anions or cations in milliequivalents per liter (fig. 1). Horizontal lines are used to separate the concentrations of various ions. Usually, six divisions are used, but more can easily be added if required.

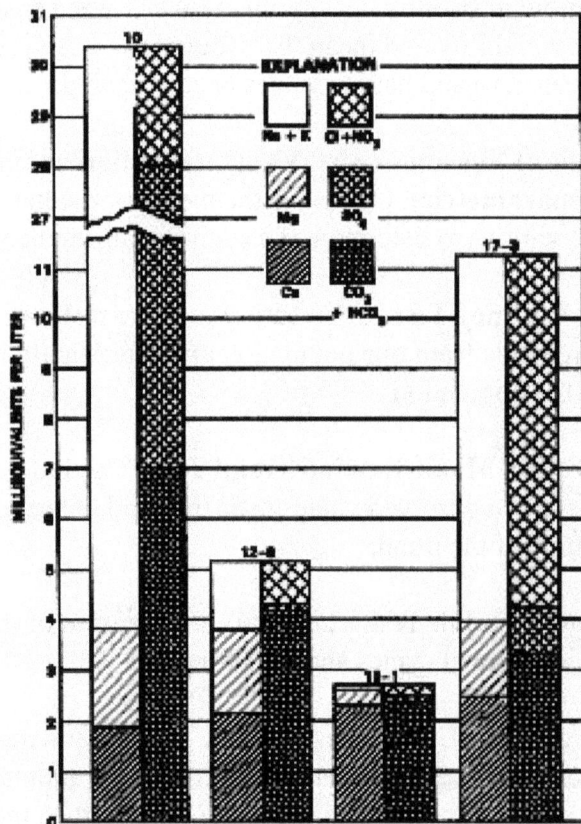

Figure 1. Example of a Collins bar chart for depicting water quality. The numbers above the bars refer to specific analysis taken from a data table. In this way, specific sampling locations can be identified in the table and correlated to the bar chart (Hem 1989).

Stiff Diagram. This is a pattern plot and can be used to evaluate the change in water quality at a location over time, or as the water passes through different geologic formations or subsurface conditions (Stiff 1951). This method uses four parallel horizontal axes extending from each side of a vertical zero axis (fig. 2). Concentrations of four cations may be plotted, one on each axis to the left of zero, and four anion concentrations in milliequivalents per liter may be plotted on each axis to the right of zero. This method gives a distinctive pattern, and is very useful in depicting water composition differences or similarities. The pattern for a particular water source tends to maintain its shape, even with concentration or dilution of the constituent. Thus, a study of water-quality patterns can often be utilized to identify different producing strata, and correlate water sources with strata over an area.

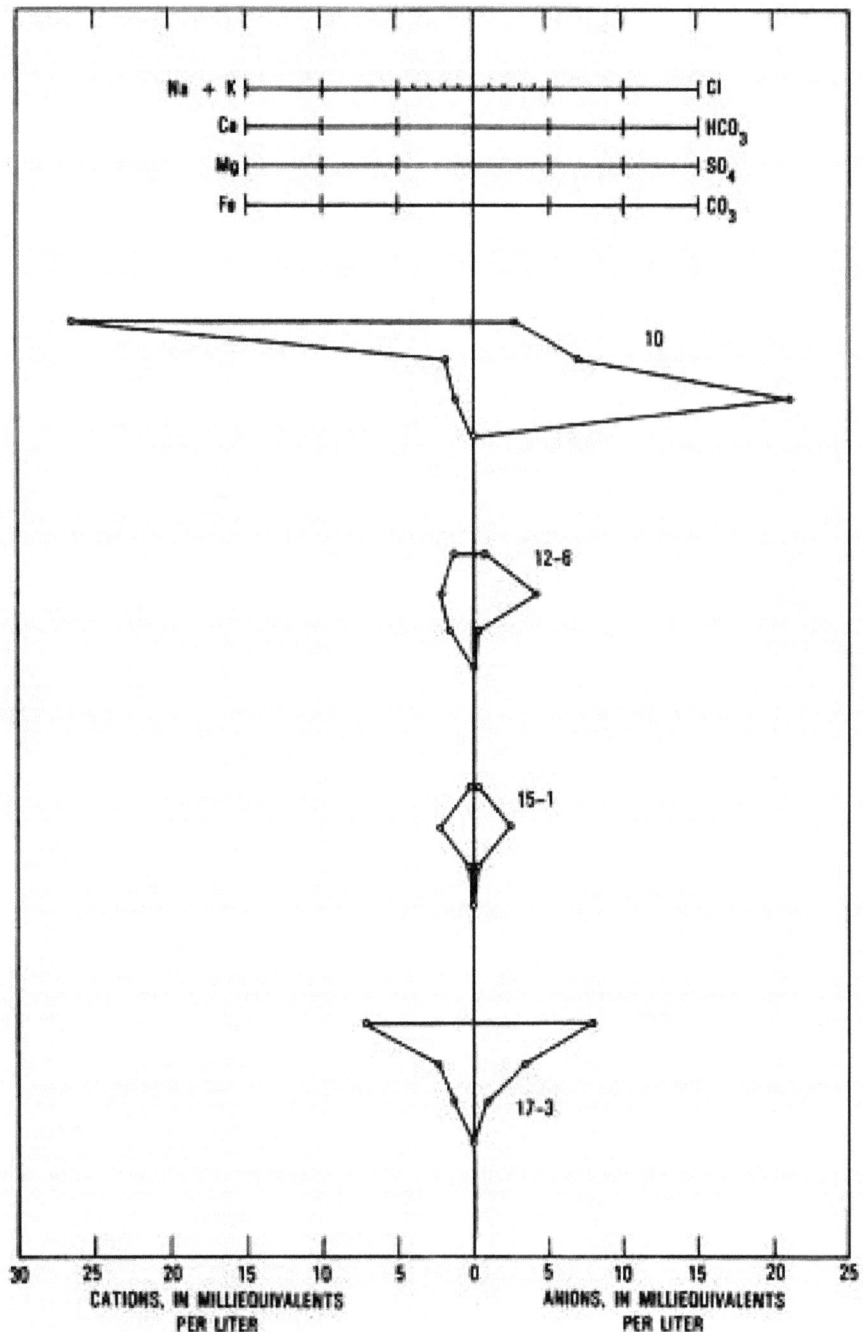

Figure 2. Examples of a Stiff diagrams for four samples. Cations are plotted as concentrations (in milliequivalents/liter) to the left of the axis, and anions are plotted to the right of the axis. The anions should always be plotted in the same sequence. Connecting the resulting points reveals an irregular pattern, as shown. Water-quality types can be readily identified by the shape of the pattern (Hem 1989).

Piper Diagram. This plot is useful for showing multiple samples and trends in major ions (Hem 1989). The central diamond-shaped field is used to show the general character of the water, and ground-water types can be quickly discriminated by position within the field. These diagrams are usually poor graphical representations to plot on maps showing water quality over a large area, because they take up a large amount of space and render the map ineffective. But they aid in interpreting the mixing of waters from different aquifers, especially when used as support with other kinds of interpretations. The circles plotted in the central field have areas proportional to dissolved-solids concentrations and are located by extending the points for the sample in the lower two triangles to the point of intersection in the diamond-shaped field (fig. 3). In the example below, the samples designated 15-1 in the lower triangles are plotted on the diamond-shaped field, extending rays parallel to the triangle axes, to the point of intersection. Distinct ground water classifications can be quickly discriminated by their position on the diamond-shaped field, as indicated in figure 4.

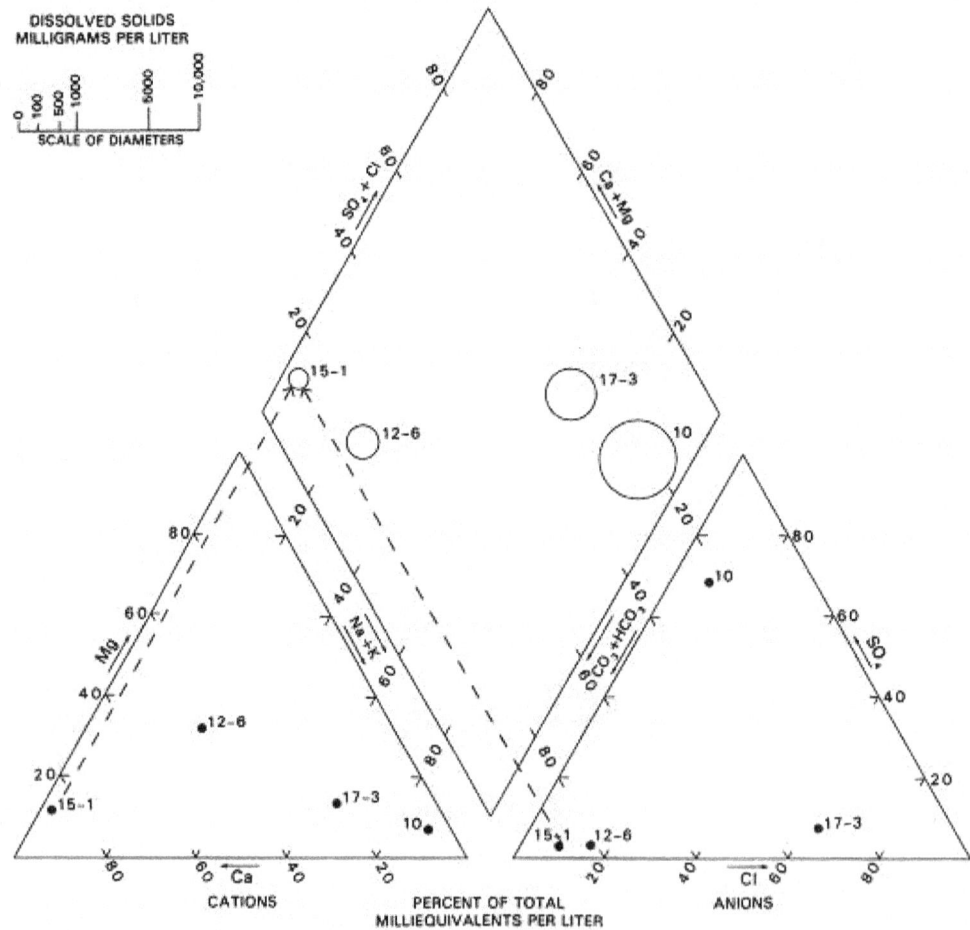

Figure 3. Example of the Piper trilinear diagram for four samples (Hem 1989).

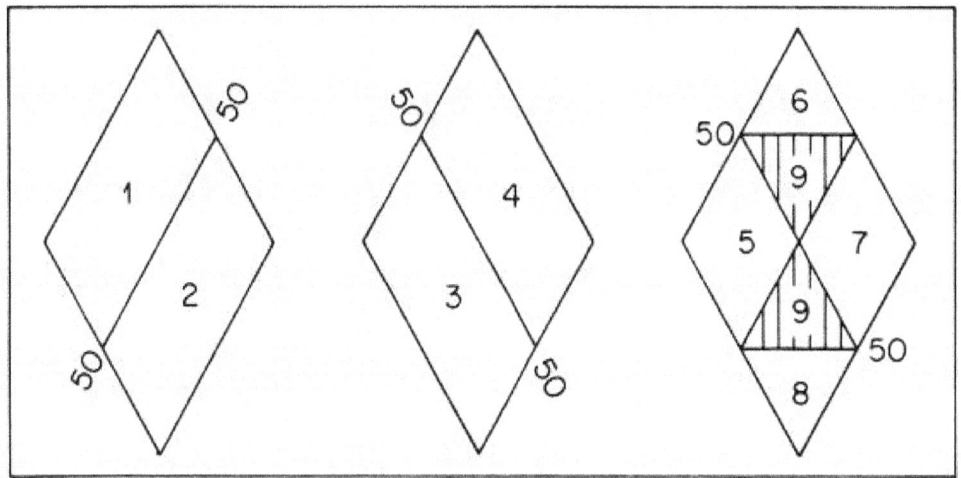

Pie Diagram. This is perhaps the most flexible method to show quality of water (fig 5). The radius of the circle is proportional to the total milliequivalents per liter. Pie charts can be conveniently plotted on base maps to show the ground-water quality for the point source; however, they are time consuming to construct, unless computer plotting software is used.

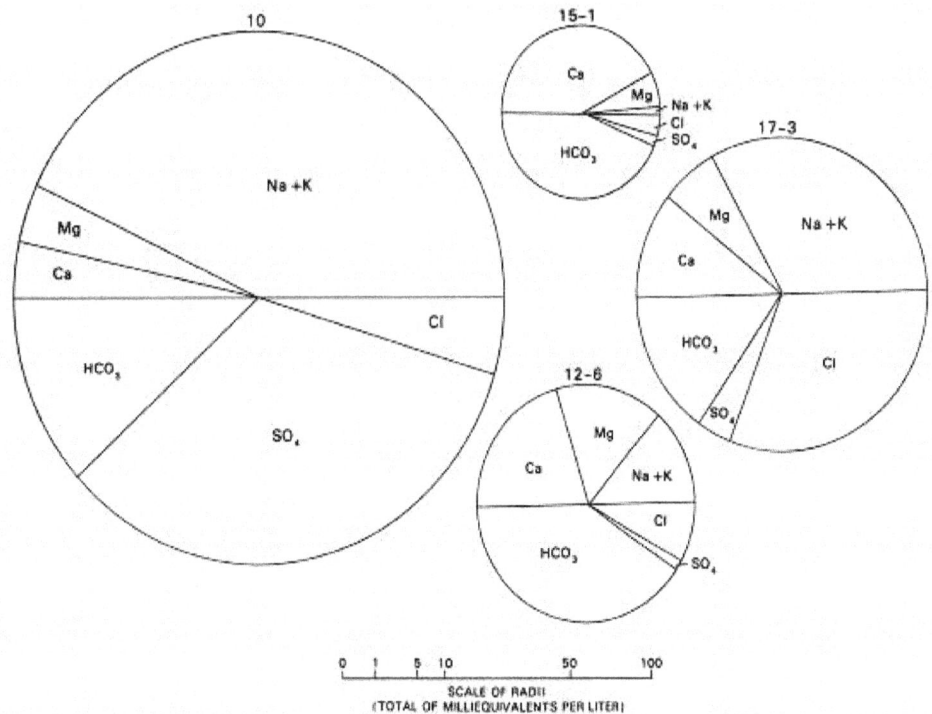

Figure 5. Examples of pie diagrams to depict water quality. The radii length on the scale indicates the concentration (in milliequivalents/liter), and the area of the circle indicates relative total ionic concentrations compared with other samples. The subdivisions of the circles represent proportions of the various ions. The numbers above the circles indicate the particular sample, taken from a table of water-quality data from the study area. By selecting samples to plot in this manner, an overall characterization of the water quality can be shown (Hem 1989).

References

Collins, W.D. 1923. Graphic representation of analysis. Industrial and Engineering Chemistry. 15: 394.

Gilbert, R.O. 1987. Statistical methods for environmental pollution monitoring. New York: John Wiley and Sons, 336 p.

Helsel, D.R.; Hirsch, R.M. 1992. Statistical methods in water resources. Amsterdam, the Netherlands: Elsevier, 529 p.

Hem, J. D. 1989. Study and interpretation of the chemical characteristics of natural water. Water-supply Paper 2254 (third edition). Washington, DC: U.S. Geological Survey. 263 p.

Piper, A.M. 1953. A graphic procedure in the geochemical interpretation of water analysis. Ground Water Note 12. Washington, DC: U.S. Geological Survey. 14p.

Sanders, T. G.; Ward, R.C.; Loftis, J.C.; Steele, T.D.; Adrian, D.D.; Yevjevich, V. 2000. Design of networks for monitoring water quality. Highlands Ranch, CO: Water Resources Publications, 336 p.

Sen, P.K. 1968. Estimates of the regression coefficient based on Kendall's Tau. Journal of the American Statistical Association. 63: 1379-1389.

Smith, R.A.; Hirsch, R.M.; Slack, J.R. 1982. A study of trends in total phosphorus measurements at NASQAN stations. Washington, DC: U.S. Geological Survey Water-supply Paper 2190, 34 pp.

Stiff, J.A., Jr. 1951. The interpretation of chemical water analysis by means of patterns. Journal of Petroleum Technology. 3 (10): 15-17.

U.S. Environmental Protection Agency. 1992. Statistical analysis of ground water monitoring data at RCRA facilities: Addendum to interim final guidance. Washington, DC: U.S. Environmental Protection Agency, http://www.epa.gov/epaoswer/hazwaste/ca/resource/guidance/sitechar/gwstats/gwstats.htm.

Appendix VII.
Geophysics

Use of geophysics can substantially reduce costs and improve the success of ground water investigations. For example, use of surface and borehole geophysical methods can provide a first estimate of the extent of a contaminant plume, thereby reducing the need for large numbers of wells to define the plume and allowing the needed wells to be optimally placed.

Surface Methods

Surface geophysical methods can be part of the geological mapping phase of the project. They can assist in the delineation of areal geology and the identification of shallow ground water conditions. They provide an indirect means of assessing a variety of hydrogeological conditions, including (1) physical properties of bedrock and unconsolidated materials, (2) delineation of subsurface lithology, (3) depth to the water table, and (4) quality of ground water. Surface geophysical methods are used to indirectly assess hydrogeological conditions and their possible controls on ground water. Three surface geophysical techniques are widely applicable to a variety of geologic settings, and are particularly useful in hydrogeologic studies. These techniques are electrical resistivity, electromagnetic conductivity, and seismic refraction and reflection. In addition, gravity and magnetic techniques are often useful in defining the geometry of geological structures in deep aquifers (Zohdy and others 1974, Bartolino and Cole 2002), and ground-penetrating radar has been used successfully to locate buried drums and to delineate detailed shallow subsurface stratigraphy and voids.

Electrical resistivity has been used effectively for near-surface geophysical studies for more than 50 years. The technique involves inducing an electric current into the ground through two electrodes and measuring the potential differences between two points on the ground with two or more additional electrodes. In essence, it is a measurement of the electrical resistance of the surficial material. This property is of great interest because the electrical resistance of the ground is related to the composition of the near-surface material, its porosity, the pore fluid conductivity, and the degree of saturation. It is used to determine lithologic changes or pore-fluid conductivity changes. For example, sands and gravels and fresh water typically have high resistivity values, but clays and contaminated water typically have low resistivity values. Success of this method depends primarily on the resistivity contrast between various geologic materials or the contrast between varying water qualities. Usually a resistivity contrast of twofold or threefold is needed to make a change in lithology, water quality, or hydrologic character detectable.

There are a variety of electrode configurations, but the most widely used are the Wenner and the Schlumberger configurations (fig. 1). The Wenner array uses four equally-spaced electrodes. The array consists of two current input

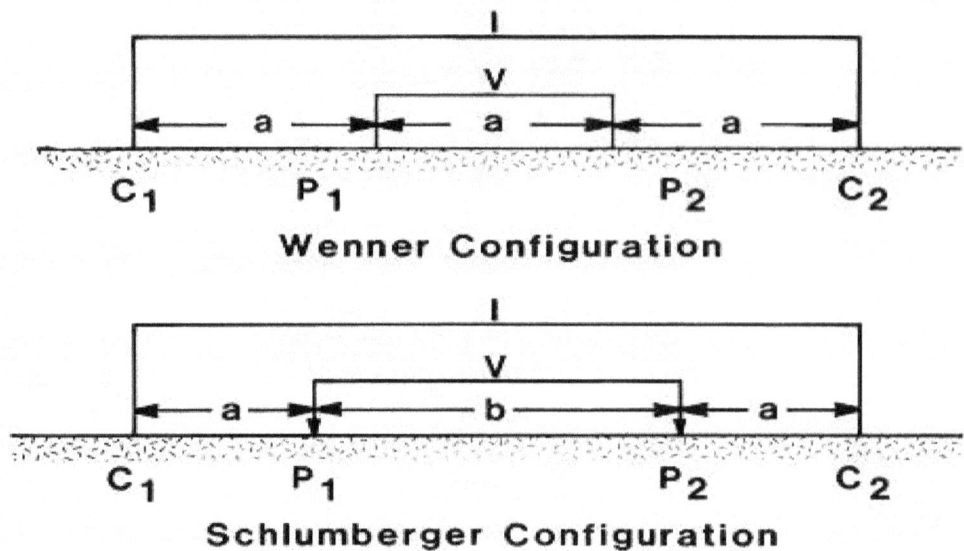

Figure 1. Common electrode arrays for surface geophysical surveys (Rehm and others 1985).

electrodes, C1 and C2, and two potential measurement electrodes, P1 and P2. The spacing between electrodes is usually referred to as the "a" spacing. The Schlumberger array differs from the Wenner array in the electrode spacings, in that the distance between a current electrode and the nearest potential electrode (a) is not equal to the distance between the two potential electrodes (b). Larger depths are penetrated by expanding the electrode array outward from the center. A more recently developed array configuration, the square array, is particularly useful for detecting fractures (Lane and others 1993).

The results must be used with caution because a resistivity sounding can have more than one interpretation. More than one combination of layers, layer thicknesses, and layer resistivities can produce the same geophysical response. The number of possible combinations that will fit the data decreases as independent geological data obtained from sources such as boreholes or outcrops reduces the number of options available for the geophysical model. For this reason, resistivity surveys must be interpreted by someone experienced in correlating geological conditions with the resistivity measurements.

An emerging technology for resistivity applications is continuous-resistivity profiling (Lane 2004). Continuous-resistivity profiling is a water-borne electrical geophysical method that is used to measure the apparent resistivity distribution of a surface water sub-bottom. This method is especially suited for delineating regions of focused ground water discharge in the sub-bottom or in the near-shore environment. It locates the freshwater/saltwater interface in the sub-bottom and images electrical properties for hydrogeological mapping of the sub-bottom. Data are collected by towing an electrode streamer behind a boat. Data collection is fast and easy, but data processing and interpretation can be time consuming.

Electromagnetic Methods

Electromagnetic (EM) techniques were originally developed for the exploration of base metals. The electromagnetic conductivity technique provides results that are similar to resistivity methods. In recent years, the technique has been applied to waste-site monitoring, particularly tracing conductive leachate plumes. The EM survey provides results that are similar to resistivity measurements. The approach involves the definition of areal anomalies in electrical conductivity that can be related to known or assumed conditions in the area. The anomalies can be the result of changes in geology, hydrology, or ground water quality.

The EM conductivity method has some distinct advantages over resistivity techniques. The equipment need not make contact with the ground, and it is very portable, making measurement taking faster. Hence, this method is very good for reconnaissance profiling. The EM transmitter and receiver coils are either held above the ground at approximately waist level or placed on the ground (fig. 2) and measurements are made as the investigator traverses across the site. The method is generally only useful at relatively shallow depths. The Geonics EM-31[34], currently the most portable instrument, can penetrate to depths of 10 to 20 feet; the Geonics EM-34 can penetrate to depths of 100 to 200 feet, depending on whether the coil is oriented vertically or horizontally (Rehm and others 1985, McNeill 1980). The basic operating principle of the EM technique is illustrated in figure 3.

Electromagnetic survey methods have been in used in hydrogeological applications since the early 1980's. The success of the technique has varied, depending on the type of application, but the method has identified leachate plumes at landfills, plumes of contaminated ground water at mine tailings sites, and water table locations at mine spoils sites (Rehm and others 1985). It should be noted, however, that the limited number of layers of contracting conductivity that can be resolved constrain the method to simple geological settings.

Figure 2. Conducting an EM survey, with the transmitter and receiver coils in the vertical orientation. (Photo by Patrick Tucci, USGS.)

[34] Mention of trade names is solely to identify equipment used and does not imply endorsement by the U.S. Department of Agriculture.

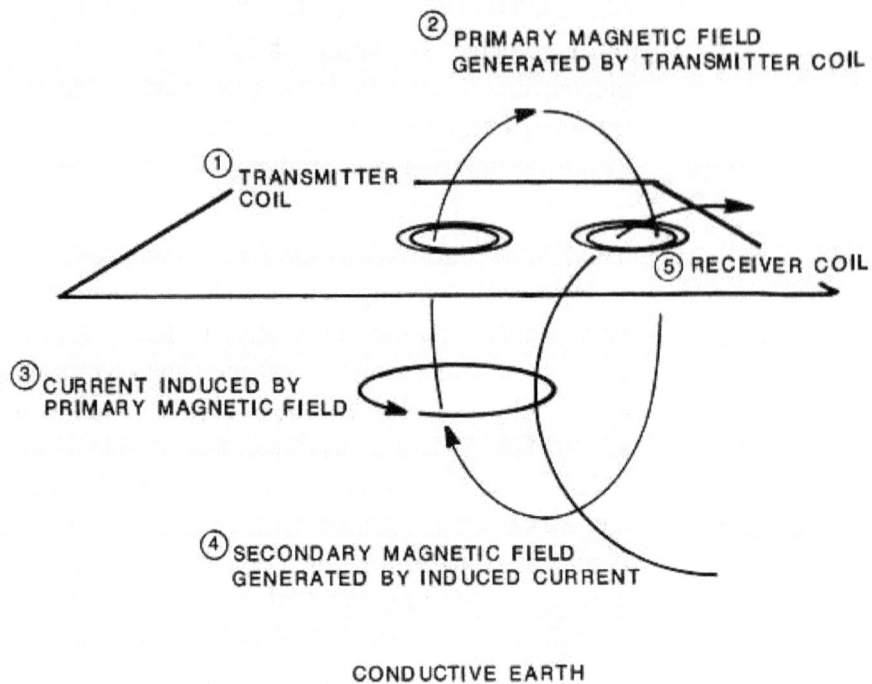

Figure 3. Operating principle of the EM technique. The transmitter coil (1) generates a magnetic field (2). This primary field induces a current (3) in a mass of conductive earth. The induced current in turn generates a secondary magnetic field (4). The receiver coil (5) senses both the primary and secondary fields. The conductivity of the earth is proportional to the ratio of the intensity of the secondary field to the intensity of the primary field (Rehm and others 1985).

Seismic Refraction Methods

Although seismic refraction methods generally have less resolution than seismic reflection methods, they have been preferred for use in shallow hydrogeological investigations for the following reasons (Zohdy and others 1974):

1. Refraction methods generally yield superior results in areas of thick alluvial or glacial fill and where large velocity contrasts exist, such as buried bedrock valleys.
2. Personnel and equipment requirements are generally simpler and less expensive for refraction surveys than for reflection surveys.

Seismic refraction techniques are designed to obtain data near the surface (typically to a depth of about 30 meters, but depths in excess of 200 meters can be achieved with more powerful seismic sources). Such techniques provide data on the refraction of seismic waves at the interface between subsurface layers and on their travel times within the layers. Properly interpreted, the refraction data make it possible to estimate the thickness and depth of geological layers (including the water table) and to assess their properties. Also, changes in the lateral facies of aquifer material can sometimes be mapped with this method (Sandlein and Yazicigil 1981)

The seismic refraction method relies upon measuring the transit time from energy source to receiver of induced vibrational energy refracted along some geological boundary (fig. 4) and assumes increasing seismic velocity with depth. Through analysis of the measured transit times as a function of source-receiver separation, it is possible, given certain assumptions, to determine the thickness and seismic velocity of all units beneath a selected source-receiver geometry. Based on the calculated velocities, it is then possible to infer lithology and physical characteristics of the area selected for analysis. It is possible that measured seismic velocities, when combined with known lithology, could be converted to porosity through a method similar to that of Wyllie and others (1958); however, it would probably fail for fine-grained sediments and, in any event, would require establishing empirical curves appropriate to the particular area.

The primary application of seismic surveys is to determine the depth to the bedrock surface or to map the elevation of the bedrock surface, because of the large velocity contrast between bedrock and unconsolidated overburden. In selected cases, refraction surveys could be employed to determine the water-table depth. Using seismic refraction methods to determine water-table depths is geophysically equivalent to the determination of bedrock depth. From a practical standpoint, the water table in a course-grained, unconfined aquifer could easily be detected; however, in a fine-grained, unconfined aquifer, the transition from saturated to unsaturated conditions is too poorly defined to be detected by refraction measurements. Haeni (1988) lists hydrogeological settings in which seismic-refraction surveys (1) can be used successfully; (2) may work, but with difficulty; and (3) will not work.

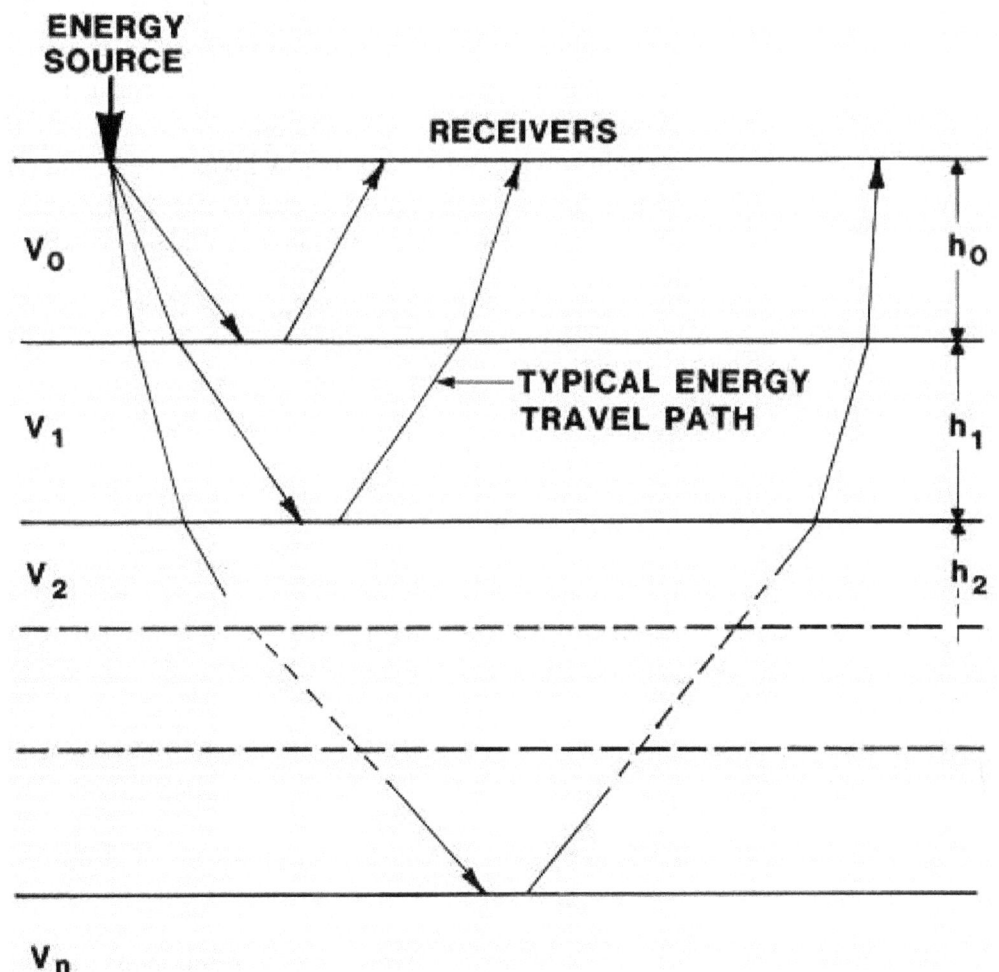

Figure 4. Typical refraction paths in seismic refraction geophysics (Rehm and others 1985).

271

Note that the method assumes increasing velocity with increasing depth. The presence of a decrease in velocity at depth can lead to significant interpretational errors. The problem is compounded by the fact that there is no way to establish from the field data whether low-velocity units at depth are affecting the results. Hence, the presence of low-velocity units at depth must be determined from logged boreholes in the area. As in electrical methods, analysis and interpretation of seismic survey data should only be done by experienced personnel with knowledge of the local geology of the area being surveyed.

Seismic Reflection Methods

Seismic reflection methods are similar to refraction methods. An acoustic signal is generated near the earth's surface, and the travel times of acoustic pulses reflected at contacts between various earth materials are measured.

Reflection surveys are generally employed to identify geological contacts at depths greater than 61 meters (200 feet). The resolution at these depths is approximately 3 meters (10 feet). The accuracy of depth determinations is limited by the uncertainty in the seismic velocities of the subsurface materials. Special equipment is available for surveys as shallow as 30 to 60 meters (100–200 feet), but refraction surveys are generally better suited than reflection surveys for use at shallower depths.

An exception to this rule is in the application of marine seismic-reflection techniques to ground water problems near surface water bodies (Haeni 1986). Detailed stratigraphic and structural information (fig. 5) can be obtained using seismic-reflection methods below lakes, rivers, and canals, where standard exploration methods cannot easily be used. The sound source and receivers (hydrophones) are towed by a boat, so that a great deal of data can be collected in a short time.

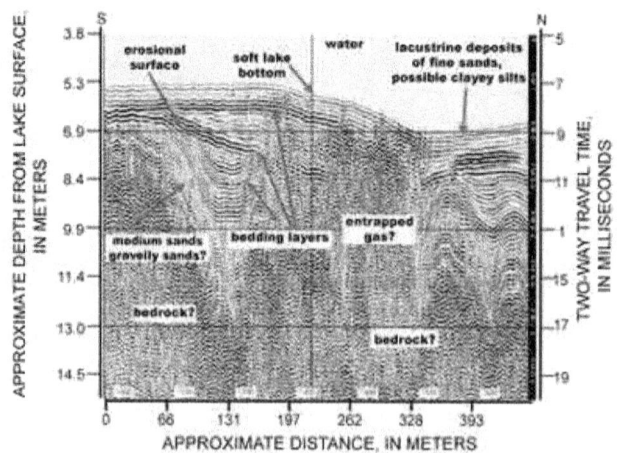

Figure 5. Detailed stratigraphic data obtained below Annabessacook Lake, Winthrop, ME, using seismic-reflection profiling.

Continuous seismic-reflection profile on Annabessacook Lake, Winthrop, Maine.

Borehole Methods

The measurement of physical earth properties using equipment lowered into drilled holes is known as geophysical well logging, borehole logging, wireline logging, or downhole logging. This type of logging requires a single hole, and thus, differs from cross-hole logging (tomography) which requires a minimum of two holes. It also differs from mud logging, core logging, or the driller's log in that no physical sample from the hole is required. Borehole logging involves an instrumentation package, known as the probe, sonde or tool, which is attached to a cable and lowered into the borehole. Normally, the probe measures or "logs" selected physical properties of the material in or near the borehole as the probe rises from the bottom of the hole. The log output typically consists of a plot of geophysical responses as a function of depth. Usually several geophysical logs are plotted simultaneously (fig. 6). The resulting downhole measurements are related to geological and hydrological conditions near the borehole.

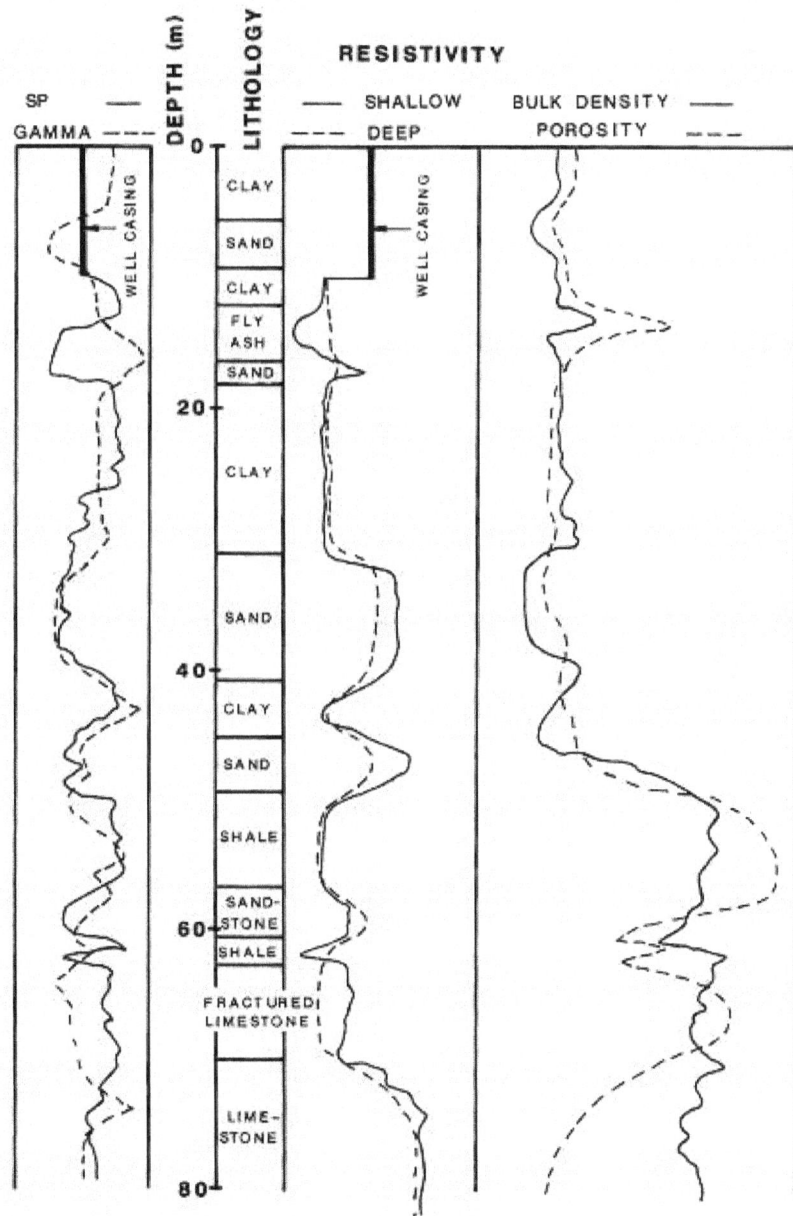

Figure 6. A typical display of several geophysical logs simultaneously displayed as output: Gamma log, spontaneous potential (SP) log, deep and shallow resistivity, bulk density, and porosity (Rehm and others 1985).

Borehole geophysical methods have great utility in ground water studies. The objective of the borehole logging effort is to provide greater detail about subsurface conditions than would be available from surface geophysical methods, drilling cuttings, or discontinuous representative or undisturbed samples. These methods are generally employed in hydrogeological applications to help meet five broad objectives:

1. To evaluate ground water quality.
2. To determine the depth to the water table.
3. To determine the depth to the bedrock surface.
4. To evaluate subsurface lithology.
5. To locate water-producing fractures.

Paillet and Crowder (1996) describe an approach for the interpretation of borehole logs in ground water studies.

Generally, the logging applications can be classified according to the parameters evaluated: measuring water quality, determining lithology, locating permeable zones, locating bedding planes and fractures, determining fluid velocity, or determining porosity. In addition to these applications, material resistivity and seismic velocity can be directly obtained to enhance the value of surface electrical or seismic measurements.

Electrical Methods

Electric logging methods form the largest single group of borehole logging techniques. Logging methods that determine electrical conductivity or resistivity in or near the borehole are the most widely used methods normally considered under the heading of electrical methods. Self-potential and induced polarization methods have more restricted applicability to hydrological investigations. Electrical conductivity or resistivity logs are conveniently divided into two classes: (1) those methods that employ electrodes in contact with borehole fluid and (2) those that rely on electromagnetic induction and require no contact with borehole fluid.

Borehole resistivity methods are generally classified according to the number of electrodes required to make a measurement. Resistance logs, also known as single-point and single-electrode logs, involve a single downhole electrode (fig. 7a). This method provides a measure of the electrical resistance associated with current flow from a point in the borehole to a point on the surface near the borehole. The log is used primarily to define contacts between materials of differing electrical properties. The primary advantage of the log is the very simple instrumentation requirements and therefore low equipment cost (Rehm and others 1985).

Normal logs, also known as two-electrode logs, use an electrode configuration like the one shown in figure 7b. The tool is usually described by the distance between electrodes A and M. The petroleum industry has standardized two-probe configurations: the short-normal at 0.41 meter (16 inches) and the long-normal at 1.63 meters (64 inches). For shallow hydrogeological studies, the distances do not appear to be standardized, but are generally within ± 50 percent of the above. The radius of investigation is approximately twice the electrode separation, or 1 meter (3 feet) for the short-normal and 3 meters (10 feet) for the long-normal (Rehm and others 1985).

The third type of resistivity log is the lateral log (fig. 7c). This log was introduced to petroleum logging to obtain the resistivity of the formation beyond the zone affected by drilling fluid. The tool is normally described by the spacing between the A and N electrodes, which in the petroleum industry is standardized at 5.69 meters (18.67 feet). In hydrological logging, there is apparently no standard, but approximately 1.8 meters (6 feet) is common. The radius of investigation for this log is approximately equal to the spacing between electrodes A and N (Rehm and others 1985).

Another type of electrical log, the micro or sidewall log, commonly employs both small-lateral, AN = 0.38 meter (1.5 inches), and small-normal, AM = 0.051 meter (2 inches), electrode spacings. The electrodes are carried in a pad, which must be held in contact with the sidewall. These logs are measuring properties within a few centimeters of the borehole, and thus are only used where detailed hydrologic information is needed (Rehm and others 1985).

Induction logs rely on electromagnetic (EM) radiation from the tool to induce secondary currents in the formation near the tool. The magnitude of these currents is then detected by a receiver within the tool. The induction methods will, therefore, operate in oil- or air-filled boreholes and in the presence of

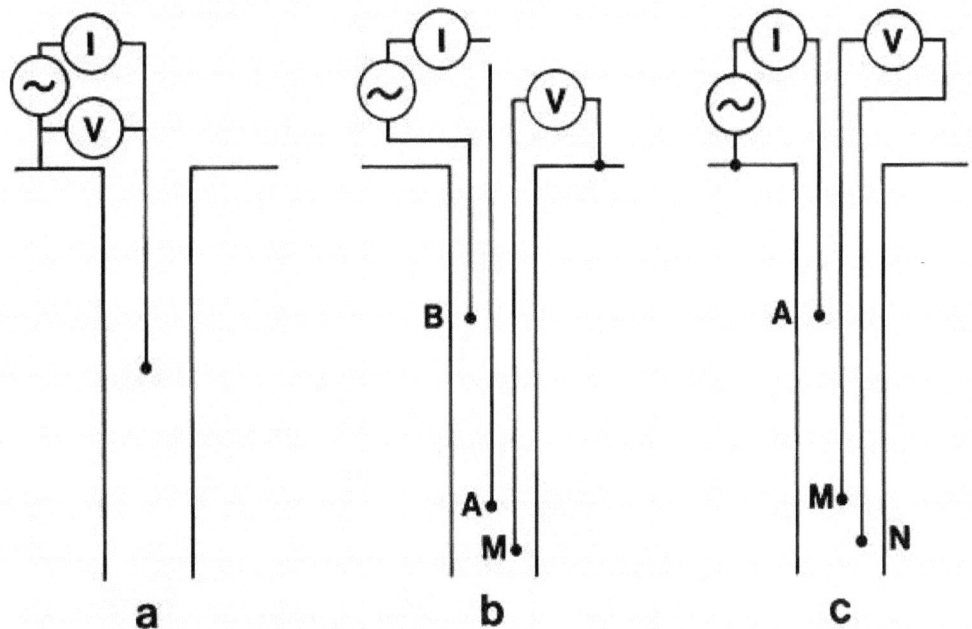

Figure 7. Electrode arrays for borehole logs: (a) resistivity log, (b) normal log, and (c) lateral log (Rehm and others 1985).

polyvinyl chloride (PVC) well casing. Highly conductive borehole fluids and steel casing, however, prevent the use of induction methods. The tools respond directly to the inverse of formation resistivity or the formation conductivity and are, therefore, known as conductivity logs (Rehm and others 1985).

From a hydrological standpoint, the primary advantage of an EM induction tool is its ability to log through PVC casing or in air-filled holes. EM induction logs, in combination with other logs, are particularly useful in the delineation of contaminant plumes that may be constrained to discrete intervals in the subsurface (fig. 8) (Williams and others 1993). EM induction probes are readily available and are commonly replacing normal-resistivity logs in ground water investigations.

SP is the potential associated with natural current flow within and near the borehole. The SP log is a measurement of these potentials over the length of the borehole. The SP voltages result primarily from conductivity differences between the drilling fluid and formation waters or from actual flow of drilling fluid into the formation. The former are known as electrochemical potentials and the latter as electrokinetic, or streaming, potentials (Rehm and others 1985).

The SP log is used primarily for determining the contacts between materials with different electrical properties, differentiating between permeable and non-permeable materials (such as sand vs. clay) and determining formation-water resistivity. In hydrological investigations, the formation waters and drilling fluid will probably display little difference in resistivity. Under these

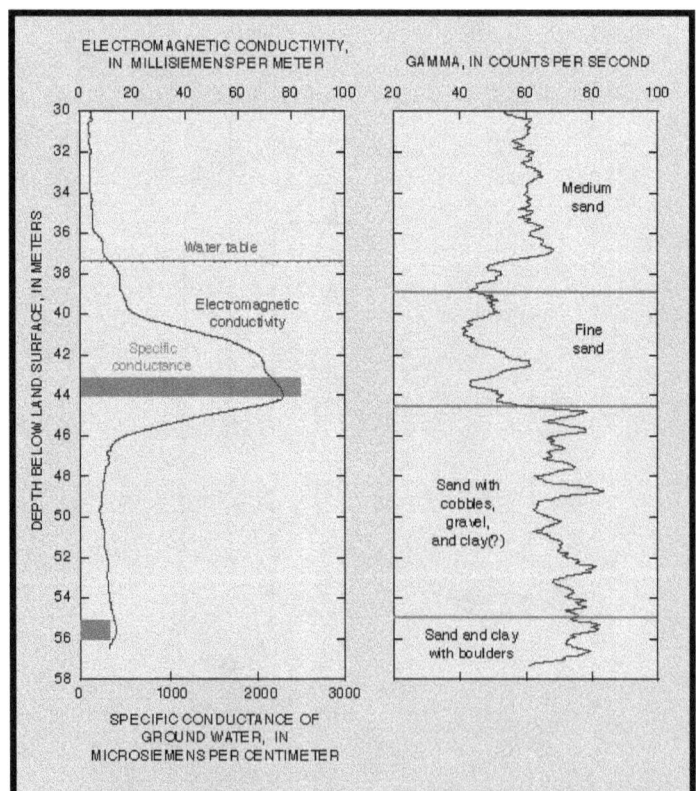

Figure 8. Electromagnetic-induction log delineates a leachate plume in a sand-and-gravel aquifer downgradient of a municipal landfill. The most highly contaminated part of the plume is at a depth of 41 to 45 meters (Williams and others 1993).

conditions, the electrochemical potential will be small or zero and, hence, of limited value. Streaming potentials may be generated as ground water flows from aquifers into the borehole or as borehole fluids flow into permeable materials (Rehm and others 1985).

Nuclear Methods

Nuclear logs are the second largest subset of logging technology. While many special purpose nuclear logs exist, the most common and widely applied nuclear logs are the natural gamma, the gamma-gamma, and the neutron log. These are all discussed in detail by Pirson (1963), Kelly (1969), Hilchie (1982), and Keys and MacCary (1976). The specific application of nuclear logging methods to hydrological problems is reviewed by the International Atomic Energy Agency (1968).

Natural-gamma logs measure the naturally occurring gamma radiation along the length of the borehole. The log can be obtained in cased or uncased holes, and in air- or fluid-filled holes. These logs may be taken in a broad energy window (natural-gamma logging) or in several narrow energy windows (spectrometric-gamma logging). Most radioactive elements are associated with clay minerals. The natural-gamma log is, therefore, primarily a clay-lithology log. The spectrometric-gamma log is intended primarily as a uranium exploration tool. The primary purpose of a natural-gamma log in the context of hydrogeological investigations is to identify clay layers penetrated by boreholes.

Gamma-gamma logs, also known as density logs, measure the effect of material near the borehole on gamma radiation emanating from a source within the logging tool. While the effect can be measured in various ways, the measurements all relate to the electron density of material near the borehole. Because electron density is related to bulk density, the log is basically a bulk-density log. If the densities of the rock matrix and pore fluid are known, the bulk density may be converted to porosity. In either case, the radius to which the measurement extends is approximately 0.3 meter (1 foot) in open holes and 0.15 meter (0.5 feet) in cased holes.

Neutron logs measure the response of material near the borehole to neutrons emitted from a source within the tool. The response may be measured in various ways, but all common measurements are related to the presence of hydrogen in the formation. Excluding potential problems with bound water, the response of the instrument is determined entirely by the amount to water in the porous medium. Hence, the tool is primarily a porosity log. The radius of influence for this tool is approximately 0.3 meter (1 foot) in open holes and a few centimeters in cased holes.

Both gamma-gamma and neutron geophysical probes contain radioactive sources, whose use is regulated by the U.S. Nuclear Regulatory Commission as well as various State agencies. Special permits are needed to store and use these sources, and extensive training in their safe use is required. If these types of logs are needed, it is best to contract with a commercial well-logging company. Natural-gamma logs do not require nuclear sources, and their use is not regulated.

Flow Logs

Flow logs measure fluid movement within the borehole as a function of depth. Flow logs fall into two classes: (1) those that directly measure fluid motion with a mechanical impeller and (2) those that measure fluid motion indirectly by measuring heat flow away from a thermal source or by electromagnetic methods. More recent designs for borehole flowmeters include acoustic-doppler and optical methods. The direct measuring devices are primarily designed for use in well production and well completion problems, and are useful for relatively large flow rates. The indirect measuring devices are better suited to lower flow rates that cannot be measured by direct measuring devices (< 2 m/min), but they require flow rates greater than 0.03 m/min (U.S. Geological Survey 1998).

Flow measurements can be useful in determining vertical direction of ground water flow (up or down), flow from a particular zone of interest (such as a fracture), or the interaction between vertically connected aquifers. They also can measure the change in vertical flow as a function of depth. Methods have recently been developed to use borehole flowmeter data to estimate hydraulic conductivity in fractured-rock aquifers (Paillet 1998)

Other Borehole Logs

Several other types of borehole logs can also provide important information for ground water investigations.

- Caliper logs provide information on the diameter of the borehole, and can be used to indicate large open fractures within the borehole or screened intervals in cased wells. They can also be used to locate constrictions in the borehole that could prevent use of large-diameter probes or "washouts" in the borehole that could influence interpretation of other logs. Caliper logging is generally fairly inexpensive, but can only be used in open boreholes in bedrock.
- Acoustical logs provide information on the velocity of sound waves in the formation. Such information is useful in interpreting surface seismic surveys, and can provide indirect information on formation density. Special acoustical tools, called televiewers, provide images that indicate the size and orientation of fractures in the borehole (fig. 9). Recent advances in televiewer logs allow processing of the data to produce "virtual cores" from the data.
- Fluid-conductance and temperature logs are useful in obtaining information on zones where water enters or leaves the borehole, or for locating zones of high electrical conductivity ground water.
- Borehole radar is a relatively new, but expensive, technique that is particularly useful in detecting subsurface fractures. The technique can often see well beyond the immediate borehole, and can be used in a cross-hole technique to determine interconnected fractures.

- Borehole television cameras (optical televiewers) or other optical imaging devices are readily available, and are useful in detecting fractures (fig. 9) and conditions in the well or borehole prior to sending down more expensive probes.

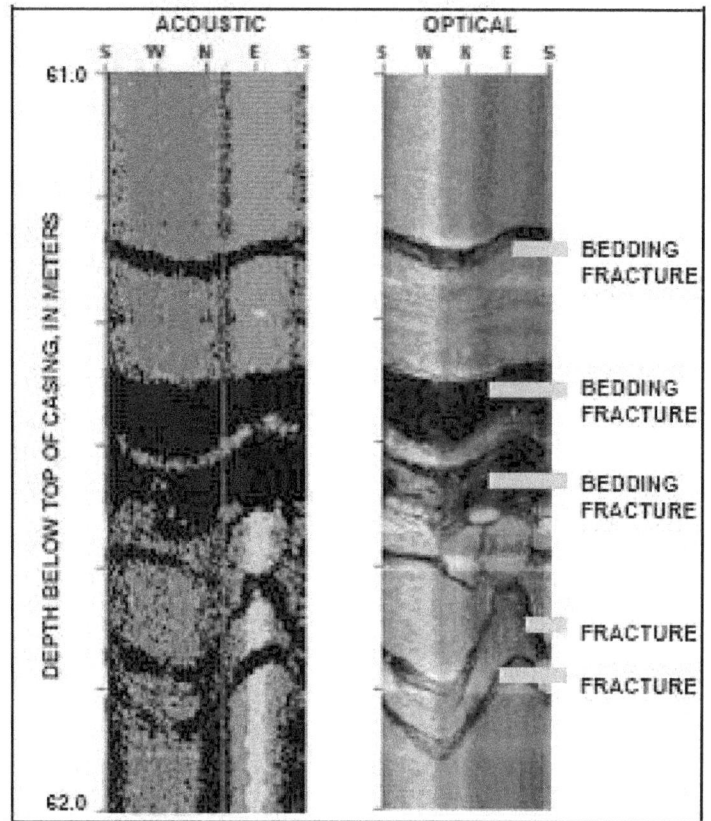

Figure 9. Acoustic and optical televiewer images of a transmissive zone in a borehole (after Williams and others 2002).

Integrated Methods

An integrated surface- and borehole-geophysical approach has been termed the "toolbox" approach (Haeni and others 2001). It is particularly useful in fractured-rock hydrological settings commonly found in the national forests. Surface geophysical methods provide site reconnaissance suitable for the development of initial conceptual models of ground water flow in the formation and location of test holes. Conventional borehole-geophysical logs, borehole imaging, and advanced single- and cross-hole geophysical methods can be interpreted to identify the location and physical characteristics of fractures, and, potentially, their hydraulic properties. Integration of surface- and borehole-geophysical data with geological, hydrological, and geochemical data provides a means for developing a comprehensive interpretation of the hydrogeological conditions at a site and a conceptual understanding of ground water flow (Haeni and others 2001). It is important to recognize that the implementation of a comprehensive integrated approach will be time consuming and resource intensive, making its application appropriate only in select circumstances.

References

Bartolino, J.R.; Cole, J.C. 2002. Ground water resources of the Middle Rio Grande Basin, New Mexico. Circular 1222. Washington, DC: U.S. Geological Survey. 130 p. [http://water.usgs.gov/pubs/circ/2002/circ1222].
Haeni, F.P. 1986. Application of continuous seismic-reflection methods to hydrologic studies. Ground Water. 24 (1): 23-31.

Haeni, F.P., 1988, Application of seismic-refraction techniques to hydrologic studies. *In* Techniques of Water-resources Investigations. Book 2, Chapter D2. Washington, DC: U.S. Geological Survey, 86 p. [http://water.usgs.gov/pubs/twri/twri2d2].

Haeni, F.P.; Lane, J.W. Jr.; Williams, J.W.; Johnson, C.D. 2001. Use of a geophysical toolbox to characterize ground water flow in fractured rock. *in* Kueper, B.H., Novakowski, K.S., and Reynolds, D.A. (eds). Fractured Rock 2001. Conference Proceedings: Toronto, Ontario, Canada. March 26-28, 2001. CD-ROM. [http://water.usgs.gov/ogw/bgas/publications/FracRock01_haeni/FracRock01_haeni.pdf].

Hilchie, D.W. 1982. Applied openhole log interpretation for geologists and engineers. Golden, CO: Hilchie, Inc.

International Atomic Energy Agency. 1968. Guidebook on nuclear techniques in hydrology. Vienna: International Atomic Energy Agency Technical Report 91.

Kelly, D.R. 1969. A summary of geophysical logging methods. Pennsylvania Department of Environmental Resources, Bureau of Topographic and Geologic Survey, 83 p.

Keys, W.S.; MacCary, L.M. 1976. Application of borehole geophysics to water-resources investigations. *In* Techniques of Water-resources Investigations. Book 2, Chapter E1. Washington, DC: U.S. Geological Survey, 126 p.

Lane, J.W.; Haeni, F.P.; Watson, W.M. 1993. Use of a square-array direct-current resistivity method to detect fractures in crystalline bedrock in New Hampshire. Ground Water. 33 (3): 476-485.

Lewis, M.R.; Haeni, F.P. 1987. The use of surface geophysical techniques to detect fractures in bedrock: An annotated bibliography. Circular 987. Washington, DC: U.S. Geological Survey. 14 p.

Lieblich, D.A.; Haeni, F.P.; Cromwell, R.E. 1992. Integrated use of surface-geophysical methods to indicate subsurface fractures at Tibbetts Road, Barrington, New Hampshire. Water-resources Investigations Report 92-4012. Washington, DC: U.S. Geological Survey. 33 p.

Paillet, F.L. 1998. Flow modeling and permeability estimation using borehole flow logs in heterogeneous fractured formations: Water Resources Research. 34 (5): 997-1010.

Paillet, F.L.; Crowder, R.E. 1996. A generalized approach for the interpretation of geophysical well logs in ground water studies: Theory and application. Ground Water. 34 (5): 883-898.

Pirson, S.J. 1963. Technical guide for well log analysis for oil and gas formation evaluation. Upper Saddle River, NJ: Prentice Hall, 326 p.

Rehm, B.W.; T.R. Stolzenburg; D.G. Nichols. 1985. Field Measurement Methods for Hydrogeologic Investigations: A Critical Review of the Literature. EPRI EA-4301. Electric Power Research Institute, Palo Alto, CA.

Sandlein, L.V.A.; Yazicigil, J.H. 1981. Surface geophysical methods for ground water monitoring Part I. Ground Water Monitoring Review. 1(3):42-46.

U.S. Geological Survey. 1998. Advances in borehole geophysics for ground water investigations. Fact Sheet 002-98. Washington, DC: U.S. Geological Survey. 4 p. [http://ny.water.usgs.gov/pubs/fs/fs00298/FS002-98.pdf].

Williams, J.H.; Lapham, W.W.; Barringer, T.H. 1993. Application of electromagnetic logging to contamination investigations in glacial sand-and-gravel aquifers: Ground Water Monitoring Review. 13 (3): 129-138.

Williams, J.H.; Lane, J.W.; Singha, K.; Haeni, F.P. 2002. Application of advanced geophysical logging methods in the characterization of a fractured-sedimentary bedrock aquifer, Ventura County, California. Washington, DC: U.S. Geological Survey Water-Resources Investigations Report 00-4083, 28 p. [http://ny.water.usgs.gov/pubs/wri/wri004083].

Wyllie, M.R.J.; Gregory, A.R.; Gardner, G.H.F. 1958. An experimental investigation of factors affecting wave velocity in porous media. Geophysics. 23:459-493.

Zohdy, A.A.R.; Eaton, G.P.; Mabey, D.R. 1974. Application of surface geophysics to ground water investigations. *In Techniques of Water-resources Investigations. Book 2, Chapter D1. Washington, DC: U.S. Geological Survey, 116 p. [http://water.usgs.gov/pubs/twri/twri2-d1].